Analog Signal Processing with Laplace Transforms and Active Filter Design

Don Meador
DeVry Institute of Technology
Kansas City, MO

DELMAR

THOMSON LEARNING™

Australia • Canada • Mexico • Singapore • Spain • United Kingdom • United States

Analog Signal Processing with Laplace Transforms and Active Filter Design, 2e

By Don A. Meador

Business Unit Director:
Alar Elken

Senior Acquisitions Editor:
Gregory L. Clayton

Development Editor:
Michelle Ruelos Cannistraci

Executive Editor:
Sandy Clark

Editorial Assistant:
Jennifer A. Thompson

Executive Production Manager:
Mary Ellen Black

Production Manager:
Larry Main

Executive Marketing Manager:
Maura Theriault

Senior Project Editor:
Christopher Chien

Art/Design Coordinator:
David Arsenault

Cover Image:
PhotoDisc, Inc.

Marketing Coordinator:
Karen Smith

NOTICE TO THE READER

Publisher does not warrant or guarantee any of the products described herein or perform any independent analysis in connection with any of the product information contained herein. Publisher does not assume, and expressly disclaims, any obligation to obtain and include information other than that provided to it by the manufacturer.

The reader is expressly warned to consider and adopt all safety precautions that might be indicated by the activities herein and to avoid all potential hazards. By following the instructions contained herein, the reader willingly assumes all risks in connection with such instructions.

The publisher makes no representation or warranties of any kind, including but not limited to, the warranties of fitness for particular purpose or merchantability, nor are any such representations implied with respect to the material set forth herein, and the publisher takes no responsibility with respect to such material. The publisher shall not be liable for any special, consequential, or exemplary damages resulting, in whole or part, from the readers' use of, or reliance upon, this material.

To Joan

Contents

1 INTRODUCTION   1

1-1	What's in This Book	1
1-2	Laplace Transforms	1
1-3	Notations	2
1-4	Calculations	2
1-5	How To Study	2
1-6	Using MATLAB and the TI-89	3

2 ANALOG SIGNAL ANALYSIS   7

2-0	Introduction	7
2-1	Step Function	8
2-2	Ramp Function	12
2-3	t^n Function	14
2-4	Impulse Function	15
2-5	Exponential Function	18
2-6	Sinusoidal Function	22
2-7	Shifted Function	33
2-8	Putting It All Together	35
2-9	Derivatives and Integrals of Waveforms	51
2-10	Periodic Waveform Values	59
2-11	Using MATLAB	68
2-12	Using the TI-89	74
	Problems	79

3 LAPLACE TRANSFORMS   93

3-0	Introduction	93
3-1	Laplace Transforms by Table	94
3-2	Inverse Laplace Transform by Table	97

3-3   Laplace Operations   98

3-4   Inverse Laplace Transform Using Traditional Techniques   121

3-5   Streamlined Inverse Laplace Transforms   142

3-6   Using Laplace Transforms to Solve Differential Equations   160

3-7   Using MATLAB   164

3-8   Using the TI-89   167

Problems   171

4 CIRCUIT ANALYSIS USING LAPLACE TRANSFORMS   179

4-0   Introduction   179

4-1   Initial Voltages and Currents   180

4-2   Laplace Impedance   183

4-3   Laplace Circuit   189

4-4   Solving Simple Circuits   191

4-5   Solving Multiple-Source Circuits   197

4-6   Solving Dependent Source Circuits   207

4-7   Solving Thévenin and Norton Circuits   210

4-8   Circuit Order   216

4-9   Using MATLAB   223

4-10   Using the TI-89   226

Problems   229

5 SINUSOIDAL STEADY STATE   237

5-0   Introduction   237

5-1   Transfer Functions   238

5-2   Pole-Zero Plot and Stability   241

5-3   Steady-State Frequency Response and the Bode Plot   246

5-4   Using MATLAB   278

5-5   Using the TI-89   282

Problems   288

6 INTRODUCTION TO FILTERS   295

6-0   Introduction   295

6-1   Filter Graphs   295

6-2   Filter Definitions   302

6-3   Op-Amps   307

Problems   314

7 NORMALIZED LOW-PASS FILTER   319

7-0   Introduction   319
7-1   Topology   320
7-2   Coefficient Matching   323
7-3   Biquads   329
7-4   Low-Pass Filter Approximations   332
7-5   Using MATLAB   358
7-6   Using the TI-89   362
       Problems   368

8 PRACTICAL FILTERS FROM THE GENERIC LOSS FUNCTIONS   373

8-0   Introduction   373
8-1   Frequency Shifting   373
8-2   Impedance Shifting   376
8-3   Gain Shifting   380
8-4   High-Pass Filter   384
8-5   Band-Pass Filter   389
8-6   Notch Filter   398
8-7   Using MATLAB   405
8-8   Using the TI-89   410
       Problems   414

Appendix A   TRANSFORM TABLES   418

A-1   Transform Pairs   418
A-2   Transform Operations   418
A-3   Transform Identities   419
A-4   Extended Transform Pairs   420

Appendix B   LAPLACE DERIVATIONS   421

B-1   Deriving Laplace Transform Pairs   421
B-2   Deriving Laplace Transform Operations   424
B-3   Deriving Complex Poles Formula   428

Appendix C   BASIC DC CIRCUIT ANALYSIS EQUATIONS   433

C-1   Identifying Series Circuits   433
C-2   Identifying Parallel Circuits   433

C-3	Series Voltage Sources	433
C-4	Parallel Current Sources	434
C-5	Ohm's Law	434
C-6	Voltage and Current Measurements	434
C-7	Voltage Divider Rule	435
C-8	Current Divider Rule	436
C-9	Kirchhoff's Voltage Law	436
C-10	Kirchhoff's Current Law	438

Appendix D SEMILOG GRAPHS 440

D-1	How to Read a Log Scale	440
D-2	Calculating Distances on a Log Scale	441
D-3	Calculating Roll-Off Rates on Semilog Graphs	444
D-4	Construction of Semilog Graph Paper	447

Appendix E OP-AMP TOPOLOGIES 450

E-1	Noninverting Single-Feedback Topology	450
E-2	Inverting Single-Feedback Topology	452
E-3	Noninverting Dual-Feedback Topology	454
E-4	Inverting Dual-Feedback Topology	458
E-5	Twin-T Topology	461

Appendix F FILTER TABLE CALCULATIONS 464

F-1	General Procedure	464
F-2	Butterworth Filter Equation	467
F-3	Chebyshev Filter Equation	470
F-4	Elliptic Filter Equation	473

Appendix G ROAD MAP 482

BIBLIOGRAPHY 488

ANSWERS TO ODD-NUMBERED PROBLEMS 489

INDEX 511

Preface

INTENDED AUDIENCE

This book is written for an introductory course on analog signal processing. The material is written at a level appropriate for students attending a four-year electronic engineering or electronic technology college. A course based on this text would appear after the completion of courses on dc circuit analysis, ac steady-state time-domain circuit analysis, and basic op-amp theory. Completion of a course in calculus would be helpful but is not absolutely essential since the sections requiring calculus may be bypassed without losing the overall mechanics of working with Laplace transforms and active filter design.

Understanding analog signal processing is becoming more important with each new development in digital technology and computer science. In digital technology, there have been giant strides in the development of pre-designed digital subsystems making complex digital systems relatively easy to design by snapping the subsystems together. Many new techniques and powerful computer languages are constantly being developed to bring to life these digital systems. Most often these digital systems monitor and control something in our analog world. Therefore, as digital systems become more advanced, it becomes possible to control more advanced analog systems. This creates the need to better understand analog signals and to better control them through filtering. This text is intended to fulfill that need by showing how to mathematically describe analog signals and filter them.

TEXTBOOK ORGANIZATION

This text includes material on the basic equations for complex analog waveforms, Laplace transforms, Laplace circuit analysis, transfer functions for analog circuits, pole-zero plots, frequency response of analog circuits, filter specifications, frequency response characteristics of op-amps, and the design of Butterworth, Chebyshev, and elliptic active filters. All the examples are worked in detail showing step-by-step how to apply the techniques discussed in the text. In an effort to use current computing technologies, new sections on how to work problems using MATLAB and the TI-89 hand calculator are added.

Chapter 1—Introduction gives a brief overview of the text. It defines the function notation used in Laplace, defines the accuracy technique used for the solutions of the examples and problems, and lists the basics for how to study this subject successfully.

Chapter 2—Analog Signal Analysis is an introduction to time domain analog signals. This chapter shows how to draw the graph of a signal given the equation, and how to write the equation of a signal from an oscilloscope display.

Chapter 3—Laplace Transform covers the Laplace transforms. The Laplace transforms are presented without discussing how they apply to either electronic or mechanical systems so that this chapter could be used as a reference in any area of study using Laplace transforms.

Chapter 4—Circuit Analysis Using Laplace Transforms and **Chapter 5—Sinusoidal Steady State** cover the application of Laplace transforms in electronics. These chapters go through basic circuit analysis, superposition, Mesh current, Node voltage, dependent source circuits, Thévenin, Norton, transfer functions, pole-zero plots, and Bode plots.

Chapter 6—Introduction to Filters, Chapter 7—Normalized Low-Pass Filter and **Chapter 8—Practical Filters from the Generic Loss Functions** cover the topic of active filter design. These chapters begin by defining low-pass, high-pass, band-pass, and notch filter characteristics. Then a discussion is given covering how to model a modern op-amp along with how to determine its limits in a filter design. Following this, the design of a normalized low-pass active filter using the Butterworth, Chebyshev, and elliptic approximations are given. Next, methods are presented that show how to change a normalized filter into a filter at a desired frequency having realistic component values. Finally, mathematical methods of how to convert a low-pass function into a high-pass, band-pass, and notch filter are shown.

GENERAL FEATURES OF THE TEXT

Laplace transforms are covered from a systems approach by keeping the in-depth mathematics in the appendices. By putting derivations in the appendices, the text avoids long dissertations on the mathematics, and gives more detail on the system used to work the examples. Thus more space is devoted for showing step-by-step solutions in the examples which is one of the text's strong points. This makes it possible to easily learn the system, and at the same time, by using the appendices, allows an in-depth study of topics when desired. This arrangement makes this text very useful for self-study or for a reference source.

Techniques and methods used in the text have been chosen so that they apply to a wide variety of problems. Instead of showing several ways to work a special case problem with techniques that only apply to the special case problem, techniques are employed that apply equally well to both special case problems and general case problems. This makes for better retention of the material with a clearer idea of how to solve a wider variety of problems.

The parallel learning curve method introduced in the first edition has been expanded in this edition. Briefly, this is a method that uses a technique from a known or previously covered topic and applies it to new material. A couple of examples of the parallel learning curve will help explain this teaching/learning method.

One example of the parallel learning curve is how the structure for finding the Laplace transform when several transform operations are involved is presented so that it parallels the structure used for writing computer code. When writing computer code, indentations are used to separate out cohesive sections such as an if-then statement or nested if-then statements. These indentations are applied similarly in this text to separate the different Laplace operations being used. Students having experience with indentations to separate sections of software code, will immediately be able to understand the structure of how to separate each Laplace operation as it is being used. The parallel transfer of knowledge from writing computer code to finding Laplace transforms requiring multiple operations is quickly transferred to the student for a more complete comprehension of this topic.

Another example of the parallel learning curve is in the presentation of active filters. Traditionally, the Butterworth filter and Chebyshev filter are presented using different design approaches. In this text both design approaches are identical. In this way, once the Butterworth filter is learned, the Chebyshev filter can be learned in a very short time because the only difference between the two filters is the coefficient table.

NEW FEATURES IN THE SECOND EDITION

The three main features added to this edition are MATLAB™, the TI-89, and the compartmentalization of the topics.

MATLAB and the TI-89 appear in separate sections in order to allow flexibility in the presentation of the material in a classroom situation. These tools provide a modern approach to solve complex signal processing problems and to help analyze results. The material can be referenced when the instructor feels it is appropriate, or the instructor can reference their own particular computer or calculator. In support of using other calculating devices, the MATLAB sections and the TI-89 sections use identical examples. This is done in order to demonstrate that when the subject matter is understood, any current calculating device can be effectively applied to this material.

MATLAB is a registered trademark of The MathWorks, Inc.

For MATLAB product information, please contact:

The MathWorks, Inc.
3 Apple Hill Drive
Natick, MA 01760-2098 USA
Tel: 508-647-7000
Fax: 508-647-7101
E-mail: info@mathworks.com
Web: www.mathworks.com

Most of the chapters and sections have been compartmentalized as much as possible to make this text flexible and at the same time have a flow from beginning to end. In this way, the depth, the amount of coverage, and, in some cases, the order of coverage can be adjusted to fit the course, class, and amount of time available. In any course, there are topics that must be covered and topics that are nice to be able to cover if time allows. Often a text will contain topics that are integrated such that sections unnecessary to a course's goals must be covered in order to understand the topics that are in the curriculum. An attempt has been made to remedy this problem by making each topic a separate compartment that can be moved around and/or bypassed as needed. In order to effectively use the compartmentalization, a map of the many different roads through this text is given in Appendix G.

Supplements:

Instructor's Guide. An Instructor's Guide contains solutions to all end-of-chapter questions and problems. ISBN: 0-7668-2819-0

Online Companion. Visit the textbook's own companion website at www.electronictech.com, featuring matlab files and text updates.

ACKNOWLEDGMENTS

I would like to thank my wife, Joan, for the many hours of working on this text with me and the display of much appreciated patience. Thank you to my Mom and Dad, Don and Donna Meador, for raising me in a loving family. I would also like to thank Greg Burnell for encouraging me to write the first edition, Greg Clayton for making this second edition possible by his persistence and encouragement, and Michelle Ruelos Cannistraci for her guidance in developing the manuscript. Thanks to David Arsenault for working with me on the artwork and for the front cover. Thanks to Pam Rockwell for taking the extra time in editing the equations. Thanks to Silvia Freeburg for making sure that everything was correctly assembled for the printing of this textbook.

Thanks to all the faculty and staff at DeVry Institute of Technology in Kansas City for the knowing and unknowing help and support. I would also like to thank the following reviewers for their work on this text:

Don Abernathy—DeVry Institute of Technology, Irving, TX

Sang Lee—DeVry Institute of Technology, Addison, IL

Jim Pannell—DeVry Institute of Technology, Irving, TX

Carlo Sapijaszko—DeVry Institute of Technology, Orlando, FL

Parker Sproul—DeVry Institute of Technology, Phoenix, AZ

Introduction

OBJECTIVES

Upon successful completion of this chapter, you should be able to:

- Recognize the function notation difference between a time domain function and an *s*-domain function.
- State why full computer/calculator accuracy is desirable.
- State the basic rules that need to be employed to study successfully.

1-1 WHAT'S IN THIS BOOK

This book covers the mathematical tools required to work electronic circuits having complex waveforms and reactive components. At the center of these mathematical tools are Laplace transforms. Using Laplace transforms, we are able to analyze a circuit having complex waveforms at multiple frequencies for the complete response (from $t = 0$ to $t = \infty$). In addition, these tools apply to steady-state responses of single-frequency sinusoidal circuits.

The first half of the book is devoted to learning these mathematical tools, and the second half is devoted to applying them to active filters. In Chapter 2 we will learn how to describe complex waveforms. These complex waveforms will not be emphasized in later chapters so that the mathematical tools can be clearly seen. While we are learning Laplace transforms, we will be using impractical circuits simply to learn the techniques—playing games with electronic circuits. Then in the second half, we will apply these techniques to a practical application—active filter design.

1-2 LAPLACE TRANSFORMS

The Laplace transform is a technique that transforms a differential equation from the time domain (equations expressed as a function of time) to the *s*-domain (equations expressed as a function of a complex variable *s*). This makes the process of solving the differential equation an algebraic process. Solving a differential equation is somewhat difficult, and in electronics, simultaneous differential equations are common. When the simultaneous differential equations are converted into simultaneous algebraic equations, the process of solving for the unknown quantities is much simpler.

In the beginning, for electronics, it is not important to understand exactly what Laplace transforms are. We are more interested in how to use the tool than how the tool was "developed and manufactured." However, when Laplace transforms are introduced in the following chapters, we begin with how the Laplace transforms are derived, but we should not be overly concerned with this. Keep in mind that the main goal is to learn how to use Laplace transforms. Understanding Laplace transforms will be much easier when you know how they are used.

1-3 NOTATIONS

In this book we must be very careful to recognize function notation. We will be concerned primarily with two types of functions, $f(t)$ and $F(s)$. These are read "f of t" and "F of s." They are not "f times t" or "f times s," but represent a function of time and a function of s just as $\sin(t)$ is a function of t and $\log(x)$ is a function of x. These two functions are the same except that one is in the time domain and the other is in the s-domain (or Laplace domain). The functions "$f(t)$" and "$F(s)$" and "$h(t)$" and "$H(s)$" are generic names and do not refer to a specific function. The functions "$e(t)$" and "$E(s)$" usually refers to a source voltage, and the functions "$v(t)$" and "$V(s)$" usually refer to a voltage drop across a component.

The typical electronic notations are used in this book. These notations are typical of most textbooks, but some may seem unusual, depending on the texts used for basic circuit analysis. You should use the notations presented in this book while learning the technique to avoid confusion and the use of "translation sheets."

1-4 CALCULATIONS

An important aspect of this book is the way the examples are calculated. Full computer/ calculator accuracy is used to calculate the answers, then this value is rounded to five significant digits. If an intermediate result is shown, subsequent calculation will be based on the full computer/calculator number rather than the five significant digits shown. For calculations based on tabulated values, only the values in the tables will be used, but after that point full computer/calculator accuracy will be maintained.

You should develop the habit of carrying the full-digit accuracy of the computer/calculator to prevent becoming obsolete in accuracy. As time progresses, the accuracy of manufactured components increases and the need for carrying more significant digits in calculations increases. If you learn to carry the full accuracy of the computer/calculator, your calculations will never become obsolete.

1-5 HOW TO STUDY

The purpose of this book is learning. Therefore, this section is the most important in the book. If you fail here, you will learn nothing in the following chapters. The rules are simple to follow but must be followed consistently. Learning Laplace transforms enough to work the problems in this book means very little. The main idea is to know Laplace transforms as well as you know Ohm's law. Then a door to many new worlds will be opened.

RULE 1: Study in short, frequent intervals of time.

Laplace transforms are easy to watch and understand when someone else is working the problems, but difficult when you wait too long to start working problems. You must work problems as soon as you are exposed to any new idea, no matter how small. More can be learned in one hour spread over several days than in two hours of unbroken time. The longer hours at a stretch will only make you feel noble—you won't retain much.

RULE 2: Always write the equation being used in its original form with variables instead of numbers.

There are always equations that must be memorized. The easiest way is to write the equation in its original form first (without copying from the text or notes). Second, use algebra to solve for the variable required. Last, substitute in the numbers. The second and last step may be reversed, depending on personal preference. Many of the equations will be similar, and this procedure will help to keep you from combining different equations.

RULE 3: Work problems without the text or notes.

When you have to use references to work a problem, the only thing you learn is that you do not know how to work the problem. You must work the problems without the text. When you are not sure what to do, try anything, but try something. (A good thing to try is to make a list of known and unknown variables and values.) When you have an answer, then and only then should you use your notes and the text to determine whether you did something wrong.

RULE 4: Keep in mind that the purpose of working a problem is not just to get the correct answer, but it is to practice a technique so that you can successfully find the answers for more complex problems.

When working a problem, very little will be learned by looking up the answer first and trying different combinations of the given numbers until the answer is found. The best way is to look at the techniques and concepts discussed in the text, apply them to the problem, and then look up the answer. In this way, you will develop an in-depth understanding that will allow you to extend your knowledge beyond the text.

1-6 USING MATLAB AND THE TI-89

Computer software packages and hand calculator/computers have greatly facilitated our ability to solve more complex problems with greater speed and accuracy. This is especially evident when pursuing a mathematically intensive career. With either of these computational devices, we can modify a problem's parameters, allow the device to calculate the answer, and use the time saved from hand calculating to ponder the results. However, we must remember that a genuine insight into a problem comes from knowing why and how much these parameters affect the results. Otherwise, we are doomed to randomly modifying parameters hoping to stumble onto a desired result. When the mechanics of the parameters are understood through hand calculating simple problems, a greater realization of how to control the parameters of a more complex problem will be gained. This does not mean that using a computational device will circumvent the learning of how parameters affect the result of a problem, but it does mean that the underlying concepts must be understood to gain the full benefits of using any computational device. In other words, using a computational device does not necessarily make you understand a problem. You must understand the fundamental concepts of the problem before you will be able to reap the tremendous benefits available through the use of computer software packages or hand calculator/computers.

Figure 1-1

In this text, MATLAB (®) from The MathWorks, Inc.[1] will be used as an example of a computer application software package, and the TI-89 shown in Fig. 1-1 from Texas Instruments, Inc. will be used as an example of a hand calculator/computer device. MATLAB was chosen because of its wide acceptance and usage. The TI-89 was chosen because it is easy to use in a classroom situation, is programmable, and is capable of symbolic manipulation. There are many other computer software packages and many different hand calculator/computers that are suitable for use in conjunction with this text. Because of this, all references to MATLAB and the TI-89 are restricted to sections at the ends of the chapters. In this way, any other software package or hand calculator/computer may be easily integrated into this text.

There are too many direct applications of MATLAB and the TI-89 to be practically presented. Therefore, the examples chosen are intended to either demonstrate a powerful use that can be employed or inspire the pursuit of more advanced applications.

Both the MATLAB and the TI-89 sections are presented with the assumption that the reader has the software package or hand calculator/computer and its instruction manual. The solutions are approached from a "how to get the answer" point of view so that MATLAB or the TI-89 can be used with minimal knowledge. However, brief explanations are

---

[1]MATLAB is a registered trademark of The MathWorks, Inc.

given so that a more thorough justification can be pursued in the reference manuals. In this way, either tool may be quickly applied to similar problems.

In most cases, these chapter-end sections will use the same examples to show how either MATLAB or the TI-89 can solve similar problems. Of course, since MATLAB is computer-based, its graphic displays will look much better than the TI-89's dot matrix display. However, the TI-89 is much more portable and far easier to use on a classroom desk.

MATLAB and the TI-89 are very powerful tools that will remove the drudgery of hand calculating and pry open the door of knowledge to gain a greater insight into the materials presented in this text than has ever been possible before.

2

Analog Signal Analysis

OBJECTIVES

Upon successful completion of this chapter, you should be able to:

- Write the time domain mathematical expression and draw the graph of an impulse, step, ramp, constant-amplitude sinusoidal, and exponential-amplitude sinusoidal function.
- Define the derivative and integral relationship between the impulse, step, ramp, and t^n functions.
- Calculate and describe the relationship between the damping constant, α, and the time constant, τ, in an exponential function.
- Calculate the exact location of the maximums and minimums of the exponential-amplitude sinusoidal function.
- Time shift a function to any desired time.
- Write the time domain equation from a display of an analog signal containing impulses, steps, and ramps.
- Write the time domain equation from a display of an exponential-amplitude sinusoidal function with a dc component given three significant points on the graph.
- Express the equation of a function containing repeating terms in a summation form.
- Solve for the derivative or the integral of a complex time domain waveform.
- Express and solve the derivative form or the integral form of the equations that describe the relationship between voltage and current of a capacitor and an inductor.
- Multiply and simplify time domain equations containing shifted step function factors.
- Calculate the average, effective, and rms value of a periodic waveform.

2-0 INTRODUCTION

Not only does the Laplace transform provide a method to analyze complex circuits but it also provides a method to analyze circuits excited by complex waveforms. A method is therefore required to describe complex waveforms. In this chapter, we define basic functions used to describe complex waveforms mathematically.

2-1 STEP FUNCTION

Ideally, a switch will supply a voltage or current to a circuit instantaneously. This instantaneous transition is shown in Fig. 2-1. The mathematical function we will use is called the *unit step function* and is denoted by

$$u(t) \tag{2-1}$$

Multiplying the unit step by a constant is used to stimulate a dc source of any value being switched on. The general form of a function with magnitude K being switched on at $t = 0$ is

$$f(t) = K\,u(t) \tag{2-2}$$

An examination of a simple function and how it contrasts with the $u(t)$ function will be helpful for a full understanding of this function. We will use the function

$$g(t) = 3t + 4 \tag{2-3}$$

for our discussion. The part of $g(t)$ enclosed in parentheses is called the *argument*. For each value of t (the argument), there exists a unique solution of $g(t)$. For example, if the argument is 2, then $g(2) = 3(2) + 4 = 10$, and with an argument of -3, $g(-3) = -5$. In contrast, $u(t)$ does not have a unique value for each t.

There are only two possible values for $u(t)$, 0 or 1. Whenever the $u(t)$ argument is greater than zero, the value of $u(t)$ is 1. When the $u(t)$ argument is less than zero, the value of $u(t)$ is 0. Therefore, $u(t)$ cannot be defined by a polynomial but must be defined piecewise as follows:

$$u(t) = \begin{cases} 1 & \text{for } t > 0 \\ 0 & \text{for } t < 0 \end{cases} \tag{2-4}$$

We also need to know how to handle $u(t)$ at $t = 0$ since it is undefined at this time. Figure 2-1 shows that the function $u(t)$ has two values at $t = 0$. We can split $t = 0$ into two parts. The time at an infinitely small distance to the right of $t = 0$ is written as

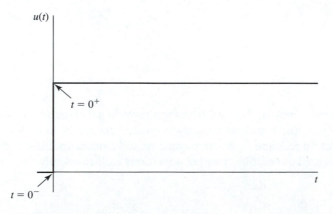

Figure 2-1

Unity step function $u(t)$. The value of $u(t)$ at $t = 0^-$ is 0, and the value of $u(t)$ at $t = 0^+$ is 1.

$t = 0^+$, and the time at an infinitely small distance to the left of $t = 0$ is written as $t = 0^-$. Therefore,

$$u(0^+) = 1$$

$$u(0^-) = 0$$

We can now define two rules for $u(t)$ at $t = 0$.

RULE 1: When we are working with the function to the right of $t = 0$, we use $u(0^+)$ for $u(t)$ at $t = 0$.

RULE 2: When we are working with the function to the left of $t = 0$, we use $u(0^-)$ for $u(t)$ at $t = 0$.

There are two important items to keep in mind about the argument of $u(t)$. First, the argument may be a function of t rather than just t. In this case, the algebraic distributive law cannot be applied since this is a function notation rather than an algebraic expression. Second, the step function's value goes through the zero-to-one transition when the value of the argument is zero. If the argument is a function of time, then the zero-to-one transition occurs when that function of time in the argument is zero.

EXAMPLE 2-1

Draw a graph of the functions $4\,u(t - 3)$ and $2\,u(5t - 25)$.

Solution: The argument of $4\,u(t - 3)$ is set equal to zero and we solve for the value of t that causes the argument to be zero.

$$t - 3 = 0$$

$$t = 3$$

(Notice that $t > 3$ causes the argument to be positive.) Therefore, we may write

$$4\,u(t - 3) = \begin{cases} 4 & \text{for } t > 3 \\ 0 & \text{for } t < 3 \end{cases}$$

The graph is shown in Fig. 2-2a.

The argument of $2\,u(5t - 25)$ is set equal to zero and t is solved.

$$5t - 25 = 0$$

$$t = 5$$

In this case, when t is greater than 5, the argument of $2\,u(5t - 25)$ is greater than zero. The result is shown in Fig. 2-2b. Note that the magnitude of the function is 2 since we cannot use the distributive law on a function notation.

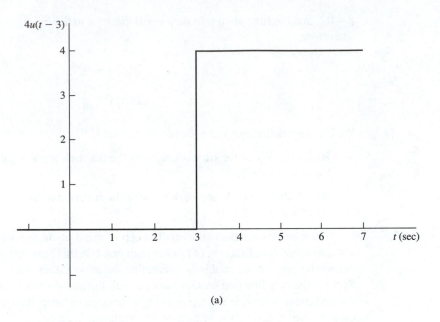

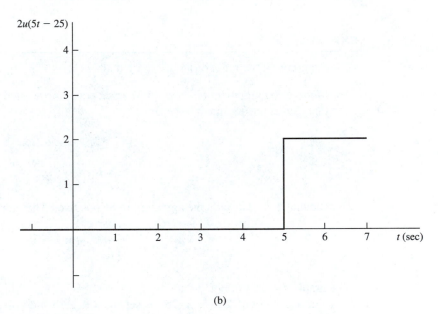

Figure 2-2

(a) Graph of $4\,u(t-3)$; (b) graph of $2\,u(5t-25)$.

The step function is sometimes referred to as the *switch* function. When we multiply the step function by another function, the product of the two functions will be zero until the argument of the step function is greater than zero. This gives us a way to represent mathematically, for instance, a sinusoidal wave "switched on" at any particular time.

Figure 2-3 shows an example of how $u(t)$ is used to switch on a sinusoidal function at $t = 0$. A continuous sinusoidal function $\sin(\omega t)$ is shown in Fig. 2-3a and a step function

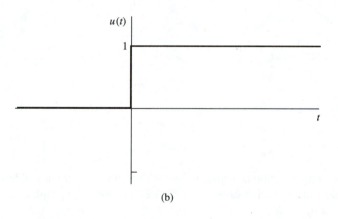

$\sin(\omega t)$

1

t

(a)

$u(t)$

1

t

(b)

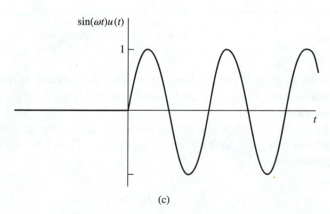

$\sin(\omega t)u(t)$

1

t

(c)

Figure 2-3

(a) Graph of sin (ωt); (b) graph of the switch function $u(t)$; (c) graph of a sinusoidal function switched on at $t = 0$.

$u(t)$ is shown in Fig. 2-3b. These functions are then multiplied together. To the left of $t = 0$, the step function's value is zero, and, therefore, the result of the multiplication is zero. To the right of $t = 0$, the step function's value is 1, and, therefore, the result of the multiplication is the value of the sinusoidal function. Figure 2-3c shows the completed graph of $g(t) = \sin(\omega t)\, u(t)$.

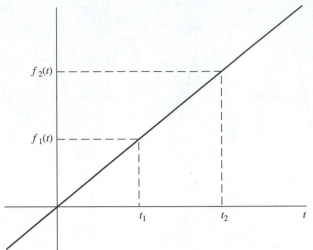

Figure 2-4

Graph of a straight line used to make a ramp function.

2-2 RAMP FUNCTION

The ramp function is a straight line with a value of zero at and before the origin increasing or decreasing with a slope of K. Figure 2-4 shows the graph of a straight line. The slope is found by

$$K = \frac{\Delta f(t)}{\Delta t} = \frac{f_2(t) - f_1(t)}{t_2 - t_1} \tag{2-5}$$

Since the slope is K and the intercept is zero, the function is written $f(t) = Kt$. If this function is voltage, the units are volts/second, and if the function is a current, the units are amperes/second.

Since the ramp function is zero before the origin, the straight-line equation is multiplied by $u(t)$ in order to form the ramp function. Therefore, the general equation for the ramp function is

$$f(t) = Kt\,u(t) \tag{2-6}$$

EXAMPLE 2-2

Draw the graph for $g(t) = 2t\,u(t)$ for $-1 < t < 3$ sec.

Solution: The result is shown in Fig. 2-5.

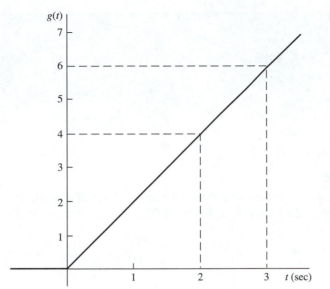

Figure 2-5

EXAMPLE 2-3

Write the equation for Fig. 2-6.

Solution: Since this is a straight line, the slope can be determined at any two points using Eq. (2-5).

$$K = \frac{f_2(t) - f_1(t)}{t_2 - t_1}$$

$$= \frac{2 - 1}{4 - 2} = \frac{1}{2}$$

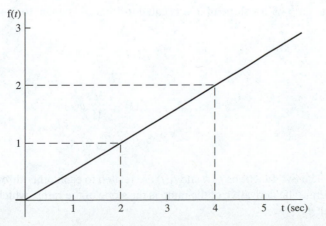

Figure 2-6

From Eq. (2-6),

$$f(t) = Kt\, u(t)$$

$$= 0.5t\, u(t)$$

2-3 t^n FUNCTION

The step function and the ramp function are part of a family of functions. The family is related by the derivative and integral. The general form for this family is

$$f_n(t) = \frac{At^n}{n!}\, u(t) \tag{2-7}$$

where n = the degree in the family
$\quad\quad A$ = a multiplying constant

(The subscripted function notation on $f_n(t)$ is not really necessary, but it helps emphasize the family relationship.)

For a step function with a height of 3, n is equal to 0 and A is equal to 3.

$$f_n(t) = \frac{At^n}{n!}\, u(t)$$

$$f_0(t) = \frac{3t^0}{0!}\, u(t)$$

Note that zero factorial is by definition equal to 1.

$$= 3\, u(t)$$

For a ramp with a slope of 3, n is equal to 1 and A is equal to 3.

$$f_n(t) = \frac{At^n}{n!}\, u(t)$$

$$f_1(t) = \frac{3t^1}{1!}\, u(t)$$

$$= 3t\, u(t)$$

The two functions $f_0(t)$ and $f_1(t)$ are related to each other through the derivative and the integral. The derivative of the ramp is the slope of the ramp, which is a constant value starting at $t = 0$.

$$f_0(t) = \frac{d}{dt} f_1(t)$$

$$= \frac{d}{dt} At\, u(t)$$

$$= A\, u(t)$$

The integral of the step must therefore be the ramp function. This is an important point that we will use later.

2-4 IMPULSE FUNCTION

In Section 2-3 we did not take the derivative of the step function. This was because $u(t)$ has a discontinuity at $t = 0$. The derivative of $u(t)$ is the impulse function and is denoted by

$$\delta(t) \tag{2-8}$$

where δ is the Greek lowercase letter delta. The general form with a multiplying constant is

$$K\delta(t) \tag{2-9}$$

This impulse function has zero width, infinite height, and a finite area. In order to understand this function, we will start with an imperfect step function and find the derivative.

In the next few paragraphs, we will work with the derivative in an electronic context. An electronic differentiator's output may not actually be the derivative of the input signal. For example, the input signal to the differentiator circuit may be a current and the output a voltage proportional to the derivative of the current. In this case, the units of the differentiator's output will not be the same units as the mathematical derivative. We read the output voltage of the differentiator and supply the correct units if the mathematical units are required.

Figure 2-7 shows an imperfect step function and its derivative. The derivative has a value only when the original function has a nonzero slope. In this case, the only time the

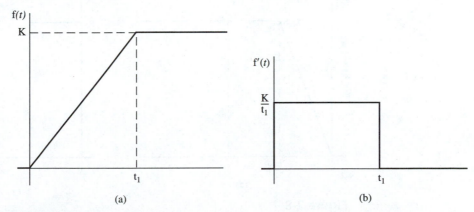

Figure 2-7

(a) Imperfect step function; (b) derivative of the imperfect step function.

derivative has a value is between $t = 0$ and $t = t_1$ seconds. The derivative, shown in Fig. 2-7b, has a value in this time interval equal to the slope of the ramp and is therefore K/t_1. The area under the derivative curve in Fig. 2-7b is

$$\text{area} = t_1 \left(\frac{K}{t_1} \right)$$

$$= K$$

Figure 2-8a is closer to being a perfect step function. The derivative in Fig. 2-8b has a larger amplitude, but the area under Fig. 2-8b is the same area that we found in Fig. 2-7b.

$$= t_2 \left(\frac{K}{t_2} \right)$$

$$= K$$

As we keep decreasing the amount of time allowed for the ramp function, the derivative will increase in height, but the area will remain constant.

$$\text{area} = \lim_{t \to 0} \frac{K}{t} t = \lim_{t \to 0} K = K$$

$$\text{height} = \lim_{t \to 0} \frac{K}{t} = \infty$$

Thus we have defined the impulse function of area K. The impulse function has infinite height, zero width, and finite area.

In electronic circuits it is unlikely that we will have a voltage or current of infinite height. It is also unlikely that we will be able to have a perfect step function. In reality, these two functions will be an imperfect step and a narrow pulse with a large amplitude, but to simplify the mathematics we will approximate these as a perfect step and an

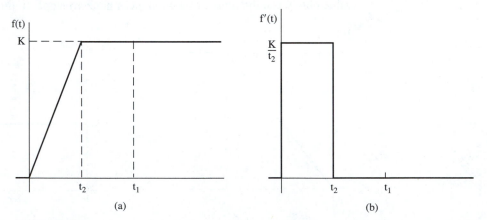

Figure 2-8

As we get closer to a perfect step function, the height of the derivative increases toward infinity, but the area under the derivative curve remains constant.

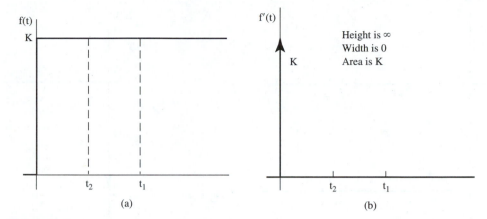

Figure 2-9

(a) Perfect step function; (b) impluse function that is the derivative of a step function.

impulse. ("Narrow" in this context means narrow when compared to the shortest time constant of the circuit.) The area of the impulse will equal the area of the narrow pulse. The impulse may be assigned any time within the limits of the original pulse since to make this approximation, the pulse must be narrow.

Figure 2-9 shows the step function, $K\,u(t)$, and its derivative, the impulse function. In equation form,

$$f(t) = K\,u(t) \tag{2-10a}$$

$$f'(t) = \frac{d}{dt}\,f(t) = K\,\delta(t) \tag{2-10b}$$

Since the height of the impulse is infinity, an arrow is used. We may draw the height of the arrow to correspond to the area, but we must never forget that the height of the impulse is infinity. Another way to show the impulse function is to draw all impulses the same height and write the value of the area next to the impulse symbol.

EXAMPLE 2-4

Graph the derivative of the function shown in Fig. 2-10.

Solution: From the graph,

$$g(t) = -2\,u(t)$$

Therefore,

$$g'(t) = -2\,\delta(t)$$

The graph of the derivative is shown in Fig. 2-11.

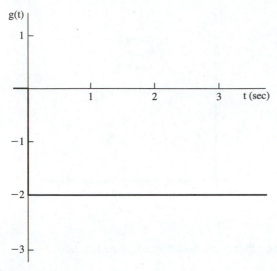

Figure 2-10

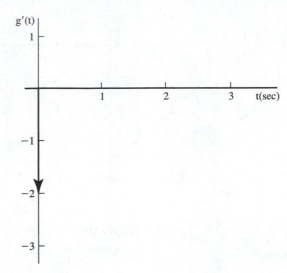

Figure 2-11

2-5 EXPONENTIAL FUNCTION

There are two forms of the exponential:

$$e^{-\alpha t} \qquad\qquad\qquad \textbf{(2-11a)}$$

$$e^{-t/\tau} \qquad\qquad\qquad \textbf{(2-11b)}$$

where $e = 2.7183\ldots$
  $\alpha =$ the damping constant
  $\tau =$ the time constant

The relationship between α and τ is found by setting the two exponentials equal to each other.

$$e^{-\alpha t} = e^{-t/\tau} \qquad\qquad\qquad \textbf{(2-12a)}$$

$$\ln(e^{-\alpha t}) = \ln(e^{-t/\tau})$$

$$\alpha = \frac{1}{\tau} \qquad\qquad\qquad \textbf{(2-12b)}$$

Figure 2-12 shows a graph of the exponential function. We will always assume that at five time constants the value of the function is 0. This is a universally accepted assumption. Five time constants is the value of time that causes the exponential function to equal e^{-5} (the exact value is $e^{-5} = 0.0067379\ldots$), and one time constant is the value of time that causes the exponential function to equal $e^{-1} (= 0.36788\ldots)$

Notice the small glitch in the drawing of Fig. 2-12 at five time constants. This is where we assume the value of the function goes to zero. On an oscilloscope display of a real

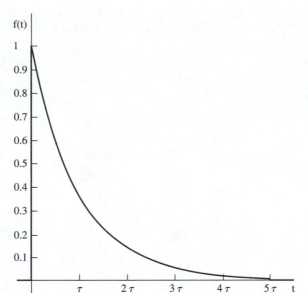

Figure 2-12

Exponential function. After five time constants the value is always assumed to be zero.

signal, the value at this point will not be zero but will continue to approach zero. The drawing has this glitch to indicate the difference between our assumption and reality.

EXAMPLE 2-5

For the function given, find (a) the amount of time for two time constants, (b) the time when the function goes to zero, and (c) α and τ. (d) Graph the function.

$$g(t) = 14e^{-20t}\, u(t)$$

Solution:

(a)   For two time constants the exponential will be e^{-2}.

$$e^{-20t} = e^{-2}$$

$$\ln(e^{-20t}) = \ln(e^{-2})$$

$$-20t = -2$$

$$t = 0.1 \text{ sec}$$

(b)   The function will be assumed to be zero at five time constants; therefore,

$$e^{-20t} = e^{-5}$$

$$t = 0.25 \text{ sec}$$

(c) To find α, we set the function equal to the general form of the exponential.

$$e^{-20t} = e^{-\alpha t}$$

$$-20t = -\alpha t$$

$$\alpha = 20$$

To find τ, we could set the function equal to the general form that uses τ, or we could use Eq. (2-12b).

$$\alpha = \frac{1}{\tau}$$

$$\tau = \frac{1}{\alpha} = \frac{1}{20} = 0.05$$

(d) The graph is shown in Fig. 2-13. Since $u(t)$ is 0 for $t < 0$, the function will be zero for $t < 0$. Since e^{-20t} is 0 for $t > 0.25$ sec, the function will be 0 for $t > 0.25$ sec. We can find the value of the function at any time by substituting in the value of t desired.

Figure 2-13

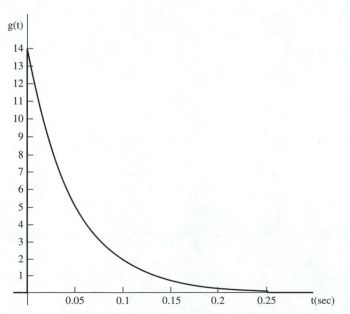

Often, we will need to write the equation for an exponential function from an oscilloscope display. To write the equation, we need to determine the multiplying constant and the damping constant α. The multiplying constant can be found by observing the magnitude of the signal at $t = 0$, and the damping constant can be found if one other point on the curve is known. The following example demonstrates this.

EXAMPLE 2-6

Determine the equation for the exponential function shown in Fig. 2-14.

Solution: The value of K is found at $t = 0$

$$v(t) = Ke^{-\alpha t}u(t)$$

$$8 = Ke^{0+}u(0^+)$$

$$K = 8$$

The function can then be written as

$$v(t) = 8e^{-\alpha t}$$

Finally, α is determined using the point at $t = 5$ msec.

$$v(t) = 8e^{-\alpha t}$$

$$2.2920 = 8e^{-\alpha\,0.005}$$

$$\frac{2.2920}{8} = e^{-\alpha\,0.005}$$

$$\ln\left(\frac{2.2920}{8}\right) = (-\alpha\,0.005)\ln(e)$$

$$\frac{\ln\left(\frac{2.2920}{8}\right)}{0.0005} = -\alpha$$

$$\alpha = 250$$

The equation from the graph can then be expressed as

$$v(t) = 8e^{-250t}u(t)$$

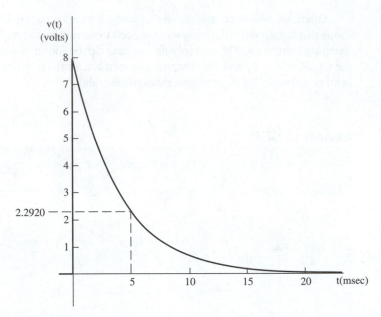

Figure 2-14

2-6 SINUSOIDAL FUNCTION

2-6-1 Constant-Amplitude Sinusoidal Function

The general form of the sinusoidal function is

$$A \sin(\omega t + \theta) \qquad (2\text{-}13)$$

where A = the peak amplitude
ω = the angular frequency, in radians per second
θ = the phase angle, in degrees or radians

The angular frequency, ω, is sometimes expressed in terms of hertz.

$$\omega = 2\pi f \qquad (2\text{-}14)$$

where f is the frequency in hertz (Hz).

The phase angle, θ, is usually expressed in degrees when phasors are being used, but ωt is always expressed in radians. When evaluating the sine function at a specific time with θ in degrees, we must first convert θ to radians before we can add it to ωt. We may therefore find it more convenient to express θ in radians. The relation between radians and degrees is

$$\pi \text{ radians} = 180° \qquad (2\text{-}15)$$

EXAMPLE 2-7

Put the following function (a) in a form suitable for evaluation on a calculator using radians and (b) in a form using degrees. (c) Evaluate the function at $t = 0.5$ sec.

$$g(t) = 10 \sin(5t + 35°)$$

Solution:

(a) The $5t$ is in radians; therefore, convert the 35°:

$$35°\left(\frac{\pi \text{ rad}}{180°}\right) = \frac{7\pi}{36} \text{ rad}$$

The function can then be expressed as

$$g(t) = 10 \sin\left(5t + \frac{7\pi}{36}\right)$$

(b) The 35° part is in degrees; therefore, convert $5t$.

$$(5t \text{ rad})\left(\frac{180°}{\pi \text{ rad}}\right) = \frac{900}{\pi} t°$$

$$g(t) = 10 \sin\left(\frac{900}{\pi} t° + 35°\right)$$

(c) In both cases, the value of $g(t)$ at $t = 0.5$ sec will be the same. Being sure that the calculator is in the correct mode (radians or degrees), we should get

$$g(0.5) = 10 \sin\left[5(0.5) + \frac{7\pi}{36}\right]$$

or

$$g(0.5) = 10 \sin\left[\frac{900}{\pi}(0.5)° + 35°\right]$$

In both cases,

$$g(0.5) = 0.30723$$

There are two methods of evaluating a sinusoidal function. The first way is to choose a value of time and solve for the value of the function at that time. This method was used in Example 2-7. The second method is to choose a known value of the sinusoidal function and solve for the time when this value will occur. In the second method, we choose a location on the sinusoidal function where we know the function's value. Figure 2-15 shows the most

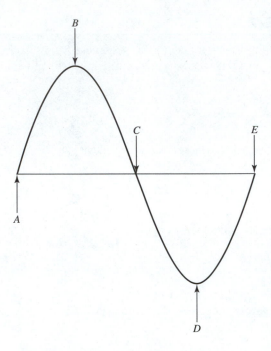

Figure 2-15

Points used to construct Table 2-1.

frequently used locations of $\sin(\omega t + \theta)$. At each location we know what $\omega t + \theta$ is equal to, as shown in Table 2-1. We can now combine these two methods to draw the sinusoidal function. This is best shown by example.

EXAMPLE 2-8

Find (a) the angular frequency, (b) the frequency in hertz, and (c) the period of the waveform. (d) Sketch the graph.

$$v(t) = 5 \sin(4t - 45°)$$

Solution:

(a) The angular frequency is

$$\omega = 4 \text{ rad/sec}$$

Table 2-1

Location	Equation (rad)
A	$\omega t + \theta = 0$
B	$\omega t + \theta = \pi/2$
C	$\omega t + \theta = \pi$
D	$\omega t + \theta = 3\pi/2$
E	$\omega t + \theta = 2\pi$

(b) The frequency is found from the relationship

$$\omega = 2\pi f$$

Solving for f yields

$$f = \frac{\omega}{2\pi} = \frac{4}{2\pi}$$

$$= 0.63662 \text{ Hz}$$

(c) The period is the reciprocal of the frequency and therefore

$$T = \frac{1}{f}$$

$$= \frac{1}{0.63662}$$

$$= 1.5708 \text{ sec}$$

(d) To sketch the graph, first convert the 45° to radians.

$$(45°)\left(\frac{\pi}{180°}\right) = \frac{\pi}{4} \text{ rad}$$

Therefore,

$$v(t) = 5 \sin\left(4t - \frac{\pi}{4}\right)$$

Next we find the time of the peaks and x-axis crossings. Using the second method and Table 2-1 gives

$$4t - \frac{\pi}{4} = 0 \qquad \Rightarrow t = 0.19635 \text{ sec}$$

$$4t - \frac{\pi}{4} = \frac{\pi}{2} \qquad \Rightarrow t = 0.58905 \text{ sec}$$

$$4t - \frac{\pi}{4} = \pi \qquad \Rightarrow t = 0.98175 \text{ sec}$$

$$4t - \frac{\pi}{4} = \frac{3\pi}{2} \qquad \Rightarrow t = 1.3744 \text{ sec}$$

$$4t - \frac{\pi}{4} = 2\pi \qquad \Rightarrow t = 1.7671 \text{ sec}$$

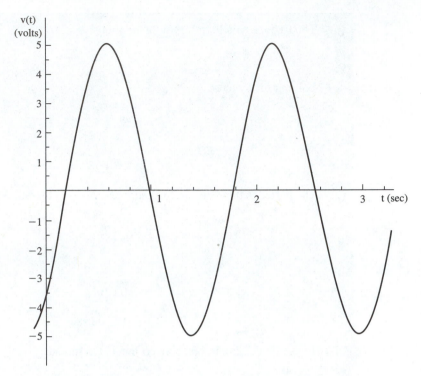

Figure 2-16

Using the first method, we can find the value of the sinusoidal function at $t = 0$.

$$v(0) = 5 \sin\left[4(0) - \frac{\pi}{4}\right] = -3.5355$$

The function is graphed in Fig. 2-16.

2-6-2 Exponential Amplitude Sinusoidal Function

The amplitude of the sinusoidal function does not have to be a constant. The amplitude may be a time-dependent function. The most important time-dependent function used for the sinusoidal amplitude is the exponential function. The general form is

$$f(t) = Ae^{-\alpha t} \sin(\omega t + \theta) \tag{2-16}$$

Figure 2-17 shows the general shape of the function. The time-dependent amplitude is $Ae^{-\alpha t}$ and is shown by the dashed line. The sine function is therefore confined within the limits of $+Ae^{-\alpha t}$ and $-Ae^{-\alpha t}$. The amplitude and consequently the function will be zero when the exponential function reaches five time constants ($e^{-\alpha t} = e^{-5}$).

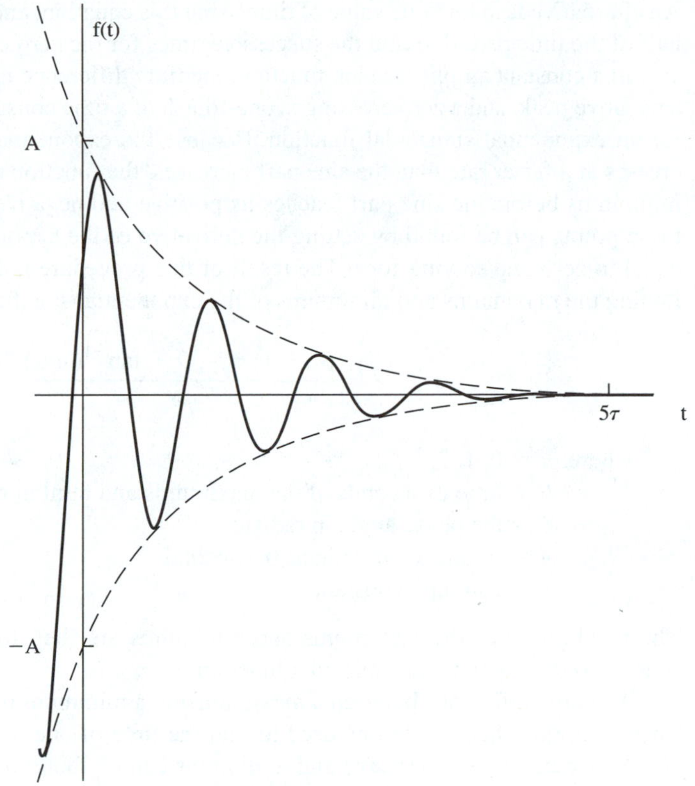

Figure 2-17

Graph of the general function $Ae^{-\alpha t}\sin(\omega t + \theta)$.

When drawing the exponential sinusoidal function or determining the equation from a graph, it is important to know where the peaks and zero crossings occur. The zero crossings are caused by the sine function becoming zero, which happens when the argument of the sine is an integer multiple of π. Therefore, we simply set the argument of the sine function equal to integer multiples of π and solve for the time that they occur.

$$(\omega t + \theta) = (n - 1)\pi$$

$$t = \frac{(n - 1)\pi - \theta}{\omega} \tag{2-17}$$

where $n = 0, 1, 2, \ldots$
t = time in seconds of the zero crossings
θ = the phase angle in radians
ω = frequency in radians per second

In this equation, the value of n could be any negative or positive integer, but since we are mainly interested in the function for positive times, we will only be interested in values of n that give us positive values of time.

This equation works well for programing a computer, but when working a problem using a hand calculator, it may be tedious to substitute into this equation for each value of n.

An alternative is to find one value of time using this equation, and then add the time of one-half of the time period to find the successive times for the zero crossings.

In a constant amplitude sine function, the time difference between a positive peak, or a negative peak, and a zero crossing is one-fourth of a time constant, but this is not the case for an exponential sinusoidal function. Because the exponential part of the function decreases at a faster rate than the sine part increases, the function reaches its maximums and minimums before the sine part reaches its positive and negative peaks. The exact time of these points can be found by setting the derivative of the exponential sinusoidal function equal to zero and solving for t. The result of this procedure is the following equation for finding the maximums and minimums of the exponential sine function:

$$t = \frac{(n-1)\,\pi - \theta + \tan^{-1}(\omega/\alpha)}{\omega}$$

(2-18)

where $n = 0, 1, 2, \ldots$
  t = time in seconds of the maximums and minimums
  θ = the phase angle in radians
  ω = frequency in radians per second
  α = damping constant

The magnitudes of the maximums and minimums are then found by substituting these times into the exponential sinusoidal function.

The time difference between a maximum and a minimum is one-half of a time period. Therefore, this equation can be used to find the time of one maximum or one minimum, and then successive maximums and minimums can be found by adding the time for one-half of a time period. Keep in mind that Eq. (2-17) must still be used to find the zero crossings since they are not exactly midway between the maximums and minimums.

EXAMPLE 2-9

Find the time that the function decays to zero. Find the value of the function at $t = 0$. Make a table with one column for times of the zero crossings, the positive peaks, and negative peaks, and a second column for the corresponding values of the function at these times. Make a sketch of the waveform.

$$f(t) = 4e^{-0.5t} \sin\left(0.4\pi t + \frac{\pi}{3}\right) u(t)$$

Solution: First we find when the function goes to zero. This is controlled by the exponential part of the function. The function will go to zero when the exponential part reaches five time constants. Therefore,

$$e^{-0.5t} = e^{-5}$$

Therefore,

$$t = 10 \text{ sec}$$

Next, the value of the function at $t = 0$ is found by substituting into the given equation.

$$f(t) = 4e^{-0.5(0)}\sin\left(0.4\pi(0) + \frac{\pi}{3}\right) = 3.4641$$

To make a table we will first find the times of the zero crossings using Eq. (2-17). Then we will find the positive peaks and negative peaks using Eq. (2-18). These results will be organized in order of time in the table.

We may use Eq. (2-17) once and add one-half of a time period to find subsequent zero crossings. One time period is found as follows:

$$2\pi f = 0.4\pi$$

$$f = \frac{0.4\pi}{2\pi} = 0.2$$

$$T = \frac{1}{f} = \frac{1}{0.2} = 5 \text{ sec}$$

Therefore, one-half time period is 2.5 seconds.

We may use any value of n and add or subtract 2.5 to find the other zero crossings. We need to find all the values between $t = 0$ and $t = 10$ since this is where the function may be nonzero.

Using $n = 0$ in Eq. (2-17), we find a time

$$t = \frac{(n-1)\pi - \theta}{\omega} = \frac{(-1)\pi - (\pi/3)}{0.4\pi} = -3.3333 \text{ sec}$$

Adding 2.5 sec to this time, we can find the times for all the zero crossings between $t = 0$ and $t = 10$ sec.

$$t = -3.3333, -0.83333, 1.6667, 4.1667, 6.6667, 9.1667, 11.667$$

If the time calculated by Eq. (2-17) had been positive, we would have subtracted 2.5 seconds until we found a negative number, to make sure we had all the positive values of time for the zero crossings.

The value of times for the maximums and minimums are found in a similar manner using Eq. (2-18). Substituting in $n = 0$, we find

$$t = \frac{(n-1)\pi - \theta + \tan^{-1}(\omega/\alpha)}{\omega} = \frac{(-1)\pi - (\pi/3) + \tan^{-1}(0.4\pi/0.5)}{0.4\pi} = -2.3847$$

Adding 2.5 sec to find the other maximums and minimums, we find

$$t = -2.3847, 0.11532, 2.6153, 5.1153, 7.6153, 10.115$$

Next we build a table. The times for the zero crossings, maximums, and minimums are put in chronological order to form one column of a two-column table. The other column will be the values of the function at these times. Table 2-2 shows the results of these calculations.

Figure 2-18 shows a finished graph of this function. To sketch the graph, we first start with an outline of the exponential part drawn above and below the *x*-axis, as shown in the figure. This is the envelope of the function, and the function will always be between these limits. Next we locate the value at $t = 0$ and the positive peaks, negative peaks, and zero crossings of the sinusoidal part. For a sketch we do not need to locate the exact magnitude of the function at the peaks. We just draw a line up to the exponential envelope.

In a rough sketch of an exponential sinusoidal function, we normally make the peaks of the sine part of the function touch the exponential envelope for the minimums and maximums of the function. In actuality, the peaks do not occur here. They occur to the left of these locations.

A problem with sketching an exponential sinusoidal function arises when more than a few cycles occur before the function goes to zero. When this happens, the exact locations of peaks and zero crossings are not found. The exponential envelope is marked and lines are scribbled back and fourth between the positive and negative limits. Even though this seems very primitive, it will give a good idea of what the graph actually looks like, and the point of drawing a complex function by hand is to get an idea of what the function will like look on an oscilloscope. When all of the graph must be accurate, we use a computer or a calculator with graphing capabilities, since this would be much faster and much simpler than trying to draw a completely accurate graph by hand.

Table 2-2

t (in seconds)	$f(t) = 4e^{-0.5t} \sin\left(0.4\pi t + \dfrac{\pi}{3}\right) u(t)$
−3.3333	0 because $u(t) = 0$
−2.3847	0 because $u(t) = 0$
−0.83333	0 because $u(t) = 0$
0.11532	3.5084
1.6667	0 because the sine part is zero
2.6153	−1.0052
4.1667	0 because the sine part is zero
5.1153	0.28799
6.6667	0 because the sine part is zero
7.6153	−0.082509
9.1667	0 because the sine part is zero
$t \geq 10$	0 because $e^{-0.5t}$ is $\geq$ 5 time constants

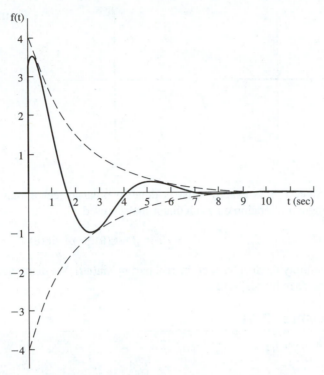

Figure 2-18

2-6-3 Splitting and Combining Sinusoidal Functions

It is often convenient to be able to split a sinusoidal function having a phase angle into the summation of a sine function without a phase angle and a cosine function without a phase angle. It is also convenient to be able to combine sinusoidal functions into a single sine function with a phase angle. A method similar to polar-to-rectangular conversion and rectangular-to-polar conversion may be used to do this process. However, instead of converting the magnitudes of the sinusoidal functions to rms values as is typically done, we will leave the magnitudes in peak values to make the math more direct. In this method, we must remember that the frequencies of the sinusoidal function must be the same and the peak magnitudes must be constants.

Figure 2-19a shows $6 \sin(\omega t)$ drawn as a phasor (using the peak value). Figure 2-19b shows $6 \sin(\omega t + 90°)$, which may also be expressed as $6 \cos(\omega t)$. With this idea we may think of the x-axis as the sine axis and measure positive angles counterclockwise from the x-axis. We may think of the y-axis as the cosine axis measuring positive angles counterclockwise from the y-axis.

To split $A \sin(\omega t + \theta)$, we project the vector lengths onto the x-axis and the y-axis as shown in Fig. 2-20. The vector length projected onto the x-axis is the magnitude of the sine part, and the vector length projected onto the y-axis is the magnitude of the cosine part. Therefore,

$$A \sin(\omega t + \theta) = [A \cos(\theta)]\sin(\omega t) + [A \sin(\theta)]\cos(\omega t) \qquad \textbf{(2-19)}$$

Figure 2-19

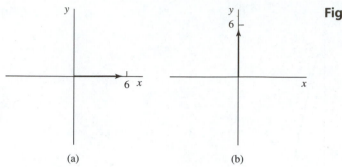

(a)   (b)

Since this is exactly what is done in polar-to-rectangular conversions, a polar-to-rectangular operation on a calculator may be used.

$$A \; \underline{/\theta} = A \cos(\theta) + jA \sin(\theta) \qquad (2\text{-}20)$$

We multiply the first term or the real part by $\sin(\omega t)$. We then remove the j and multiply the second term by $\cos(\omega t)$.

EXAMPLE 2-10

Split into sine and cosine parts.

$$5 \sin(3t + 70°)$$

Solution:

$$5 \; \underline{/70°} = 1.7101 + j4.6985$$

Therefore,

$$5 \sin(3t + 70°) = 1.7101 \sin(3t) + 4.6985 \cos(3t)$$

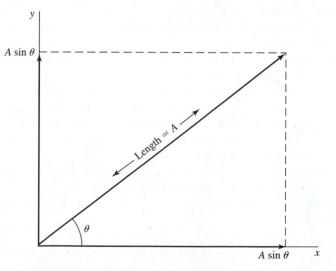

Figure 2-20

$A \sin(\omega t + \theta)$ being split into a sine and a cosine component.

EXAMPLE 2-11

Combine into a single sine function.

$$4e^{-3t}\sin(2t) - 3e^{-3t}\cos(2t)$$

Solution: If the exponentials or the frequency of both sinusoidal functions had been different, we would be unable to combine the sine and cosine functions. Since in this case the powers of the exponentials and the frequency of the sinusoidal are the same, we may combine the functions. First factor out the exponential,

$$e^{-3t}[4\sin(2t) - 3\cos(2t)]$$

and work with just the sinusoidal parts. We put the function in a polar form and convert to rectangular form.

$$4 - j3 = 5 \underline{/-36.870°}$$

This leads us to the single sine function with a phase angle.

$$5e^{-3t}\sin(2t - 36.870°)$$

2-7 SHIFTED FUNCTION

All functions do not start at $t = 0$. We need a way to shift or delay a function. In this section, we develop the shifting theorem. Figure 2-21 shows the function $f_1(t) = 2t\,u(t)$ shifted by 2 seconds.

Table 2-3 shows the value at $f_1(t) = 2t\,u(t)$ and the values from the graph shown in Fig. 2-21. Notice that the values for the graph are identical to the function $f_1(t)$ except they occur 2 seconds later. We can accomplish this by making all the t's in the $f_1(t)$ function equal to $(t - 2)$. Therefore, we can write the equation for Figure 2-21 as

$$f_2(t) = f_1(t - 2) = 2(t - 2)\,u(t - 2)$$

Notice that $u(t - 2)$ will be zero until we reach $t = 2$ sec. Substituting into $f_2(t)$, we will get the values shown in Table 2-3.

The shifting theorem can be summarized as follows: To shift a function to start at $t = a$ seconds, we replace each t with $t - a$. Therefore, the function $f_1(t)$,

$$f_1(t) = f(t)\,u(t) \tag{2-21a}$$

becomes a new function, $f_2(t)$,

$$f_2(t) = f_1(t - a) = f(t - a)\,u(t - a) \tag{2-21b}$$

If a is positive, the function will be shifted to the right of $t = 0$ and delay the function by a seconds. If a is negative, the function will be shifted to the left.

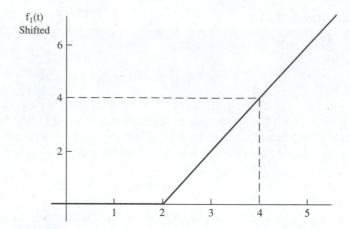

Figure 2-21

The function 2t $u(t)$ shifted to the right by 2 seconds.

EXAMPLE 2-12

Shift the following functions so that they start at 4 seconds.

(a) $t\,u(t) + 4\,u(t)$
(b) $4e^{-2t}\sin(7t + \pi/2)\,u(t)$

Solution:
(a) $(t - 4)\,u(t - 4) + 4\,u(t - 4)$
(b) $4e^{-2(t-4)}\sin[7(t - 4) + \pi/2]\,u(t - 4)$

When working with shifted functions, we must have them in a proper form. This means that all occurrences of the t in a term of the function must be identical to the form of the argument of the $u(t)$ multiplier. For example, if the step function part of a term is $u(t - 5)$, then all occurrences of t in this term must be in the form of $(t - 5)$. Part (b) of the next example demonstrates this.

Table 2-3

t	$f_1(t)$	Graph values
0	0	0
1	2	0
2	4	0
3	6	2
4	8	4
5	10	6

EXAMPLE 2-13

Shift each function back to the origin.

(a) $7(t - 3)^2\, u(t - 3)$
(b) $5\sin(2t - 10)\, u(t - 4)$

Solution:

(a)   Looking at the step function, we see that the function was delayed by 3 seconds. We want to shift the function to the opposite direction; therefore, let $t = t + 3$.

$$7[(t + 3) - 3]^2\, u[(t + 3) - 3] = 7t^2\, u(t)$$

(b)   Again we begin by looking at the step function, and we see that this function has been shifted by 4 seconds. The sine function must first be put in the proper form of a shifted function.

$$\sin(2t - 10)$$

$$\sin(2t - 8 - 2)$$

$$\sin[2(t - 4) - 2]$$

Therefore, the function in a proper form is

$$5\sin[2(t - 4) - 2]\, u(t - 4)$$

We now substitute $t = t + 4$.

$$5\sin\{2[(t + 4) - 4] - 2\}\, u[(t + 4) - 4] = 5\sin(2t - 2)\, u(t)$$

2-8 PUTTING IT ALL TOGETHER

Most complex waveforms can be expressed using the impulse function, step function, ramp function, sinusoidal function, and the shifting theorem. In this section, we use all of these concepts together to describe complex waveforms.

When dealing with complex waveforms, we either need to graph a function from a mathematical expression, or we need to write a mathematical expression from a given graph. In each case, there are some general guidelines to follow. We will look first at graphing a function from a mathematical expression.

There are two methods for graphing a mathematical expression. The first is to graph each term separately and then add the functions point by point until you know what the total function looks like. The other method is to evaluate the function at points of time where a new term in the mathematical expression becomes active. The first method is used when the function is complex or we are unsure of how to graph it. The second method is used more frequently since it is faster to use.

EXAMPLE 2-14

Graph the function using the two methods just discussed.

$$f(t) = 5\,u(t) - 7\,u(t - 2) + 3\,u(t - 3)$$

Solution: Using the first method, we graph each term as shown in Fig. 2-22a–c. We then add these at each location in time and plot the sum as shown in Fig. 2-22d. The second method is basically the same as the first method except that we are seeing things change mathematically instead of graphically. The significant times and values are shown in Table 2-4. We simply plug into the equation at the values of time shown in the table. Remember that when the argument of the step function is positive, its value is equal to 1, and when the argument of the step function is negative, its value is zero. As you become more experienced, this will become a mental process not requiring a written table.

Table 2-4

t	$5\,u(t) - 7\,u(t - 2) + 3\,u(t - 3)$	$f(t)$
$t = 0^-$	$0 - 0 + 0$	0
$t = 0^+$	$5 - 0 + 0$	5
$0 < t < 2$	$5 - 0 + 0$	5
$t = 2^-$	$5 - 0 + 0$	5
$t = 2^+$	$5 - 7 + 0$	-2
$2 < t < 3$	$5 - 7 + 0$	-2
$t = 3^-$	$5 - 7 + 0$	-2
$t = 3^+$	$5 - 7 + 3$	1
$t > 3$	$5 - 7 + 3$	1

To simplify the calculations when several terms are active at the same time, we may combine them into a single effective term (or a few effective terms). Notice in Table 2-4 that after $t = 2$ sec, we could think of the $5\,u(t) - 7\,u(t - 2)$ as a single function of $-2\,u(t - 2)$. After $t = 3$ sec, we could combine the first three terms and think of them as $1\,u(t - 3)$. This is very helpful in functions of several terms since we will in effect be working with a smaller number of terms at any given time.

EXAMPLE 2-15

Graph the function

$$v(t) = 3t\,u(t) - 2(t - 2)\,u(t - 2)$$

Solution: The first term of the function is drawn in Fig. 2-23a, and the second term is drawn in Fig. 2-23b. These two graphs will be added point-by-point to form the complete function shown in Fig. 2-23c.

In the time interval from $t = 0$ to $t = 2$, we see that the graph of $3t\,u(t)$ increases at a rate of 3 and that the graph of $-2(t - 3)\,u(t - 3)$ is zero. Therefore, in this section of time, the final graph in Fig. 2-23c shows a ramp with a slope of 3.

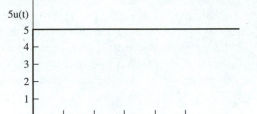

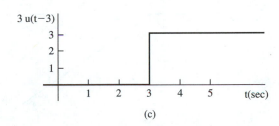

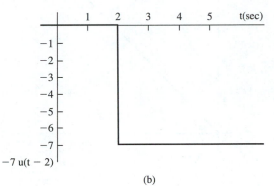

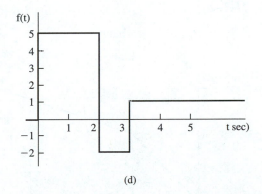

Figure 2-22

(a–c) Each term in Example 2-14 graphed separately; d—complete function.

After $t = 2$ sec, the $-2(t - 3)\,u(t - 3)$ term becomes active. In this interval of time, we see that as Fig. 2-23a increases by 3 for each unit of time, Fig. 2-23b decreases by 2. This in effect gives us a slope of 1. Notice that even though we have added a negative slope, the result of the graph is a positive slope. In general, when adding ramps the slope of the final graph will be the algebraic summation of the active slopes.

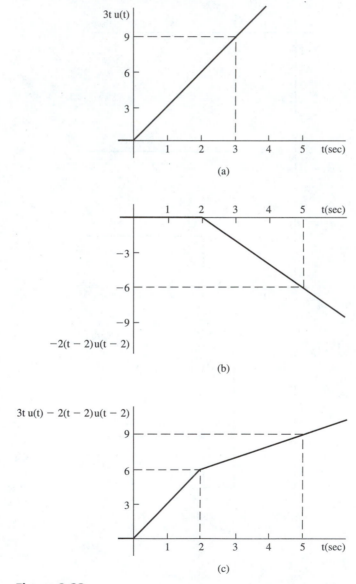

Figure 2-23

The addition of two ramps with unequal slopes.

EXAMPLE 2-16

Graph the function.

$$g(t) = 4t\,u(t) - 4(t - 2)\,u(t - 2) - 8(t - 5)\,u(t - 5) + 8(t - 6)\,u(t - 6)$$

Solution: From $t = 0$ until $t = 2$ sec the first term is the only active term. At $t = 2$ sec the second term also becomes active. Since both terms are ramps and have equal but opposite slopes, the second term counteracts further changes caused by the first term. The function therefore remains at the level it obtained at $t = 2$ sec. This can be seen by graphing the two functions separately (Fig. 2-24a and b) and adding them (Fig. 2-24c).

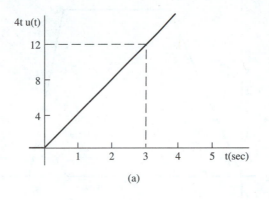

(a)

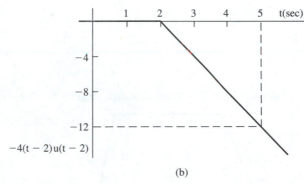

(b)

Figure 2-24

Two ramps of equal but opposite slopes cancel each other out.

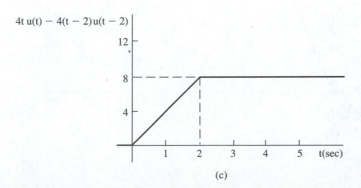

(c)

At $t = 5$ sec, the third term adds a negative ramp. Therefore, at $t = 5$ sec we get a ramp in the negative direction, as shown in Fig. 2-25a. At $t = 6$ sec, another ramp starts. It has the same magnitude slope as the ramp that starts at $t = 5$ sec, but this ramp is positive. Therefore, these last two ramps counteract each other and the function remains at 0 for $t > 6$. The final result is shown in Fig. 2-25b.

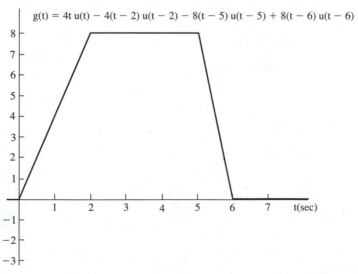

Figure 2-25

(a)

(b)

Now let's do the reverse and write the mathematical expression from the graph. To write the mathematical expression, we look at each point where the graph changes its current direction or level. At each point we try different functions until we find the one that gives us the desired result, but it is not all guesswork. We can make some general rules about common function changes that will occur.

The first common change is a step change. Figure 2-26 shows a step change occurring at $t = a$. When there are two values at one time, there will always be a step function involved. The change in the height is $f(a^+) - f(a^-)$ whether it is a positive or negative change. Therefore, the term to add to the function is

$$[\, f(a^+) - f(a^-)]\, u(t - a) \tag{2-22}$$

Another common change is the slope of a ramp. Figure 2-27 shows a function with a positive change in the slope at $t = a$. The slope of A becomes a slope of B when a term with a slope of K becomes active in the function.

$$B = A + K$$

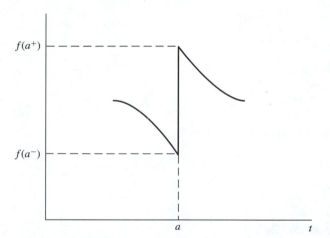

Figure 2-26

Step change in an arbitrary function occurring at $t = a$ sec.

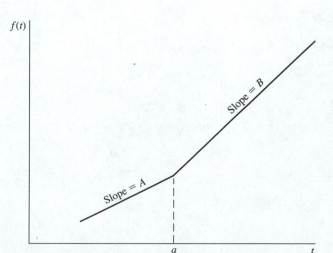

Figure 2-27

Slope change in a function occurring at $t = a$ sec.

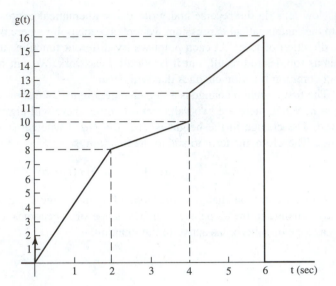

Figure 2-28

Therefore,

$$K = B - A \tag{2-23a}$$

The term to add to the function is then

$$K(t - a)\,u(t - a) \tag{2-23b}$$

EXAMPLE 2-17

Write the function for Fig. 2-28.

Solution: Starting at $t = 0$ there is an impulse with an area of 2; therefore, the first term is

$$2\,\delta(t)$$

For $t = 0$ to 2 sec there is a ramp. Using Eq. (2-5), we can find the slope.

$$K = \frac{\Delta f(t)}{\Delta t} = \frac{8 - 0}{2 - 0} = 4$$

Therefore, this term is

$$4t\,u(t)$$

For $t = 2$ to 4 sec the slope reduces. Using Eq. (2-5), the slope reduces to

$$K = \frac{\Delta f(t)}{\Delta t} = \frac{10 - 8}{4 - 2} = 1$$

From Eq. (2-23a),

$$K = B - A$$

$$= 1 - 4$$

$$= -3$$

This term becomes

$$-3(t - 2)\, u(t - 2)$$

At $t = 4$ sec we have two terms to add to our function: a step change and a ramp change. First we will find the step change using Eq. (2-22).

$$[f(a^+) - f(a^-)]\, u(t - a)$$

$$(12 - 10)\, u(t - 4)$$

$$2\, u(t - 4)$$

Also at $t = 4$ the slope changes to

$$K = \frac{\Delta f(t)}{\Delta t} = \frac{16 - 12}{6 - 4} = 2$$

and the previous slope was 1. Using Eq. (2-23a) gives us

$$K = B - A$$

$$= 2 - 1$$

$$= 1$$

$$K(t - a)\, u(t - a)$$

$$(t - 4)\, u(t - 4)$$

At $t = 6$ sec the function has a step and a change in the ramp slope. For the step,

$$[f(a^+) - f(a^-)]\, u(t - a)$$

$$(0 - 16)\, u(t - 6)$$

$$-16\, u(t - 6)$$

The new ramp has a slope of 0 and the old slope was 2; therefore,

$$K = B - A$$

$$= 0 - 2$$

$$= -2$$

$$K(t - a)\, u(t - a)$$

$$-2(t - 6)\, u(t - 6)$$

The complete equation is the sum of the terms.

$$g(t) = 2\, \delta(t) + 4t\, u(t) - 3(t - 2)\, u(t - 2) + 2\, u(t - 4)$$
$$+ (t - 4)\, u(t - 4) - 16\, u(t - 6) - 2(t - 6)\, u(t - 6)$$

EXAMPLE 2-18

Determine the equation for the exponential sinusoidal function shown in Fig. 2-29.

Solution: This is an exponential sinusoidal function with a step and therefore has the general form of

$$f(t) = K_1\, u(t) + K_2 e^{-\alpha t} \sin(\omega t + \theta)\, u(t)$$

Since the second term may have a nonzero value at $t = 0$, we cannot use Eq. (2-22) to find the value of K_1. However, we can determine its value when the second term dies out, which will occur as t approaches ∞ because the value of $e^{-\alpha t}$ approaches 0, causing the second term to be equal to 0. The value of the function, as seen from Fig. 2-29, approaches 10 as t approaches ∞. From this reasoning,

$$f(t) = K_1\, u(t) + K_2 e^{-\alpha t} \sin(\omega t + \theta)\, u(t)$$

$$f(\infty) = K_1\, u(\infty) + K_2 e^{-\infty} \sin[\omega(\infty) + \theta]\, u(\infty)$$

$$10 = K_1\,(1) + 0$$

$$K_1 = 10$$

This means that the exponential sinusoidal function has been shifted up by 10. From this, we realize that a crossing of 10 corresponds to a zero crossing of the exponential sinusoidal part, as described in Section 2-6-2.

In the sine part of the term, we may find ω by measuring the distance between either a maximum and a minimum or between two consecutive crossings of the final value of the function. The time difference between either of these two locations will be one-half the time period of the frequency of the sine function. We cannot use the distance between a maximum or minimum and a crossing because the distance between these

Figure 2-29

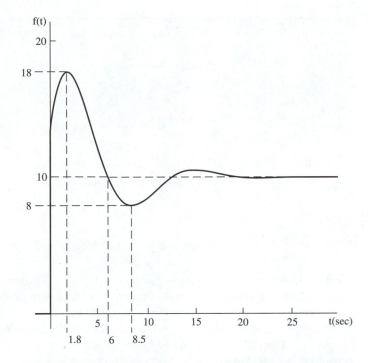

points is not one-fourth the period of the frequency. Since the times are shown for a consecutive maximum and a minimum, we double the difference between these two points to find the time period.

$$T = 2(t_{\text{at min}} - t_{\text{at max}})$$

$$= 2(8.5 - 1.8)$$

$$= 13.4 \text{ sec}$$

The value of ω is then found by

$$\omega = \frac{2\pi}{T}$$

$$= \frac{2\pi}{13.4}$$

$$= 0.46889 \text{ rad/sec}$$

We may now look for the phase angle, θ. We know that the zero crossing of an exponential sinusoidal function is caused by ($\omega t + \theta$) becoming either π or 2π, and we know that the zero crossing corresponds to the crossing of the steady-state value in this function. By observing the shape of the curve in Fig. 2-29, we see that the crossing of the steady-state value at $t = 6$ sec is where ($\omega t + \theta$) = π. We already know the value of ω, so we solve for θ.

$$(\omega t + \theta) = \pi$$

$$[0.46889(6) + \theta] = \pi$$

$$\theta = \pi - 0.46889(6)$$

$$\theta = 0.32823 \text{ rad}$$

When the values that have been found so far are substituted in, we see that there are two more quantities to find to complete the equation.

$$f(t) = K_1 \, u(t) + K_2 e^{-\alpha t} \sin(\omega t + \theta) \, u(t)$$

$$= 10 \, u(t) + K_2 e^{-\alpha t} \sin(0.46889t + 0.32823) \, u(t)$$

From here we need to generate two equations to solve for the two unknowns. The two values must be at locations where the sine part is nonzero in order to keep the term with the unknown variables from dropping out. If the value at $t = 0$ had been given on the graph, then K_2 could have been solved directly, since the exponential part would be equal to 1 at this time, and then one other point could have been used to solve for α. In this case, we are not given the value of the function at $t = 0$, and two simultaneous equations must be generated at $t = 1.8$ sec and at $t = 8.5$ sec. The first equation is found at the $t = 1.8$ sec location, and it is solved for K_2.

$$f(t) = 10 \, u(t) + K_2 e^{-\alpha t} \sin(0.46889t + 0.32823) \, u(t)$$

$$18 = 10 \, u(1.8) + K_2 e^{-\alpha 1.8} \sin[0.46889(1.8) + 0.32823] \, u(1.8)$$

$$K_2 = 8.6804 \, e^{+\alpha 1.8}$$

The same procedure is followed using the $t = 8.5$ sec location.

$$f(t) = 10 \, u(t) + K_2 e^{-\alpha t} \sin(0.46889t + 0.32823) \, u(t)$$

$$8 = 10 \, u(8.5) + K_2 e^{-\alpha 8.5} \sin[0.46889(8.5) + 0.32823] \, u(8.5)$$

$$K_2 = 2.1701 \, e^{+\alpha 8.5}$$

The right sides of the two equations are set equal to each other since they are both equal to K_2, and then the equation is solved for α.

$$8.6804 \, e^{+\alpha 1.8} = 2.1701 \, e^{+\alpha 8.5}$$

$$\frac{e^{+\alpha 1.8}}{e^{+\alpha 8.5}} = \frac{2.1701}{8.6804}$$

$$\ln(e^{\alpha 1.8 - \alpha 8.5}) = \ln\left(\frac{2.1701}{8.6804}\right)$$

$$\alpha 1.8 - \alpha 8.5 = -1.3863$$

$$\alpha = 0.20691$$

All that is left is to substitute $\alpha = 0.20691$ into either of the two equations to solve for K_2.

$$K_2 = 2.1701\, e^{+\alpha 8.5}$$

$$= 2.1701\, e^{(0.20691)8.5}$$

$$= 12.598$$

All of the unknown values have been solved for and they are substituted in for the final answer.

$$f(t) = 10\, u(t) + 12.598 e^{-0.20691t} \sin(0.46889t + 0.32823)\, u(t)$$

When writing mathematical expressions from a given graph, we often find a waveform that repeats. This type of waveform has an infinite number of terms, but we can express it as a summation. Summations are not difficult if a few simple rules are followed.

1. Write out the terms that complete approximately two cycles. The first term commonly will not be in the summation, and we need enough terms to see what is repeating and how it is repeating.
2. Starting with the second or third term, write down everything that does not change from term to term, leaving space for the items that are changing.
3. Fill in the blanks with some form of the counting variable, n. Start n at any convenient value, since we can change the starting value of n later. Table 2-5 shows some common forms of the counting variable and the sequence that each form follows.
4. After you supply a form of n for all the blanks that were left from step 2, try a value of n one increment before where you are currently starting n. If this gives

Table 2-5

n	$n + 1$	$(-1)^n$	$(-1)^{n+1}$	$2n$	$2(n + 1)$
0	1	1	-1	0	2
1	2	-1	1	2	4
2	3	1	-1	4	6
3	4	-1	1	6	8

you the beginning terms you left out, the entire function can be expressed by the summation alone. If this does not produce the beginning terms, they must be expressed separately before the summation.

5. Change the summation to the shortest form. This is done by examining the forms used for *n*. If there are mostly *n* + 1 terms, substitute *n* = *n* − 1 into all forms of *n*. If there are mostly *n* − 1 terms, substitute *n* = *n* + 1 for all *n*'s. If there are mostly *n* terms, leave it alone.

Following these steps will produce the best form of a summation in the shortest amount of time.

EXAMPLE 2-19

Write the equation for Fig. 2-30 in a summation form.

Solution:

Step 1.   First write the terms for the first couple of cycles. Each cycle is 2 sec.

$$2\,u(t) - 3\,u(t - 1) + 3\,u(t - 2) - 3\,u(t - 3) + 3\,u(t - 4) + \cdots$$

Step 2.   Starting with the second term, write down everything that stays constant. Also, arbitrarily start *n* at zero.

$$\sum_{n=0}^{\infty} 3\,u(t - \quad)$$

Notice that a blank has been left for the sign of the term.

Step 3.   Find the form of the counting variable, *n*, to use in the blanks. From Table 2-5 you can find the forms required to fill in the blanks.

$$\sum_{n=0}^{\infty} (-1)^{n+1}\,3\,u[t - (n + 1)]$$

Step 4.   When you let *n* = −1 it does not give the first term of the equation, and therefore you will have to write it as a separate term.

Step 5.   There are mostly *n* + 1 terms in the summation. Let *n* = *n* − 1 to shorten the expression.

$$\sum_{n-1=0}^{\infty} (-1)^{n-1+1}\,3\,u[t - (n - 1 + 1)]$$

$$\sum_{n=1}^{\infty} (-1)^{n}\,3\,u(t - n)$$

The equation in its shortest form is

$$i(t) = 2\,u(t) + \sum_{n=1}^{\infty} (-1)^{n}\,3\,u(t - n)$$

Figure 2-30

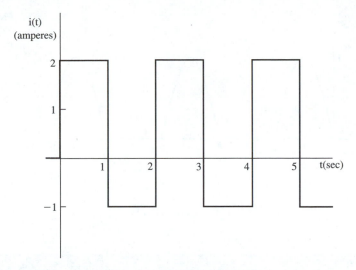

EXAMPLE **2-20**

Write the equation for the graph shown in Fig. 2-31 in a summation form.

Solution:

Step 1.   Write the terms of the first couple of cycles. First find the ω for the sine wave on which this function is based. At $t = 2$ sec, ωt is equal to one-half of a complete cycle.

$$\omega t = \frac{2\pi}{2} \qquad \text{at } t = 2 \text{ sec}$$

$$\omega(2) = \pi$$

$$\omega = \frac{\pi}{2}$$

The first term of the waveform is then

$$\sin\left(\frac{\pi}{2}t\right)u(t)$$

At $t = 2$ sec, another sine function needs to be started to cancel the first term. (There are a couple of ways to do this, but functions must always be written in a shifted form.)

$$\sin\left(\frac{\pi}{2}t\right)u(t) + \sin\left[\frac{\pi}{2}(t-2)\right]u(t-2)$$

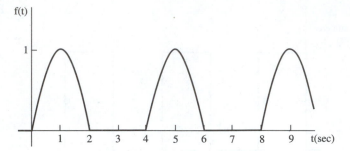

Figure 2-31

Notice that at $t = 2$ sec the second term will be 180° out of phase with the first term, and the two terms will cancel each other from $t = 0$ to ∞.

At $t = 4$ sec, we need to start a new sine wave and then cancel it at $t = 6$ sec. The third and fourth terms will therefore be

$$\sin\left[\frac{\pi}{2}(t - 4)\right]u(t - 4) + \sin\left[\frac{\pi}{2}(t - 6)\right]u(t - 6)$$

Step 2. Starting with the second term, write everything that stays constant.

$$\sum_{n=1}^{\infty} \sin\left[\frac{\pi}{2}(t - \quad)\right]u(t - \quad)$$

Step 3. Using Table 2-5, fill in the blanks.

$$\sum_{n=1}^{\infty} \sin\left[\frac{\pi}{2}(t - 2n)\right]u(t - 2n)$$

Step 4. In this particular case, when we start n at 0 we get the first term. So we just start at $n = 0$ instead of at 1, and the completed function is

$$f(t) = \sum_{n=0}^{\infty} \sin\left[\frac{\pi}{2}(t - 2n)\right]u(t - 2n)$$

Step 5. Since all the terms do not have $n + 1$ or $n - 1$ form, we will leave the summation as it is.

2-9 DERIVATIVES AND INTEGRALS OF WAVEFORMS

There are a number of circuits that integrate and differentiate an input signal. The simplest circuits are the inductor and capacitor. Assuming the initial conditions of zero, the voltage across these devices given the current is

$$v(t) = L\frac{di(t)}{dt} \qquad (2\text{-}24)$$

$$v(t) = \frac{1}{C}\int_0^t i(t)\,dt \qquad (2\text{-}25)$$

and the current through these devices given the voltage is

$$i(t) = \frac{1}{L}\int_0^t v(t)\,dt \qquad (2\text{-}26)$$

$$i(t) = C\frac{dv(t)}{dt} \qquad (2\text{-}27)$$

Since even the simplest components integrate and differentiate, we need to know what to expect when a function is integrated or differentiated. We may integrate or differentiate functions either mathematically or graphically. Either method may be superior, depending on the particular case. First we will look at the mathematical way of integrating and differentiating, then the graphic method.

There are three areas that may not have been covered in beginning calculus that are common to complex functions in electronics. First are graphs with discontinuities in them. The discontinuity will be either a step or an impulse. These will be handled by evaluating the function on the left of the discontinuity, the right of the discontinuity, and then at the discontinuity. On either side of a discontinuity, normal calculus procedures are used.

We have already covered, in concept, the integral and derivative of the step and impulse function in Sections 2-3 and 2-4. Section 2-3 showed that the integral of a step is the ramp function.

$$\int_0^t u(t)\,dt = t\,u(t) \qquad (2\text{-}28)$$

In Section 2-4, we defined the impulse by finding the slope (the derivative) at the step function's discontinuity.

$$\frac{d}{dt}u(t) = \delta(t) \qquad (2\text{-}29)$$

Since the derivative of the step is the impulse, the integral of the impulse must be the step function.

$$\int_0^t \delta(t)\, dt = u(t) \quad , \tag{2-30}$$

Notice that we instantaneously sum up the area contained in the impulse at $t = 0$. The derivative of the impulse normally does not show up in practical circuits. This concept is more theoretical and will not be covered here.

Second is removing $u(t)$ from an integral. We will always be integrating from $t = 0$ to t. Since we are concerned with the function to the right of $t = 0$, we may use $t = 0^+$ for the lower limit of integration. In this interval, $t = 0^+$ to t and $u(t)$ has a constant value of 1 and may be factored out of the integral like any other constant.

Third is finding the integral and derivative of a shifted function. Since we always write our shifted function in the form of $f(t - a)\, u(t - a)$, we may integrate the function in its unshifted form, $f(t)\, u(t)$, and then shift the result back to time $t = a$.

EXAMPLE 2-21

Show that the preceding paragraph is true for the following function:

$$\int_0^t 4(t - 2)\, u(t - 2)\, dt$$

Solution: The unshifted form of $4(t - 2)\, u(t - 2)$ is $4t\, u(t)$, which we will integrate. Since $u(t)$ has a constant value of 1 from $t = 0^+$ to ∞, we may remove $u(t)$ from the integral as we could any other constant.

$$\int_0^t 4t\, dt\, u(t) = \frac{4t^2}{2}\bigg|_0^t u(t)$$

$$= 2t^2\, u(t)$$

Then we shift the function to start at $t = 2$ sec.

$$2(t - 2)^2\, u(t - 2)$$

We will now work the problem without shifting back to the origin, and we should get the same result.

$$\int_0^t 4(t - 2)\, u(t - 2)\, dt$$

Since the function is zero until $t = 2$ sec, we are actually integrating from $t = 2^+$ to t.

$$\int_2^t 4(t - 2)\, u(t - 2)\, dt$$

Over these limits, $u(t - 2)$ has a constant value of 1 and may therefore be removed from under the integral.

$$\left[\int_2^t 4(t - 2) \, dt \right] u(t - 2)$$

$$\left(\int_2^t 4t \, dt + \int_2^t -8 \, dt \right) u(t - 2)$$

$$\left[\frac{4t^2}{2} \bigg|_2^t + (-8t) \bigg|_2^t \right] u(t - 2)$$

$$[2t^2 - 2(2)^2 - 8t + 8(2)] \, u(t - 2)$$

$$2(t^2 - 4t + 4) \, u(t - 2)$$

$$2(t - 2)^2 \, u(t - 2)$$

The derivative of a shifted function is handled in a similar fashion. Shift the function to the origin, take the derivative, and shift the result back.

When taking the derivative of a function, we must consider each term of the function as the product of two functions of t. One of the functions will typically be $u(t)$. Therefore, the derivative of each term will be

$$\frac{d}{dt} f(t) \, u(t) = f'(t) \, u(t) + f(t) \, \delta(t)$$

Since in the second term $\delta(t)$ is zero at any time except at $t = 0$, the only time $f(t)$ makes any difference is at $t = 0$. We may, therefore, use $f(0)$ in the second term.

$$\frac{d}{dt} f(t) \, u(t) = f'(t) \, u(t) + f(0) \, \delta(t) \tag{2-31}$$

EXAMPLE 2-22

Find the derivative of

$$4 \, u(t) + 2 \cos(4t) \, u(t)$$

Solution: Take the derivative of each term.

$$\frac{d}{dt} 4 \, u(t) = (0) \, u(t) + 4 \, \delta(t) = 4 \, \delta(t)$$

$$\frac{d}{dt} 2 \cos(4t) \, u(t) = -(2)(4)\sin(4t) \, u(t) + 2 \cos[4(0)] \, \delta(t)$$

$$= -8 \sin(4t) \, u(t) + 2 \, \delta(t)$$

The derivative of the function is the sum of the two previous results.

$$\frac{d}{dt} [4 \, u(t) + 2 \cos(4t) \, u(t)] = 6 \, \delta(t) - 8 \sin(4t) \, u(t)$$

EXAMPLE 2-23

Mathematically find the integral of the function in Fig. 2-32. Graph the result.

Solution: First we find the equation of the function.

$$g(t) = -10 \, u(t) + 20 \, u(t - 4) - 5(t - 8) \, u(t - 8)$$
$$+ 5(t - 10) \, u(t - 10) + 5 \, \delta(t - 12)$$

Next we shift each term back to the origin, integrate it, and shift the result back to the proper time.

$$\int_0^t -10 \, u(t) \, dt = -10t \, u(t) \implies -10t \, u(t)$$

$$\int_0^t 20 \, u(t) \, dt = 20t \, u(t) \implies 20(t - 4) \, u(t - 4)$$

$$\int_0^t -5t \, u(t) \, dt = \frac{-5}{2}t^2 \, u(t) \implies \frac{-5}{2}(t - 8)^2 \, u(t - 8)$$

$$\int_0^t 5t \, u(t) \, dt = \frac{5}{2} t^2 \, u(t) \implies \frac{5}{2} (t - 10)^2 \, u(t - 10)$$

$$\int_0^t 5 \, \delta(t) \, dt = 5 \, u(t) \implies 5 \, u(t - 12)$$

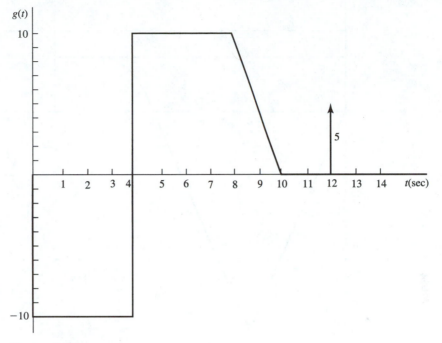

Figure 2-32

Combining all terms gives

$$\int_0^t g(t)\,dt = -10t\,u(t) + 20(t-4)\,u(t-4)$$

$$-\frac{5}{2}(t-8)^2\,u(t-8) + \frac{5}{2}(t-10)^2\,u(t-10) + 5\,u(t-12)$$

Figure 2-33 shows the graph of the function. The first term starts the graph off with a slope of −10. Since the second term, which starts at $t = 4$, is also a ramp, it combines with the initial ramp, causing a ramp with a slope of +10. Next a squared term is added. Notice the effect of combining the squared term with the current positive slope. The effect is a quadratic curve. Then at $t = 10$, the next squared term starts. When the two squared terms combine, instead of canceling each other's effect, they create a negative slope of 10 that cancels the slope that was started at $t = 4$. This negative ramp created at $t = 10$ can be verified by graphing the two squared terms on a separate axis. Finally, the last term adds a step at $t = 12$.

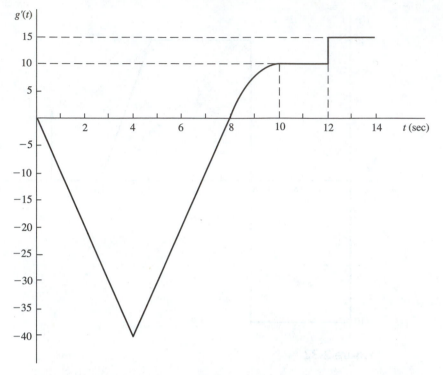

Figure 2-33

EXAMPLE 2-24

Mathematically find the derivative of the function shown in Fig. 2-34. Graph the result.

Solution: Again we write the equation of the function first.

$$g(t) = 5\,u(t) - 5\,u(t - 1) + 5(t - 2)\,u(t - 2) - 10\,u(t - 3) - 5(t - 3)\,u(t - 3)$$

Next we shift each term back to the origin, find the derivative, and then shift each term back to the proper time delay.

$$\frac{d}{dt}\,5\,u(t) = (0)\,u(t) + 5\,\delta(t) \qquad \Rightarrow 5\,\delta(t)$$

$$\frac{d}{dt} - 5\,u(t) = -(0)\,u(t) - 5\,\delta(t) \qquad \Rightarrow -5\,\delta(t - 1)$$

$$\frac{d}{dt}\,5t\,u(t) = 5\,u(t) + 5(0)\,\delta(t) \qquad \Rightarrow 5\,u(t - 2)$$

$$\frac{d}{dt} - 10\,u(t) = -(0)\,u(t) - 10\,\delta(t) \qquad \Rightarrow -10\,\delta(t - 3)$$

$$\frac{d}{dt} - 5t\,u(t) = -5\,u(t) - 5(0)\,\delta(t) \Rightarrow -5\,u(t - 3)$$

Figure 2-34

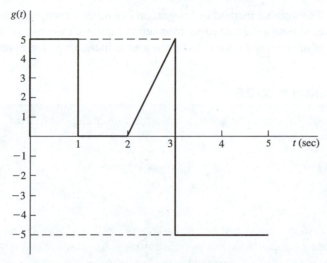

The resulting equation is

$$\frac{d}{dt} g(t) = 5\,\delta(t) - 5\,\delta(t-1) + 5\,u(t-2) - 10\,\delta(t-3) - 5\,u(t-3)$$

Figure 2-35 shows the graph of the result. Notice the $-10\delta(t)$ is still drawn from the x-axis even though the end of it could be considered starting at a level of 5. This is because the impulse actually has an infinite height and we stop it at -10 only to note its area, not its height.

Figure 2-35

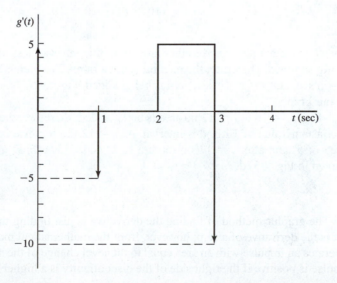

The graphic method of integration is simply a matter of calculating the area under a curve. Sliding a sheet of paper from left to right, we can observe, to the left of the paper, the rate of area being summed and if the area is increasing, decreasing, or remaining constant.

EXAMPLE 2-25

Graphically find the integral of the function in Fig. 2-36. (Note that it is the same graph used in Example 2-23.)

Solution: From $t = 0$ to $t = 4$ sec the area is

$$-10 \times 4 = -40$$

Therefore, we have summed an area of -40 when we get to $t = 4$ sec. If we slide a piece of paper from left to right across $g(t)$, we see that a constant amount of area is being added as we move the paper. Since this is a constant area being summed, we will have a ramp starting at zero area at $t = 0$ and ending with an area of -40 at $t = 4$ sec. This is shown in Fig. 2-37a.

From $t = 4$ to $t = 8$ will also add a constant amount of area as we slide our paper across the function, but this will be a positive area. The total area added when we reach $t = 8$ sec is

$$(8 - 4) \times 10 = 40$$

This is added to our present area already summed up to $t = 4$ sec, shown in Fig. 2-37b. Since the area is positive, we will have a positive ramp. This ramp will take us back to zero at $t = 8$ sec because the negative and positive areas summed up to $t = 8$ sec are equal.

From $t = 8$ to $t = 10$ the area under the curve is

$$\frac{1}{2}(10 - 8) \times 10 = 10$$

As we move our paper across the function in this interval, less and less positive area is being summed. Therefore, the integral is not a ramp. The area is mostly added closer to $t = 8$ sec, and at $t = 10$ sec, zero is being added. Figure 2-37c shows this being added to our graph.

From $t = 10$ to $t = 12$ no area is being added; therefore, the integral value stays at a constant value of 10 in this interval. At $t = 12$ the impulse instantaneously adds an area of 5, and after $t = 12$ no more area is added. The final graph of the integral is shown in Fig. 2-37d.

The graphic method of finding the derivative is just finding the slope. The only tricky part is the derivative of a step; however, from the mathematical method, we found that this generates an impulse with an area equal to the level change of the step. The direction of the impulse is positive if the right side of the discontinuity is a higher level than the left side.

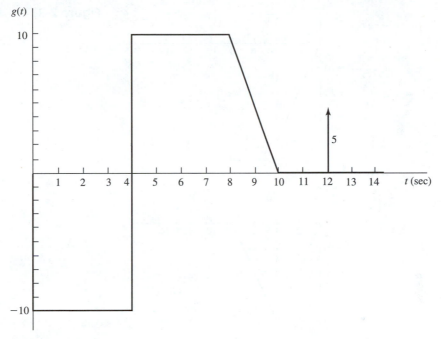

Figure 2-36

EXAMPLE **2-26**

Graphically find the derivative of Fig. 2-38a. (Note it is the same graph used in Example 2-24.)

Solution: The solution is shown in Fig. 2-38b. Let's look at what happens at each time. At $t = 0$ there is a step going up as we go from the left to the right side of $t = 0$. The change is 5, so we get an impulse at $t = 0$ with an area of 5.

From $t = 0^+$ to $t = 1^-$ the slope is zero, and therefore the derivative is zero.

Since at $t = 1$ there is a step change going down with a difference of 5, we get another impulse of area 5 but in the negative direction.

From $t = 2$ to $t = 3$ sec there is a ramp with a slope of 5. The derivative will then be 5.

At $t = 3$ there is a step down. The magnitude of this step is 10. This will yield a negative impulse going down with an area of 10.

After $t = 3$ the slope of the function is zero. The derivative will then also be zero.

2-10 PERIODIC WAVEFORM VALUES

The average, effective, and rms values are typically applied to periodic waveforms. The ultimate goal of knowing these values is to determine the actual amount of energy or heat that is dissipated by an electrical component. Knowing how to calculate these values allows us to select components with appropriate voltage, current, and power ratings.

Figure 2-37

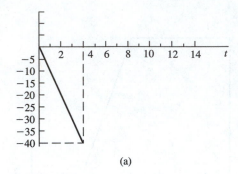

(a)

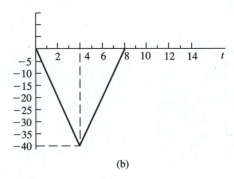

(b)

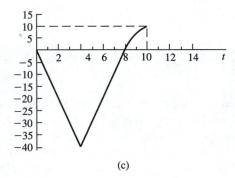

(c)

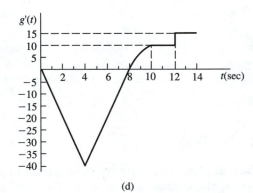

(d)

Figure 2-38

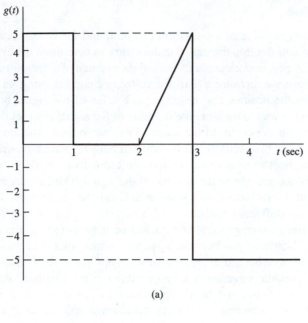

(a)

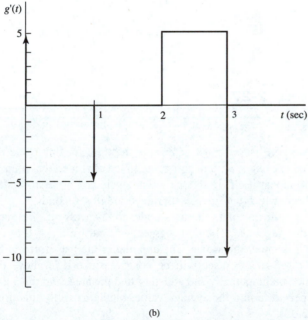

(b)

Since this section applies mainly to periodic waveforms, we should review the definition of a periodic waveform. In simple terms, a periodic waveform repeats its shape once every time period. Therefore, the following equation must be true for any and every value of time, t, for a waveform with a period of T seconds.

$$f(t + T) = f(t) \qquad\qquad (2\text{-}32)$$

2-10-1 Average Value

The average value of a waveform can be found for voltage, current, and power, and this section will develop the equation necessary to determine the average value of these waveforms. A practical example of finding the average of a waveform is often performed in the lab. When we measure a periodic voltage or current using an oscilloscope, we not only measure the positive and negative peak value of the signal but we also measure the dc value. One way to measure the dc value of the signal is to set the oscilloscope's zero volts location for the center of the screen. Then we measure the amount of voltage shift in the signal when the oscilloscope is switched between the ac measurement setting and the dc measurement setting. The amount of shift in volts is the dc value of the signal, and this dc value is the average or mean value of the signal. Thus, a pure sine wave, such as is found from the ac wall socket of a residential house, has an average or mean value of zero since it will not shift when measured in this way on an oscilloscope. We may then say that this ac signal has an average value of zero, and we may also say it contains no dc value or dc component. However, just because it has an average voltage of zero does not mean that it will not cause a resistor to dissipate heat when applied to the resistor.

A periodic waveform's average value can be calculated mathematically by summing up the area above and below zero (the *x*-axis) for one time period and dividing by the period of the waveform. Therefore, the average or dc value of a signal can be expressed as

$$f_{avg} = f_{dc} = \frac{1}{T} \int_0^T f(t)\, dt \qquad (2\text{-}33)$$

EXAMPLE 2-27

Determine the average value of the graph shown in Fig. 2-39.

Solution: First we must determine the time for one cycle of the waveform. This is done by starting at a point and going to the next identical point where the waveform begins to repeat itself. In this case, one cycle can be seen to start at $t = 0$ and finish at $t = 4$ sec, which will give us a time period of 4 seconds.

Next we calculate the area under this waveform. We may split up the cycle into two time segments. The first segment of time from $t = 0$ to $t = 2$ has a negative area since it is under the *x*-axis. The area under the waveform in the second time segment from $t = 2$ to $t = 4$ sec will be positive since it is above the *x*-axis. These can be summed up graphically and added to find the area, and this result is then divided by the time period to find the average value. Mathematically this process is performed using Eq. (2-33).

$$f_{dc} = \frac{1}{T} \int_0^T f(t)\, dt$$

$$= \frac{1}{4} \left\{ \int_0^2 f(t)\, dt + \int_2^4 f(t)\, dt \right\}$$

Figure 2-39

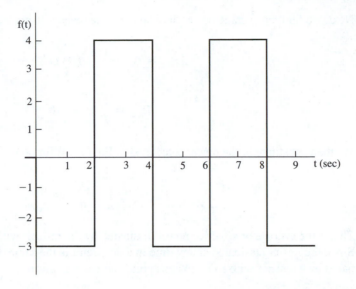

$$= \frac{1}{4}\left\{-(2)(3) + (2)(4)\right\}$$

$$= 0.5$$

The dc, or average, value for this function is therefore 0.5.

In the previous problem the integrations could have been done graphically. Typically, it is easier to use the graphic method of integration as described in Section 2-9 when the periodic function is composed of ramps, steps, and impulses, but if the periodic wave form is sinusoidal, then using calculus will be much easier.

2-10-2 Effective or rms Value

The effective, or rms, value of voltage (or current) is the amount of dc voltage (or current) that will produce the same amount of power dissipation by a resistor as the ac voltage that is currently across the resistor. The effective or rms power is the actual power dissipated by the resistor and can be found by multiplying the rms voltage across the resistor by the rms current through the resistor. Note that resistance is being used since resistance is the only component that will actually dissipate power. Let's derive the equations for the rms voltage.

We start by determining the instantaneous power dissipated by a resistor, R, with a periodic voltage, $v_{ac}(t)$, across it. The amount of power dissipated at any point in time can be calculated by the square of the voltage at any point in time divided by the resistance.

$$p(t) = \frac{v_{ac}^2(t)}{R} \tag{2-34}$$

Next, we find the average power dissipated by the resistor using Eq. (2-33).

$$f_{avg} = f_{dc} = \frac{1}{T} \int_0^T f(t)\, dt$$

$$P_{avg} = \frac{1}{T} \int_0^T p(t)\, dt$$

We then substitute in the power calculated in Eq. (2-34) for $p(t)$.

$$P_{avg} = \frac{1}{T} \int_0^T \frac{v_{ac}^2(t)}{R}\, dt \tag{2-35}$$

This is the average or effective power dissipated by the resistor with an ac voltage across it. Since the power dissipated at any time in a dc circuit is the same, the average power dissipated by a resistor with a dc voltage applied across it is given by

$$P_{avg} = \frac{v_{dc}^2}{R} \tag{2-36}$$

To find the relationship between the dc voltage in Eq. (2-36) and the ac voltage in Eq. (2-35), we set the right side of Eq. (2-35) equal to the right side of Eq. (2-36) and solve for v_{dc}, which will be the rms voltage v_{rms}.

$$P_{avg} = \frac{v_{dc}^2}{R} = \frac{1}{T} \int_0^T \frac{v_{ac}^2(t)}{R}\, dt$$

$$v_{dc}^2 = \frac{1}{T} \int_0^T v_{ac}^2(t)\, dt$$

$$v_{dc} = v_{rms} = \sqrt{\frac{1}{T} \int_0^T v_{ac}^2(t)\, dt} \tag{2-37}$$

Notice how R cancels out. This indicates that the relationship between rms value and the ac value is independent of the component that the voltage is across.

The mathematical process used in Eq. (2-37) is the reason for the subscript on v_{rms}. This subscript is used because the right side of the equation is the square **r**oot of the **m**ean (or average) of the ac voltage **s**quared.

Using a similar process, we can find the amount of dc current through a resistor that produces the same amount of power dissipation as the ac current through the resistor. This is given by

$$i_{dc} = i_{rms} = \sqrt{\frac{1}{T} \int_0^T i_{ac}^2(t)\, dt} \tag{2-38}$$

The rms value is most frequently applied in steady-state sinusoidal ac circuit analysis. Applying Eq. (2-37) and Eq. (2-38) to this type of signal reveals that the rms value can be found by dividing the amplitude (or peak value) of the sine function by the square root of 2. Once this relationship is established, Eq. (2-37) and Eq. (2-38) are never referenced again. This is not the case when the signals are complex. In that case, the equations must be applied every time a new waveform is encountered.

In Eq. (2-37) and Eq. (2-38), the waveforms are squared. When complex waveforms containing step functions are squared, multiplication of step functions having different time shifts occurs. We must, therefore, determine how to interpret the multiplication of step functions.

The multiplication of two step functions is graphed in Fig. 2-40. Figure 2-40a shows the step function $u(t - a)$, and Fig. 2-40b shows another step function, $u(t - b)$, where $a > b$. In the time period from $t = 0$ to $t = b$, the multiplication of the two functions is zero times zero, for a result of zero. In the time interval from $t = b$ to $t = a$, $u(t - a) = 0$ and $u(t - b) = 1$, and therefore the multiplication of these two functions in this time interval is zero. In the time interval for $t > a$, both functions are 1 and the resulting multiplication is 1. The graph of the multiplications of $u(t - a)$ times $u(t - b)$ is shown in Fig. 2-40c. From this we can conclude that if the $a > b$, then the result will be $u(t - a)$, and by a similar manner we can conclude that if $a < b$, the result will be $u(t - b)$. This can be expressed in the equation

$$u(t - a)\, u(t - b) = \begin{cases} u(t - a) & \text{for } a \geq b \\ u(t - b) & \text{for } a < b \end{cases} \tag{2-39}$$

By similar reasoning we can establish the result of multiplying an impulse by a step.

$$\delta(t - a)\, u(t - b) = \begin{cases} \delta(t - a) & \text{for } a \geq b \\ 0 & \text{for } a < b \end{cases} \tag{2-40}$$

EXAMPLE 2-28

The voltage across a 10-Ω resistor is shown in Fig. 2-41. Calculate the power dissipated by the resistor.

Solution: The easiest way to find the power dissipated by the resistor in this case will be to find the square of the rms voltage divided by the resistor. The rms voltage can be found by using Eq. (2-37).

$$v_{\text{rms}} = \sqrt{\frac{1}{T} \int_0^T v_{\text{ac}}^2(t)\, dt}$$

This equation indicates that the value of the resistance has no effect in determining the rms value, so the resistance value will be ignored. The equation requires us to find the square of the voltage shown in Fig. 2-41. First we write the equations for $v(t)$.

$$v(t) = 2t\, u(t) - 6\, u(t - 2) - 6\, u(t - 5) - 6\, u(t - 8) \cdots$$

Figure 2-40

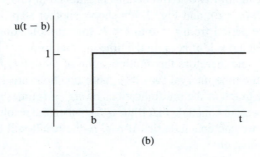

(a)

(b)

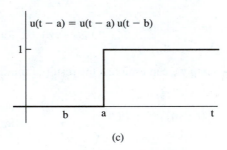

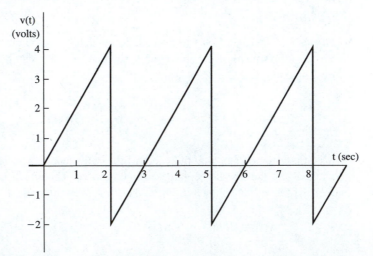

(c)

Figure 2-41

We perform the integral from $t = 0$ to T, and in this case $T = 3$ sec. Therefore, we only require the terms in $v(t)$ that are active in this time interval, since they are the only terms responsible for creating this cycle. The equation is therefore reduced to

$$v(t) = 2t\, u(t) - 6\, u(t - 2)$$

From Eq. (2-37), we see that we must find the square of the function

$$v^2(t) = [2t\, u(t) - 6u(t - 2)]^2$$

Multiplying out the right side, we have

$$v^2(t) = 4t^2\, u(t)\, u(t) - (2)[12t\, u(t)\, u(t - 2)] + 36\, u(t - 2)\, u(t - 2)$$

Simplifying the multiplications of step functions using Eq. (2-39), we find

$$v^2(t) = 4t^2\, u(t) - 24t\, u(t - 2) + 36\, u(t - 2)$$

Notice that the second term is not in a proper shifted form and must be modified as described in Section 2-7. This term written in a proper shifted form is

$$-24t\, u(t - 2) = -24(t - 2)\, u(t - 2) - 48\, u(t - 2)$$

The voltage squared can then be expressed as

$$v^2(t) = 4t^2\, u(t) - 24(t - 2)\, u(t - 2) - 48\, u(t - 2) + 36\, u(t - 2)$$

$$= 4t^2\, u(t) - 24(t - 2)\, u(t - 2) - 12\, u(t - 2)$$

Next, from Eq. (2-37) we must find the integral of $v^2(t)$.

$$\int_0^3 [4t^2\, u(t) - 24(t - 2)\, u(t - 2) - 12\, u(t - 2)]\, dt$$

Each term is removed and integrated separately. The first time is quite simple to integrate.

$$\int_0^3 4t^2\, u(t)\, dt = \left. \frac{4t^3}{3} \right|_0^3 = 36$$

The second term is a shifted function, so we will shift the function to the origin, integrate it from 0 to t, and then shift the result back to the original time.

$$\int_0^3 -24(t - 2)\, u(t - 2)\, dt \Rightarrow \int_0^t -24(t)\, u(t)\, dt = \frac{-24t^2}{2} \Rightarrow -12(t - 2)^2\, u(t - 2)$$

From this we substitute in $t = 3$ sec for the upper limit of the integral.

$$-12(t-2)^2 \, u(t-2) \Big|_{t=3} = -12$$

The last term is integrated by a similar process.

$$\int_0^3 -12 \, u(t-2) \, dt \Rightarrow \int_0^t -12 \, u(t) \, dt = -12t \Rightarrow -12(t-2) \, u(t-2)$$

Substituting in the upper limit,

$$-12(t-2) \, u(t-2) \Big|_{t=3} = -12$$

Restating the rms equations and substituting the values in for the integral, the solution can be found.

$$v_{\text{rms}} = \sqrt{\frac{1}{T} \int_0^T v_{\text{ac}}^2(t) \, dt}$$

$$= \sqrt{\frac{1}{3} \int_0^3 4t^2 \, u(t) - 24(t-2) \, u(t-2) - 12 \, u(t-2) \, dt}$$

$$= \sqrt{\frac{1}{3} (36 - 12 - 12)}$$

$$= 2 \text{ Volts}_{\text{rms}}$$

The power dissipated by the 10-Ω resistor can then be calculated.

$$P = \frac{v_{\text{rms}}^2(t)}{R} = \frac{(2)^2}{10} = 0.4 \text{ Watts}$$

2-11 USING MATLAB

Since MATLAB was originally designed for matrix algebra, its approach to calculating expressions and storing values is oriented towards matrix algebra. However, in our use of MATLAB, we will often want to use a single-row matrix, or vector, as a list of individual values. We will want each element of the matrix to be used one at a time to calculate independent values for an expression. This will be conveyed by different operation symbols.

When an expression is entered, MATLAB uses the "*" operation symbol to indicate matrix multiplication and the ".*" to indicate using the elements of the matrix independently. Similarly, the "/" and "^" symbols indicate matrix division and matrix exponent power, and the "./" and ".^" symbols indicate individual use of the elements in the matrix for division and exponent power. Since addition and subtraction are performed the same for the matrix as for the elements of the matrix independently, there is no need for a different operation symbol.

MATLAB does not consider an expression that is entered on the command line to be a symbolic expression even though it contains symbolic variables. In MATLAB, an expression is consider symbolic only if the expression is stored in and used from a variable location. In order to use symbolic expressions, the Symbolic toolbox must be installed in MATLAB. A significant difference is that when entering a symbolic expression into a variable, the operators for multiplication, division, and so on must be entered without the "." in front of them. This will be pointed out again when symbolic expressions are used in examples in later sections, since there is such a significant difference.

In this section and subsequent MATLAB sections, the following text style will be used to indicate the text that is to be typed in response to the MATLAB prompt ">>":

```
t = linspace (-10, 10, 200);
```

This style is used to make it easy to determine what is to be typed in the MATLAB Command Window.

In this section, Example 2-29 will show how to create an array of values to be evaluated by a symbolic expression and how to create the $u(t - a)$ function that will be used in subsequent examples. Example 2-30 will show how to graph the expression given in the first example. Example 2-31 will show how to graph a more complex piecewise function.

EXAMPLE 2-29

Calculate the values of $f(t)$ in Table 2-2 for the entries $t = -2.3847, 0.11532, 2.6153, 5.1153,$ and 7.6153. The function from the table is repeated here for convenience.

$$f(t) = 4e^{-0.5t} \sin\left(0.4\pi t + \frac{\pi}{3}\right) u(t)$$

Solution: MATLAB has the built-in function for $\sin(x)$, but it does not have a built-in function for $u(t)$. Therefore, we need to create this function. To do this, the M-files feature of MATLAB will be employed. An M-file is a text file having the extension ".m" that contains MATLAB commands. When the function is referenced in the Command Window, MATLAB looks up the file and executes its contents. With this very powerful feature, almost any function can be created.

Any text editor that is capable of storing a text file with a ".m" extension may be used to create an M-file, but this can be accomplished more directly in MATLAB. From the MATLAB desktop, select **File -> New -> M-File.** A Text Editor Window will open that will allow the required text command to be entered. In the text box type the following two lines:

```
function a = u(b)
a = (b>=0);
```

Then save the file under the name "u." The MATLAB text editor will automatically save the file in the correct folder with the ".m" extension. Next, close the file. This will cause the text editor to close and will return us to the normal MATLAB display. With the file saved in the correct folder, we can type what and our function name "u" will appear in the list of M-files.

These two lines are all that is required to create the step function. The first line defines the function and its arguments. The "a" and the "b" are the arguments, and "u" is the name of the function. The variables that will be passed to this function do not need to have the same name as the arguments listed in the function definition because these arguments are really placeholders. In fact, we can pass an expression for the "b" argument.

The value output when the "u" matrix is called is based on the values passed in the "b" matrix. For each element in the "b" matrix that meets the specified condition, a value of 1 will be output in the corresponding element, and for each element in the "b" matrix that does not meet the specified condition, a value of 0 will be output. Since "b" is a placeholder, we may pass expressions such a $t - 1$. This will, therefore, define the function $u(t - a)$ for use in any of our MATLAB expressions. We can test this function by creating a matrix t = [-3, -2, -1, 0, 1, 2, 3] and then entering u(t - 1). This will cause the printout of [0, 0, 0, 0, 1, 1, 1]. Entering u(t - 2) will cause a printout of [0, 0, 0, 0, 0, 1, 1].

Next, we need to set up a single-row matrix, or vector, containing values of time that will be used in the calculation of our function. This is done by entering the following command in the Command Window:

```
t = [-2.3847, 0.11532, 2.6153, 5.1153, 7.6153];
```

Next we enter the following equation, being sure to use ".*" for the multiplication sign since we want to know the value of our function at each individual location.

```
ft = 4 .* exp(-0.5 .* t)
.* sin (0.4 .* pi .* t + pi/3) .* u(t)
```

Once this line is typed and the Enter key is pressed, a display of the values of the function at each time in the t matrix will be printed on the screen. In actuality, this last statement created a matrix called ft that contains the values of the function for each value of time in the t matrix.

The Command Window printout of MATLAB is shown in Fig. 2-42.

```
>>t = [-2.3847, 0.11532, 2.6153, 5.1153, 7.6153];
>>ft = 4 .* exp(-0.5 .* t) .* sin (0.4 .* pi .* t + pi/3) .* u(t)
ft =
   0 3.5084 -1.0052 0.2880 -0.0825
```

Figure 2-42

MATLAB Command Window for Example 2-29.

EXAMPLE 2-30

Display the plot of the function on the graph screen.

$$f(t) = 4e^{-0.5t}\sin\left(0.4\pi t + \frac{\pi}{3}\right)u(t)$$

Solution: In most graphing programs, individual x-y locations are sequentially plotted with a line drawn between the consecutive x-y locations to make the graph look continuous. Often, a graphing program is designed so that it automatically sets a step size along the x-axis to give the best resolution possible on the screen, and some graphing programs allow us to specify the increments. MATLAB has extensive graphing capabilities and it does allow us to specify our step size for graphing. We will use this capability when drawing the graph in this example.

When specifying the step size along the x-axis, we must be careful to have enough steps to prevent the graph from appearing too rough or jagged. We must also limit the number of steps to keep from using too much memory and unnecessarily slowing the plotting process. We need to find an appropriate number of steps that will give us an accurate representation of the function. The step size required for this is quickly learned through observation, but a good starting value typically used is 100 points.

For this problem, we will specify the x-axis, or t values, to have 100 points equally spaced. Since our function exists from 0 sec to 10 sec, we will set our graph to display from -1 sec to 11 sec so that we can observe the function slightly before it starts at $t = 0$ and slightly after the five time constants. We will create a matrix, or vector, of linearly spaced time values that we will call "t". We enter the following command line to create our linearly spaced matrix of values:

```
t = linspace (-1, 11, 100);
```

This creates a matrix with 100 values equally spaced between -1 and 11. To double the resolution, we may change the last number to 200. Notice in this command that there is a semicolon at the end. This tells MATLAB to store the values in the t matrix but not to print the values on the screen.

Our original equation is multiplied by $u(t)$. We, therefore, will need to use the $u(t)$ function that was created in Example 2-29. If it has not been created, then go back to Example 2-29 and enter it before continuing this example.

The next step is to calculate a matrix with the corresponding y-values for each value of t. Remember that we must use the ".*" notation so that the values in the t matrix will be used individually instead of as a matrix. We enter the following command line to calculate the y-axis values:

```
ft = (4 .* exp(-0.5 .* t) .* sin (0.4 .* pi .* t + pi/3))
                      .* u(t);
```

```
>>t = linspace (-1, 11, 100);
>>ft = (4 .* exp(-0.5 .*t) .* sin (0.4 .* pi .* t + pi/3)) .* u(t);
>>plot(t,ft); grid on; xlabel ('t seconds'); ylabel ('f(t)'); figure(gcf)
```

Figure 2-43

MATLAB Command Window for Example 2-30.

We now have the two matrices. The values for the *x*-axis locations are in the *t* matrix, and the values for the corresponding *y*-axis locations are in the *ft* matrix. Finally, we enter the command line to plot these *x*-*y* locations as

```
plot (t,ft); grid on; xlabel ('t seconds');
        ylabel ('f(t)'); figure (gcf)
```

This line is really five command lines separated by semicolons. The first command, `plot (t,ft)`, actually plots the graph, and the second command, `grid on`, causes the graph to have a dashed grid. The next two commands print the text indicated along the graph's edges to label the x-axis and the y-axis. The last command, `figure (gcf)`, causes the Figure Window to come to the front and display the graph. These commands could be put on separate lines, but it is more convenient to combine them on a single line.

Figure 2-43 shows the Command Window for this example, and Fig. 2-44 shows the Figure Window from MATLAB.

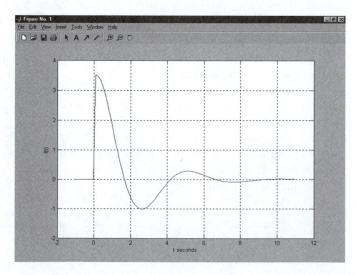

Figure 2-44

EXAMPLE 2-31

For the equation shown, display the plot of the function in the Figure Window of MATLAB.

$$f(t) = -2t\,u(t) + 6\,u(t-2) + 6\,u(t-5)$$

Solution: First we must determine an appropriate range of values for the *x*-axis. By examining the function, we see that it does not have any active terms until 0 sec and that the last term in the function becomes active at 5 sec. We therefore need to include on our graph the transition just before 0 sec and enough of the graph after 5 sec to determine how it will continue as it approaches infinity. Therefore, in this case, a reasonable range is from -1 second to 10 sec. These values may be easily modified and the function regraphed if they do not give us a good enough representation of the function.

With the range decided, we can create a matrix of 100 equally spaced time values. Enter the following command line to accomplish this:

```
t = linspace (-1, 10, 100);
```

Since functions with discontinuities change abruptly, the number of points may have to be increased. If the steps between the time locations are large, a step function will not have a sharp edge in its transition. Using the $u(t)$ function created in Example 2-29, we enter the command to calculate the values of our function as

```
ft = -2 .* t .* u(t) + 6 .* u(t - 2) + 6 .* u(t - 5);
```

Finally, we write the commands to plot the result in the Figure Window.

```
plot (t,ft); grid on; xlabel ('t seconds');
        ylabel ('f(t)'); figure (gcf)
```

The MATLAB Command Window is shown in Fig. 2-45 and the Figure Window for the graph of this function is shown in Fig. 2-46. Notice in Fig. 2-46 that the graph clearly shows the step starting before $t = 2$ sec. This is caused by making the steps too far apart. This inaccuracy will be dramatically improved by changing the number of steps from 100 to 200, recalculating the *ft* matrix, and plotting the function again.

```
>>t = linspace(-1, 10, 100);
>>ft = -2 .* t.* u(t) + 6 .* u(t - 2) + .* u(t-5);
>>plot(t,ft); grid on; xlabel('t seconds'); ylabel('f(t)'); figure (gcf)
```

Figure 2-45

MATLAB Command Window for Example 2-31.

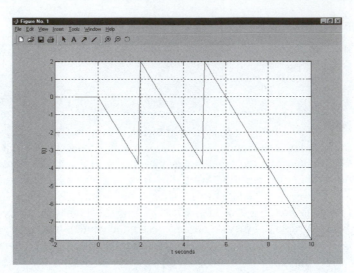

Figure 2-46

The previous examples demonstrate how MATLAB can be used to graph the waveforms of this section. Waveforms too complex to draw by hand are simple using MATLAB. By using the symbolic capabilities, we could find the integral and derivative of many analog signals. There are many other ways that MATLAB can be used with the material presented in the chapter.

2-12 USING THE TI-89

In this section and the following sections for the TI-89 calculator, two basic methods will be used to show what to enter into the calculator. One method shows the specific keystrokes and the other method shows how the entered keystrokes will appear on the Entry line at the bottom of the calculator's screen. The keystrokes method will be used for complicated keyboard entries and entries that do not display a change on the Entry line. As an example, the keystrokes method of storing the expression $x + 7$ into the memory location y is shown by

$$x \boxed{+} 7 \boxed{STO\blacktriangleright} y \boxed{ENTER}$$

and the Entry line method is shown by

$$x + 7 \rightarrow y$$

The next few examples demonstrate the power of symbolic entry for calculations and for graphing. Example 2-32 shows how to calculate the values of an expression by entering the algebraic expression and assigning the variable to a value. This example also shows how to create the $u(t)$ function that will be used in subsequent examples. Example 2-33 shows how to graph the equation from the first example on the Graph screen, and Example 2-34 shows how to graph a more complex piecewise function on the Graph screen.

EXAMPLE 2-32

Calculate the values of $f(t)$ in Table 2-2 for the entries $t = -2.3847$, 0.11532, 2.6153, 5.1153, and 7.6153. The function from the table is repeated here for convenience.

$$f(t) = 4e^{-0.5t} \sin\left(0.4\pi t + \frac{\pi}{3} \right) u(t)$$

Solution: The function has a $u(t)$ factor in it. Because the TI-89 does not have this function, we must create it. We begin by opening the Program Editor for creating a function. From the Home screen, enter the following keystrokes:

APPS 7:Program Editor 3:New. . .

Figure 2-47a shows how the calculator's screen should appear after entering these keystrokes. In the dialog box, enter the following keystrokes:

(▶) 2:Function ENTER (▼) (▼) u ENTER ENTER

Figure 2-47b shows how the screen should appear just before pressing the Enter key twice. Note that the folder is set to "ASP" but any folder will do. After the Enter key has been pressed twice, the Program Editor will appear and the calculator is ready to accept the programming code. Enter the code so that the display appears as shown in Fig. 2-47c.

The function will be automatically saved when exiting the Program Editor. The Program Editor is exited by pressing

2nd [QUIT]

The function that was just entered defines the switch function $u(t)$. The first line in Fig. 2-47c specifies the form that must be used to call this function. The "u" is the function name, and the "t" is the argument that will be passed to this function. The "t" is not a variable. It is really a placeholder for whatever expression we want to put in this location. The next line, "Func," tells the calculator that this program will be used as a function similar to how other functions, such as the sine function, are used in the calculator. The "If t < 0" will test whatever expression is passed for "t" to determine if its value is less than zero. If it is, then the next line of code, "Return 0," will be executed. This will return a value of 0 for the function's value. If the value of the expression for "t" is not less than zero, then the "Return 0" line will be skipped and the "Return 1" line will be executed. This will cause the value of 1 to be returned for the function's value.

Along with the $u(t)$ function, the expression we want to graph also contains a sinusoidal function. Whenever a sinusoidal function is involved, we must make sure that the calculator is in the correct mode of either radians or degrees. In this case, we will need the calculator in the radian mode. To put the calculator in the radian mode, from the Home screen enter the following keystrokes:

MODE (scroll to Angle) (▶) 1:RADIAN ENTER ENTER

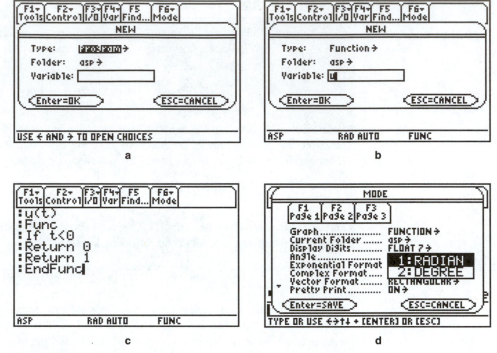

Figure 2-47

The calculator's display should look like Fig. 2-47d just before the Enter key is pressed twice.

Next, enter the function on the Entry line as follows:

$$4*e\hat{\ }(-0.5*t)*\sin(0.4*\pi*t + \pi/3)*u(t)|t = -2.3847$$

The "|" symbol is called the "with" symbol and, in this case, it is telling the calculator to replace the "t" in the expression with the value of -2.3847. After this is entered on the Entry line and the Enter key is pressed once, the value of 0 will be displayed. This is because when the expression is evaluated with $t = -2.3847$, the $u(t)$ will be equal to zero since the time is negative. If the expression did not have the $u(t)$ value, the calculator would have displayed the value -12.245.

After the Enter key is pressed and the value of 0 is displayed for $t = -2.3847$, press the ⊙ arrow. Then press the ⊙ arrow enough times to place the cursor just to the right of the equal symbol, and then press the CLEAR key only once. This will remove the -2.3847 so that the next value for t may be entered. This process is continued until all the values of t are calculated. Figure 2-48 shows the screen after the last value has been calculated.

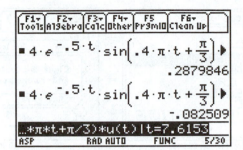

Figure 2-48

EXAMPLE 2-33

Display the plot of the following function on the Graph screen:

$$f(t) = 4e^{-0.5t} \sin\left(0.4\pi t + \frac{\pi}{3}\right) u(t)$$

Solution: Before graphing the function, we need to put the calculator into the function-graphing mode by entering the following keystrokes from the Home screen:

MODE (you will be on the "Graph" entry line) ▶ 1:FUNCTIONS
ENTER ENTER

Figure 2-49a shows the calculator's screen just before the Enter key is pressed twice.

There are two ways to graph the function. One way is from the Entry line on the Home screen and the other way is using the Y = Editor. We will use the direct entry method from the Home screen. First we will remove all the functions from the Y = Editor. (Any functions that you may want to keep may be deselected in the Y = Editor instead of removing them as the next list of keystrokes will do. See the calculator's guidebook for more information.) To clear all the functions use the following keystrokes:

APPS 2:Y=Editor F1 8:↓Clear Functions ENTER HOME

Figure 2-49b shows how the screen should appear just before the Enter and Home keys are pressed.

We must make use of the $u(t)$ function that was created in Example 2-32, since a $u(t)$ appears in the function we need to graph. If this function has not been created, go back to the previous example and enter it into the calculator before continuing with this example.

We may now enter the function to graph into the Entry line as follows:

4*e^(−0.5*t)*sin(0.4*π*t + π/3)u(t)|t = x

but do not press the Enter key just yet.

Notice on this line that we are assigning t to x. This is an easy way to change the variable instead of rewriting the expression. If the expression was not already in the History Area from the last example, we could simply enter the expression in with x's.

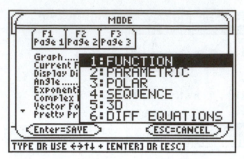 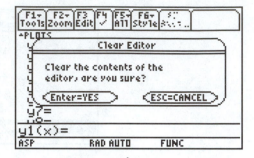

a   b

Figure 2-49

However it is done, we need to have the expression in terms of *x* because the graphics display is set up to graph functions of *x*.

Next we need to add some commands just before our expression on the Entry line that will clear the Graph screen and tell the calculator to graph our function. The following keystrokes will accomplish this:

2nd ◄ CATALOG (scroll to ClrGraph) ENTER 2nd [:]
CATALOG (scroll to Graph) ENTER

The line will now extend past what can be displayed on the Entry line, but the completed line should be as follows:

$$\text{ClrGraph:Graph } 4*e\wedge(-0.5*t)*\sin(0.4*\pi*t + \pi/3)*u(t)|t = x$$

Making sure that the calculator is in radian mode, we now press the Enter key. This will cause the calculator to display the Graph screen and plot our function. When the calculator has finished graphing the function, we may zoom in or out using the zoom commands from the toolbar at the top of the Graph screen. There are several more options shown on the toolbar at the top of the Graph screen that allow us to trace the graph line and display the exact values.

Figure 2-50 shows the display zoomed in and a cursor located at $x = 1$ and $y = 1.80296$. When you are done examining the graph, press the Home key to return to the Home display.

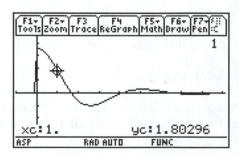

Figure 2-50

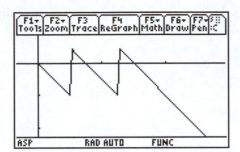

Figure 2-51

EXAMPLE **2-34**

For the function shown, display the plot on the Graph screen.

$$f(t) = -2t\,u(t) + 6\,u(t-1) + 6\,u(t-5)$$

Solution: We begin by making sure the calculator is in the proper mode for graphing. If you have not changed the settings from the last example, the following keystrokes are unnecessary, but if you do not have the calculator set up, then you must enter the following keystrokes:

MODE (you will be on the "Graph" entry line) ⊙ 1:FUNCTIONS
ENTER ENTER
APPS 2:Y=Editor ENTER F1 8:↓Clear Functions ENTER HOME

Writing our equation to plot the graph only requires one line entered on the Entry line.

ClrGraph:Graph $-2*t*u(t)+6*u(t-2)+6*u(t-5)|t=x$

Figure 2-51 shows a zoomed-in view of the graphed function.

The previous examples demonstrate how the TI-89 can calculate and graph analog signals, and from this we can surmise that the calculator is capable of calculating and graphing signals that are too complex to do by hand. In addition, since this calculator is capable of performing symbolic calculus, it can be used to calculate the relationship between the voltage and current on capacitors and inductors.

PROBLEMS

Section 2-1

1. Draw a graph of the functions.
   (a) $v(t) = 5\,u(t)$
   (b) $i(t) = -2\,u(t)$

2. Draw a graph of the function.

$$f(t) = 4\,u(t-3)$$

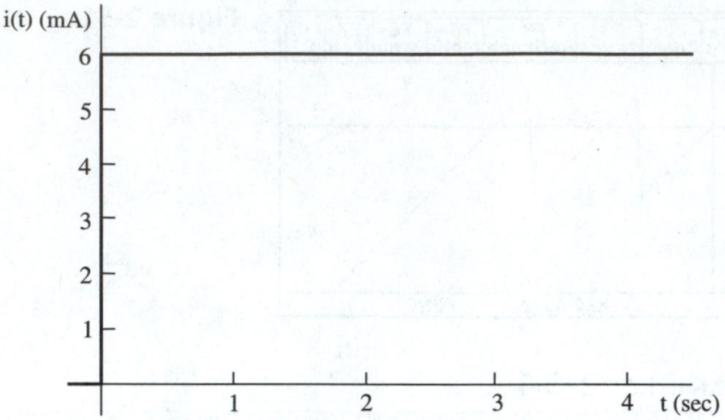

i(t) (mA)

Figure 2-52

3. Write the equation for Fig. 2-52.

Section 2-2

4. Draw a graph of the function.

$$f(t) = \frac{3}{5} t \, u(t)$$

5. Draw a graph of the function.

$$i(t) = (t - 2) \, u(t - 2)$$

6. Write the equation of Fig. 2-53.

Section 2-3

7. Show that the derivative of $f_3(t)$ as defined by Eq. (2-7) is equal to $f_2(t)$.

8. Show that the integral of $f_2(t)$ as defined by Eq. (2-7) is equal to $f_3(t)$. Assume zero for the initial condition.

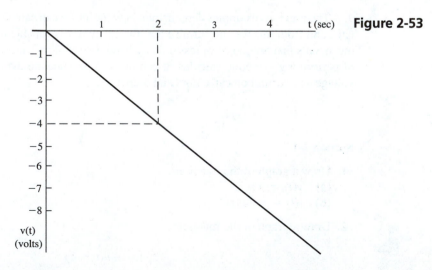

Figure 2-53

Section 2-4

9. Draw the graph of $f(t) = 4\,\delta(t)$.

10. Approximate the function as an impulse and draw the graph.

$$f(t) = 10\,u(t) - 10\,u(t - 0.01)$$

Section 2-5

11. For the function shown:
    (a) Find the time constant.
    (b) Find the damping constant.
    (c) Make a table of values for the function from $t = 0$ to when the function is zero (5τ) in steps of $\frac{1}{2}$ time constant.
    (d) Using the table in part (c), draw the function.

$$v(t) = 5e^{-2\times10^3 t}\,u(t)$$

12. Write the equation for Fig. 2-54.

Section 2-6

13. Convert the following functions into a form suitable for evaluating the functions on a calculator in radians and then in degrees.
    (a) $14 \sin(6t + 30°)$
    (b) $9 \cos(5t + 15°)$

14. For the function shown:
    (a) Find the frequency in rad/sec and hertz.
    (b) Find the voltage at $t = 0$.
    (c) Find the time where the peaks and zero crossings occur for the first cycle.

$$v(t) = 4 \sin(30t + 40°)\,u(t)$$

15. For the function shown:
    (a) Find the time the function decays to zero.
    (b) Find the value of the function at $t = 0$.

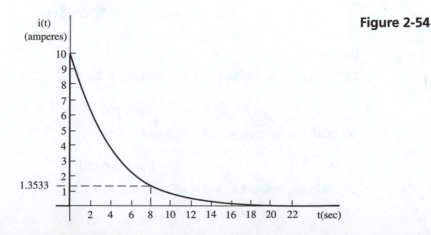

Figure 2-54

(c)   Make a table showing the times and values of the maximums, minimums, and zero crossings from $t = 0$ until the function is zero.

(d)   Draw the function from the table in part (c).

$$i(t) = 4e^{-2t} \sin(5t)\, u(t)$$

16. For the function shown:
    (a)   Find the time the function decays to zero.
    (b)   Find the value of the function at $t = 0$.
    (c)   Make a table showing the times and values of the maximums, minimums, and zero crossings from $t = 0$ until the function is zero.
    (d)   Draw the function from the table in part (c).

$$v(t) = 6e^{-20t} \sin(24\pi t + 35°)$$

17. Split the function into sine and cosine parts.

$$10 \sin(4\pi t + 57°)$$

18. Split the function into sine and cosine parts.

$$8e^{-2t} \sin(7\pi t + \pi/7)$$

19. Split the function into sine and cosine parts.

$$4 \cos(3\pi t + 40°)$$

20. Combine the function into a sine function with a phase angle.

$$4 \sin(5t) + 3 \cos(5t)$$

21. Combine the function into a sine function with a phase angle.

$$6e^{-7t} \sin(4\pi t) + 5e^{-7t} \cos(4\pi t)$$

22. Combine the function into a sine function with a phase angle.

$$4 \sin(2\pi t) + 3e^{-4t} \cos(5\pi t)$$

Section 2-7

23. Shift the function so that it will start at $t = 2$ sec.

$$4t\, u(t)$$

24. Shift the function so that it will start at $t = 5$ sec.

$$5 \sin(4\pi t)\, u(t)$$

25. Shift the function so that it will start at $t = 3$ μsec.

$$7e^{-2t} \sin(3t + 40°)\, u(t)$$

26. Shift the function back to the origin.

$$4\,\delta(t - 2)$$

27. Shift the function back to the origin.

$$4e^{(-2t+6)}\sin[4(t - 3) + 10°]\,u(t - 3)$$

28. Shift the function back to the origin.

$$(4t + 25)\,u(t - 5)$$

29. Draw a graph of the function.

$$v(t) = 5(t - 4)\,u(t - 4)$$

30. Draw a graph of the function.

$$i(t) = 10\,\delta(t - 1)$$

31. Draw a graph of the function.

$$v(t) = 5e^{-2(t-3)}\,u(t - 3)$$

32. Draw a graph of the function showing two complete cycles.

$$i(t) = 4\,\sin[5(t - 4)]\,u(t - 4)$$

Section 2-8

33. Draw a graph of the function.

$$f(t) = 2\,u(t) + 3\,u(t - 1) - u(t - 2) - 4\,u(t - 3)$$

34. Draw a graph of the function.

$$v(t) = 3\,\delta(t) + 10t\,u(t) - 30\,u(t - 3)$$

35. Draw a graph of the function.

$$i(t) = 8\,u(t) - 4t\,u(t) - 5\,\delta(t) + 8(t - 2)\,u(t - 2)$$

36. Draw a graph of the function.

$$f(t) = 5t\,u(t) - 5(t - 2)\,u(t - 2) - 10\,u(t - 4)$$

37. Draw a graph of the function.

$$v(t) = 4\,u(t) + 4e^{-0.5t}\sin(0.5\pi t + 30°)$$

38. Write the mathematical expression for Fig. 2-55.

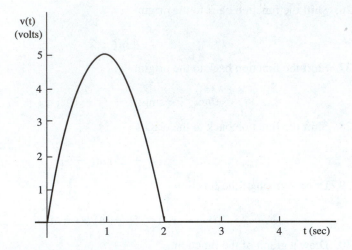

Figure 2-55

39. Write the mathematical expression for Fig. 2-56.

40. Write the mathematical expression for Fig. 2-57.

41. Write the mathematical expression for Fig. 2-58.

42. Write the mathematical expression for Fig. 2-59.

43. Write the mathematical expression for Fig. 2-60.

44. Draw a graph of the function.

$$f(t) = \sum_{n=0}^{\infty} 4(-1)^n \, u(t - 2n)$$

45. Draw a graph of the function.

$$v(t) = -5t \, u(t) + \sum_{n=0}^{\infty} 5 \, u(t - n)$$

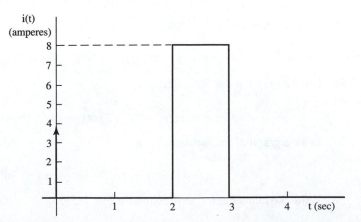

Figure 2-56

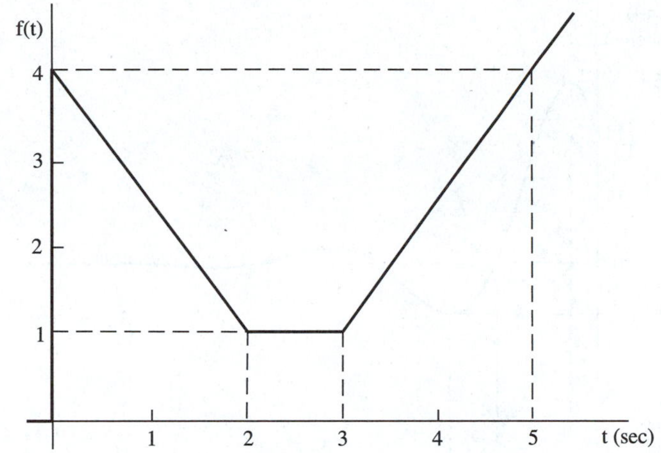

Figure 2-57

f(t) graph:

(graph showing f(t) starting at 4 at t=0, decreasing linearly to 1 at t=2, constant at 1 from t=2 to t=3, then increasing linearly to 4 at t=5 and continuing upward)

46. Draw a graph of the function.

$$i(t) = \sum_{n=0}^{\infty} 4(t - 5n)\, u(t - 5n) - u[t - (2 + 5n)]$$

$$- 4[t - (2 + 5n)]\, u[t - (2 + 5n)] + u[t - (3 + 5n)]$$

$$- 4[t - (3 + 5n)]\, u[t - (3 + 5n)]$$

$$+ 4[t - (5 + 5n)]\, u[t - (5 + 5n)]$$

47. A function generator has a square-wave output of 1 kHz. It has a positive swing of 20 V and a negative swing of 20 V. Draw a graph of the waveform and write the equation in a summation form. Assume that a positive swing starts at $t = 0$.

48. A function generator has a digital square-wave output with a variable duty cycle. Write the equation in a summation form for the output when it has a duty cycle of 70%. Assume that the signal goes to 5 V at $t = 0$, and assume that the signal is at 10 kHz.

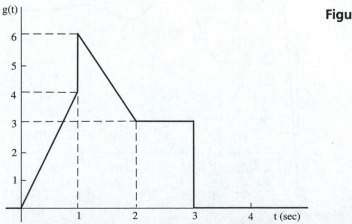

Figure 2-58

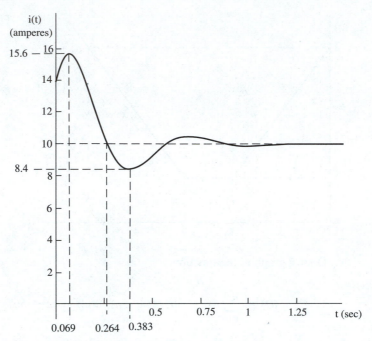

Figure 2-59

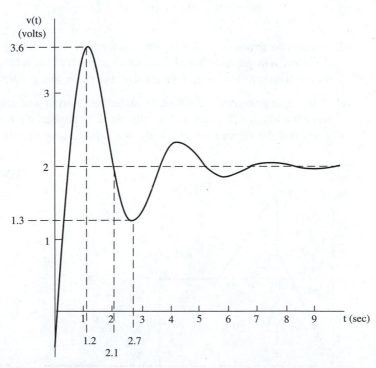

Figure 2-60

Figure 2-61

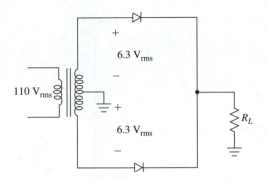

49. The full-wave rectifier shown in Fig. 2-61 has a 60-Hz 110-V rms signal applied to the input of the transformer. Assuming no voltage drop across the diodes, write the equation in a summation form for the output of the rectifier.

Section 2-9

50. Mathematically find the integral of the function.

$$f(t) = 4\,u(t) + 5\,\delta(t - 2) - 4\,u(t - 3)$$

51. Mathematically find the integral of the function.

$$f(t) = \delta(t) + 7\,u(t) - 8(t - 1)\,u(t - 1)$$

52. Mathematically find the derivative of the function.

$$f(t) = 3t\,u(t) - 3(t - 4)\,u(t - 4) - 12\,u(t - 4)$$

53. Mathematically find the derivative of the function.

$$f(t) = u(t) + 5(t - 2)^2\,u(t - 2)$$

54. Mathematically find the integral of the function.

$$f(t) = e^{-4}u(t) + \sin[3(t - 1) + 30°]\,u(t - 1)$$

55. Mathematically find the integral of the function.

$$f(t) = 5\,u(t - 1) + e^{-2(t-2)}\sin[4(t - 2)]\,u(t - 2)$$

56. Mathematically find the voltage across a 1-μF capacitor that has the current $i(t)$ applied to it.

$$i(t) = \delta(t) + t\,u(t) - (t - 2)\,u(t - 2)$$

57. Mathematically find the derivative of the function.

$$f(t) = 6e^{-3t}\,u(t) + \sin[4(t - 3) + 75°]\,u(t - 3)$$

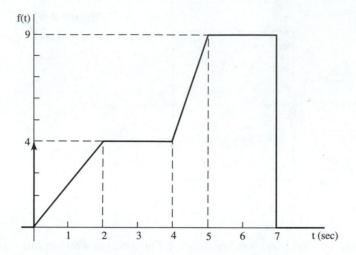

Figure 2-62

58. Mathematically find the derivative of the function.

$$f(t) = e^{-4(t-2)}\sin[3(t-2) + 40°]\,u(t-2)$$

59. Mathematically find the voltage across a 5-mH inductor that has the current $i(t)$ going through it.

$$i(t) = u(t) + 4(t-3)\,u(t-3)$$

60. Graphically find the integral of Fig. 2-62.

61. A 2-H inductor has across it the voltage shown in Fig. 2-63. Find the current through the inductor graphically and mathematically.

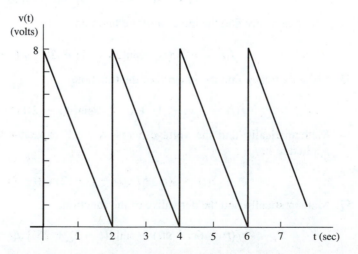

Figure 2-63

Figure 2-64

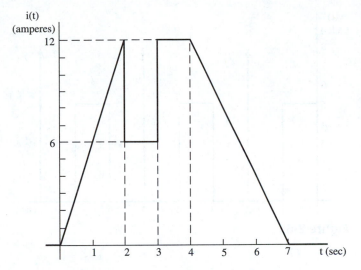

62. Graphically find the derivative of Fig. 2-64.

63. Figure 2-65 shows the current applied to a $\frac{1}{2}$-F capacitor. Write the equation for the voltage across the capacitor.

Section 2-10

64. Find the average value of the periodic waveform in Fig. 2-63.

65. Find the average value of the periodic waveform in Fig. 2-66.

66. Find the average value of the periodic waveform in Fig. 2-67.

67. Find the average value of the periodic waveform in Fig. 2-68.

Figure 2-65

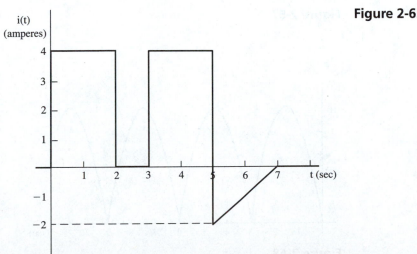

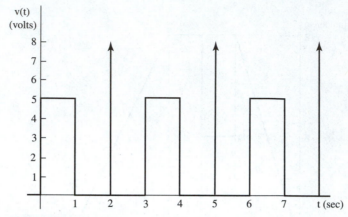

Figure 2-66

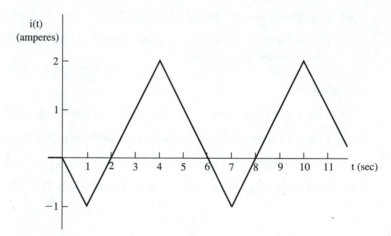

Figure 2-67

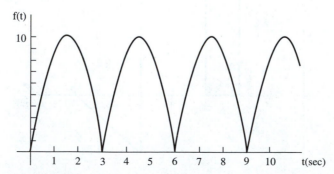

Figure 2-68

68. Simplify the following functions:
    (a) $5\,u(t-2)\,u(t-2)$
    (b) $3(t-4)\,u(t)\,u(t-4)$

69. Simplify the following functions:
    (a) $5e^{-(t-5)}\,u(t-3)\,u(t-5)$
    (b) $15(t-3)\,u(t-3)\,u(t-5)$

70. For Fig. 2-63,
    (a) Find the rms value of the voltage.
    (b) Determine the power dissipated by a 500-Ω resistor with this voltage across it.

71. For Fig. 2-67,
    (a) Find the rms value of the current.
    (b) Determine the power dissipated by a 20-Ω resistor with this current through it.

72. For Fig. 2-68,
    (a) Calculate the average value of the waveform.
    (b) Calculate the rms value of the waveform.

Laplace Transform

OBJECTIVES

Upon successful completion of this chapter, you should be able to:

- State the Laplace transform equation and derive the basic Laplace transform pairs using this equation.
- Define the standard form for expressing a Laplace function.
- Determine the Laplace transform of simple time-domain waveforms using a table of Laplace transform pairs.
- Determine the inverse Laplace transform of simple Laplace functions using a table of Laplace transform pairs.
- Determine the Laplace transform of complex waveforms that require the use of multiple Laplace transform operations at the same time.
- Create an expanded table of Laplace transform pairs by using the Laplace transform operations.
- Define and identify poles and zeros, and determine the order of each pole and zero.
- Find the inverse Laplace transform of first-order and multiple-order real poles using traditional techniques of partial fraction expansion or using the streamlined approach.
- Find the inverse Laplace transform of first-order complex poles using both the traditional complex numbers approach and the traditional real numbers approach.
- Find the inverse Laplace transform of first-order and multiple-order complex–conjugate poles using the streamlined approach.
- Solve integral–differential and differential equations for continuous systems and switched systems.

3-0 INTRODUCTION

This chapter covers finding the Laplace transform and the inverse Laplace transform of a function in depth. The depth presented here is intended not only for using Laplace transforms in electronics but also for using Laplace transforms in other areas such as control systems.

In this chapter, a traditional approach is blended with a modern streamlined approach. The traditional approach allows us to see more of why we do each mathematical step, and the streamlined approach allows us to do each mathematical step more efficiently. Both methods will work equally well, but the streamlined approach is more practical when applying Laplace transforms to electronic circuits.

Learning Laplace transforms is like learning how to ride a bicycle. You can watch someone ride for a long time but you will never learn until you get on the bicycle and try. You will not learn Laplace transforms by reading examples and watching someone work problems—you must work the problems.

This chapter will use quadratic equations in a special form, so we need to define the form before starting. The quadratic equation for the roots of

$$as^2 + bs + c$$

is usually expressed as

$$\text{roots} = \frac{-b \pm \sqrt{b^2 - 4ac}}{2a}$$

In our case, the coefficient of the squared variable is typically unity ($a = 1$). Therefore, if we let $a = 1$, the quadratic function becomes

$$s^2 + bs + c$$

and the quadratic equation becomes

$$\text{roots} = \frac{-b}{2} \pm \sqrt{\left(\frac{b}{2}\right)^2 - c}$$

If $(b/2)^2$ is $<c$, the function is complex and the roots will have a real part and an imaginary part:

$$\text{roots} = -\alpha \pm j\omega$$

When this is the case we can use a factored form that shows us the roots:

$$[(s + \alpha)^2 + \omega^2]$$

3-1 LAPLACE TRANSFORMS BY TABLE

3-1-1 The Laplace Integral

The Laplace transform is defined as

$$F(s) = \int_0^\infty f(t)\, e^{-st}\, dt \tag{3-1}$$

This can also be stated in a shorthand notation that we will use most of the time.

$$\mathscr{L}[f(t)] = F(s) \tag{3-2}$$

The time-domain function and its corresponding Laplace function are called a *transform pair.*

This is a single-sided Laplace transform since it goes from zero to infinity instead of from minus infinity to plus infinity. This means that all of our time-domain functions must exist at or after $t = 0$. This implies that all time functions are multiplied by $u(t - a)$, where a must be equal to or greater than zero.

Note that functions in the time domain are shown with lowercase characters and functions in the Laplace domain are shown with uppercase characters.

Our purpose is to learn Laplace transforms as a tool, but just to take some of the mystery out of it, we will derive a transform pair.

EXAMPLE 3-1

Find $\mathscr{L}[u(t)]$ using the Laplace integral.

Solution: Using Eq. (3-1) and given $f(t) = u(t)$,

$$F(s) = \int_0^\infty f(t) \, e^{-st} \, dt$$

$$= \int_{0^+}^\infty u(t) \, e^{-st} \, dt$$

$$= \int_0^\infty 1 e^{-st} \, dt$$

We recognize the integral form

$$\int_0^\infty e^u \, du = e^u \Big|_0^\infty$$

Letting

$$u = -st$$

we get

$$du = -s \, dt$$

Therefore,

$$\frac{-1}{s} \int_0^\infty e^{-st} (-s \, dt)$$

$$\frac{-e^{-st}}{s} \Big|_0^\infty = 0 - \frac{-1}{s}$$

The result is

$$\mathscr{L}[u(t)] = \frac{1}{s}$$

Appendix A shows the Laplace transform of several functions that we will need. Each of the transform pairs shown in Section A-1 in Appendix A are derived in Appendix B using Eq. (3-1), as demonstrated in the previous example. Appendix A also contains a table of transform operations and a list of transform identities that we often need to reference as we learn Laplace transforms.

Let's first look at the identities listed in Section A-3. The first entry (I-1) shows the proper way to write the Laplace transform of a function $f(t)$ as stated in Eq. (3-2) and the inverse Laplace of a function $F(s)$. The second entry (I-2) states that $f(t)$ is not equal to $F(s)$ but that

$$f(t) \Leftrightarrow F(s)$$

which is read "$f(t)$ transforms to $F(s)$."

Entry (I-3) show that a constant can be moved inside or outside the Laplace and inverse Laplace notation. Entry (I-4) shows that the Laplace transform of a sum of functions is equal to the sum of the Laplace transform of each function, and the inverse Laplace transform of a sum of functions is equal to the sum of the inverse Laplace transform of each function. Both of these identities are true because the Laplace transform is an integral function.

3-1-2 Laplace Transform by Table

The Laplace transform of a function will be found by using the transform pairs and the transform operations in Appendix A. We will begin with simple examples requiring only the transform pairs, and then in a later section we will use examples requiring the transform operations.

In the following examples, each time-domain function is replaced by its Laplace-domain counterparts listed in Appendix A, and then the Laplace function is put in a standard form. The standard form can be stated as follows:

Standard form: A single fraction having the numerator and denominator factored with real roots in the form of $(s + \alpha)$ and complex roots in the form of $[(s + \alpha)^2 + \omega^2]$. Time-shifted expressions are shown in separate terms.

EXAMPLE 3-2

Find the Laplace transform of the voltage

$$v(t) = 2\,\delta(t) + 10\,u(t) + 12t\,u(t)$$

Solution:

$$V(s) = \mathscr{L}[2\,\delta(t) + 10\,u(t) + 12t\,u(t)]$$

Using identity (I-4) gives

$$= \mathscr{L}[2\,\delta(t)] + \mathscr{L}[10\,u(t)] + \mathscr{L}[12t\,u(t)]$$

Using identity (I-3) yields

$$= 2\mathcal{L}[\delta(t)] + 10\mathcal{L}[u(t)] + 12\mathcal{L}[t\,u(t)]$$

Using transform pairs (P-1), (P-2), and (P-3), respectively, leads to

$$= 2 + \frac{10}{s} + \frac{12}{s^2}$$

Then we put the answer in a standard form.

$$V(s) = \frac{2(s + 2)(s + 3)}{s^2}$$

EXAMPLE 3-3

Find $\mathcal{L}[\sin(2t + 35°)]$.

Solution: Since we are using single-sided Laplace transforms, this expression (or any other function) must be assumed to be multiplied by $u(t)$. First we must split the sine function into a sine and cosine.

$$1\ \underline{/35°} = 0.81915 + j0.57358$$

Therefore,

$$\sin(2t + 35°) = 0.81915 \sin(2t) + 0.57358 \cos(2t)$$

Using (P-5) and (P-6) gives

$$\frac{(0.81915)(2)}{s^2 + 4} + \frac{0.57358s}{s^2 + 4}$$

$$\mathcal{L}[\sin(2t + 35°)] = \frac{0.57358(s + 2.8563)}{s^2 + 4}$$

3-2 INVERSE LAPLACE TRANSFORM BY TABLE

Finding the inverse Laplace transform by table is simply a matter of rewriting the function as a sum of Laplace expressions that appear in the table of transform pairs, and then replacing each term with the time-domain counterpart. This method can get quite complicated and difficult on complex functions. Therefore, we will limit its use to simple functions and use other, more powerful techniques for complicated expressions.

EXAMPLE **3-4**

Find

$$\mathcal{L}^{-1}\left[\frac{5}{s}\right]$$

Solution: This is just a case of factoring out the 5 so that we can use (P-2).

$$5\mathcal{L}^{-1}\left[\frac{1}{s}\right] = 5\,u(t)$$

EXAMPLE **3-5**

Find

$$\mathcal{L}^{-1}\left[\frac{s+6}{s^2+9}\right]$$

Solution: From the denominator we see that this is a sinusoidal type of function. We can just split the numerator to see that this is a sine function and a cosine function.

$$\mathcal{L}^{-1}\left[\frac{s}{s^2+3^2}\right] + \mathcal{L}^{-1}\left[\frac{(2)(3)}{s^2+3^2}\right]$$

Using (P-6) and (P-5), we have

$$\cos(3t) + 2\sin(3t)$$

and combining into a single sine function gives

$$2 + j = 2.2361\ \underline{/26.565°}$$

$$2.2361\sin(3t + 26.565°)$$

3-3 LAPLACE OPERATIONS

By using only a few transform pairs and the transform operations, we will be able to find the Laplace transform for a large variety of functions. The transform operations are very powerful and we will go through an example of each. The transform operations are shown in Section A-2 in Appendix A.

There is a general procedure that is followed when using the Laplace operations. When this procedure is understood, the application of all of the Laplace operations will be relatively easy. The procedure can be described as follows:

1. State the Laplace operation that will be used.
2. Set the time-domain side of the stated operation equal to the function for which the Laplace transform is being sought, and then solve for $h(t)\, u(t)$.
3. State a Laplace transform pair that applies to $h(t)\, u(t)$, and determine $H(s)$.
4. Perform the Laplace operation indicated in step 1.

3-3-1 Shifted Function Operation

This operation allows us to find the Laplace and the inverse Laplace transform of any time-shifted function. In the time domain, the time-shifting operation is performed by replacing "t" in $h(t)\, u(t)$ with "$t - a$" to form a time-shifted function $h(t - a)\, u(t - a)$. This same time-shifting operation can be performed in the Laplace domain by multiplying $H(s)$ by "e^{-as}" to form the Laplace time-shifted function $e^{-as}H(s)$. The reverse can also be performed. To time shift back to the origin in the time domain, we replace "$t - a$" with "t," and to time shift back to the origin in the Laplace domain, we remove "e^{-as}." This process is expressed in transform operation (O-1).

$$h(t - a)\, u(t - a) \Leftrightarrow e^{-as}H(s)$$

Since all of the other transform pairs and transform operations are for unshifted functions, we must always shift time-shifted functions to the origin first before using any entries in either table. All the other transform operations may be used in any order, but if time shifting is involved, you must shift the time-domain function to the origin as the first step and time shift it back out in the Laplace domain as the last step. This first and last rule applies to finding the inverse Laplace transform as well.

EXAMPLE 3-6

Find

$$\mathcal{L}[10\cos(8(t - 5))\, u(t - 5)$$

Solution: First we state the operation we will be using.

$$h(t - a)\, u(t - a) \Leftrightarrow e^{-as}\, H(s)$$

Next, the time-domain side of this operation is set equal to the function for which the Laplace transform is sought.

$$h(t - a)\, u(t - a) = 10\cos[8(t - 5)]\, u(t - 5)$$

Then we solve for $h(t) u(t)$. In this case, we assign $a = 5$, and then time shift both sides to the origin. Therefore,

$$h(t) u(t) = 10 \cos(8t) u(t)$$

We can now determine $H(s)$ by using transform pair (P-6).

$$\cos(\omega t) u(t) \Leftrightarrow \frac{s}{s^2 + \omega^2}$$

$$H(s) = \frac{10s}{s^2 + 64}$$

Finally, we multiply both sides by e^{-as}, where $a = 5$, to complete transform operation (O-1).

$$e^{-as} H(s) = \frac{10s\, e^{-5s}}{s^2 + 64}$$

Therefore,

$$\mathscr{L}[10 \cos[8(t - 5)]\, u(t - 5)] = \frac{10s\, e^{-5s}}{s^2 + 64}$$

EXAMPLE 3-7

Find

$$\mathscr{L}[4\, \delta(t - 1) + 7\, u(t - 2) - 2\, u(t - 3) - 8(t - 3)\, u(t - 3)]$$

Solution: Because all of the terms are time shifted, we must use the time-shifting operation (O-1).

$$h(t - a)\, u(t - a) \Leftrightarrow e^{-as} H(s)$$

Each of the four terms is set equal to the time-domain side of this operation. Notice that the first term with the impulse does not have a $u(t)$ multiplier. This term does not require a $u(t)$ multiplier because $\delta(t)$ is zero for $t < 0^-$.

Listing each term separately and solving for $h(t) u(t)$,

$$h_1(t - a)\, u(t - a) = 4\delta(t - 1) \qquad \Rightarrow h_1(t)\, u(t) = 4\delta$$

$$h_2(t - a)\, u(t - a) = 7u(t - 2) \qquad \Rightarrow h_2(t)\, u(t) = 7\, u(t)$$

$$h_3(t - a)\, u(t - a) = -2u(t - 3) \qquad \Rightarrow h_3(t)\, u(t) = -2\, u(t)$$

$$h_4(t - a)\, u(t - a) = -8(t - 3)u(t - 3) \Rightarrow h_4(t)\, u(t) = -8t\, u(t)$$

Using the appropriate transform pair, we can find the respective $H(s)$ for each term.

$$\delta(t) \Leftrightarrow 1 \qquad \Rightarrow \mathcal{L}[4\,\delta(t)] \quad = H_1(s) = 4$$

$$u(t) \Leftrightarrow \frac{1}{s} \qquad \Rightarrow \mathcal{L}[7\,u(t)] \quad = H_2(s) = \frac{7}{s}$$

$$u(t) \Leftrightarrow \frac{1}{s} \qquad \Rightarrow \mathcal{L}[-2\,u(t)] = H_3(s) = \frac{-2}{s}$$

$$t\,u(t) \Leftrightarrow \frac{1}{s^2} \qquad \Rightarrow \mathcal{L}[-8t\,u(t)] = H_4(s) = \frac{-8}{s^2}$$

Each of these terms is then multiplied by e^{-as} to time shift them back out in the Laplace domain.

$$\mathcal{L}[4\,\delta(t-1)] \qquad\qquad = e^{-as}\,H_1(s) = 4e^{-s}$$

$$\mathcal{L}[7\,u(t-2)] \qquad\qquad = e^{-as}\,H_2(s) = \frac{7e^{-2s}}{s}$$

$$\mathcal{L}[-2\,u(t-3)] \qquad\qquad = e^{-as}\,H_3(s) = \frac{-2e^{-3s}}{s}$$

$$\mathcal{L}[-8(t-3)\,u(t-3)] = e^{-as}\,H_4(s) = \frac{-8e^{-3s}}{s^2}$$

Each term is then summed together.

$$4e^{-s} + \frac{7e^{-2s}}{s} + \frac{-2e^{-3s}}{s} + \frac{-8e^{-3s}}{s^2}$$

Now we combine all the terms with the same time shift.

$$4e^{-s} + \frac{7e^{-2s}}{s} + \frac{-2(s+4)e^{-3s}}{s^2}$$

Sometimes a function is not in the proper form of a shifted function. That is, the function does not have the same form as the step function's argument. There are two ways to change a function into a shifted form. Using an algebraic method, we multiply out, add and subtract terms, and collect terms appropriately so that we get a proper shifted form. Using a graphic method, we draw the function and rewrite the equation from the graph.

EXAMPLE 3-8

Find the Laplace transform of the function $v(t)$. Show both methods to force the function into a proper shifted form.

$$v(t) = (2t - 1) u(t - 2)$$

Solution: This function is not in the proper form of a shifted function because $(2t - 1)$ is not in the form of the step function's argument, $(t - 2)$.

Algebraic method. First multiply out.

$$v(t) = 2t u(t - 2) - u(t - 2)$$

Then add terms to the function to create the form needed, and subtract the same terms so that the function's value will not change.

$$v(t) = 2t u(t - 2) - [4 u(t - 2)] - u(t - 2) + [4 u(t - 2)]$$

$$v(t) = (2t - 4) u(t - 2) + (-1 + 4) u(t - 2)$$

$$v(t) = 3 u(t - 2) + 2(t - 2) u(t - 2)$$

Graphic method. Figure 3-1a shows the function $(2t - 1)$, and Fig. 3-1b shows this function multiplied by $u(t - 2)$. We write the equation from Fig. 3-1b using methods shown in Chapter 2.

$$v(t) = 3 u(t - 2) + 2(t - 2) u(t - 2)$$

Now that we have $v(t)$ in a proper shifted form, we can write the Laplace transform. All of the terms are time shifted, so transform operation (O-1) will be used.

$$h(t - a) u(t - a) \Leftrightarrow e^{-as} H(s)$$

Setting each term equal to the time-domain side of this operation and solving for $h(t)$ $u(t)$, we have

$$h_1(t - a) u(t - a) = 3u(t - 2) \qquad \Leftrightarrow h_1(t) u(t) = 3 u(t)$$

$$h_2(t - a) u(t - a) = 2(t - 2) u(t - 2) \Leftrightarrow h_2(t) u(t) = 2t u(t)$$

Using the appropriate transform pair, we can find the respective $H(s)$ for each term.

$$u(t) \Leftrightarrow \frac{1}{s} \quad \Rightarrow \mathcal{L}[3 u(t)] = H_1(s) = \frac{3}{s}$$

$$t u(t) \Leftrightarrow \frac{1}{s^2} \quad \Rightarrow \mathcal{L}[2t u(t)] = H_2(s) = \frac{2}{s^2}$$

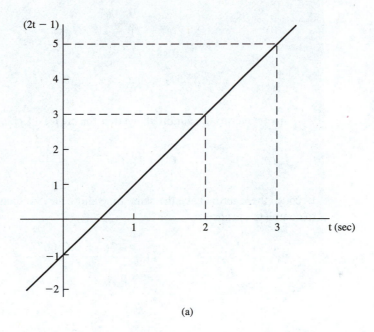

(a)

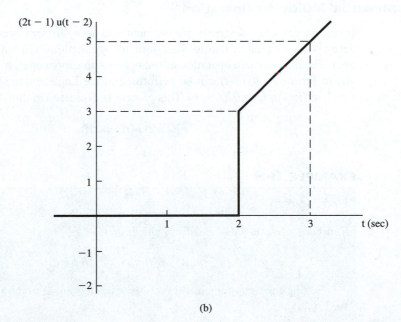

(b)

Figure 3-1

Each of these terms is then multiplied by e^{-as} to time shift it back out in the Laplace domain.

$$\mathcal{L}[3\,u(t-2)] = e^{-as}\,H_1(s) = \frac{3e^{-2s}}{s}$$

$$\mathcal{L}[2(t-2)\,u(t-2)] = e^{-as}\,H_2(s) = \frac{2e^{-2s}}{s^2}$$

The two terms are summed together for the total Laplace voltage.

$$V(s) = \frac{3e^{-2s}}{s} + \frac{2e^{-2s}}{s^2}$$

Because these terms have the same time shift, they are combined into a single fraction to complete the answer.

$$V(s) = \frac{3(s + 0.66667)e^{-2s}}{s^2}$$

3-3-2 Exponential Multiplier Operation

In electronics a lot of signals are transient and decay to zero over a period of time. This decay is exponential by nature. Therefore, many functions will have an exponential multiplier. The time-domain operation of multiplying an exponential, e^{-bt}, times a function $h(t)$ $u(t)$ to form $e^{-bt}\,h(t)\,u(t)$ can be performed in the Laplace domain by setting s equal to $s + b$ in $H(s)$ to form $H(s + b)$. This process is expressed in transform operation (O-2).

$$e^{-bt}\,h(t)\,u(t) \Leftrightarrow H(s + b)$$

EXAMPLE 3-9

Find $\mathcal{L}[3te^{-2t}]$.

Solution: We may use transform operation (O-2).

$$e^{-bt}\,h(t)\,u(t) \Leftrightarrow H(s + b)$$

Setting the time-domain side of this operation equal to the function, we can solve for $h(t)\,u(t)$.

$$e^{-bt}\,h(t)\,u(t) = 3t\,e^{-2t}$$

$$h(t)\,u(t) = 3t$$

We notice that there is no $u(t)$ on the right side of this equation, but a $u(t)$ may be multiplied times the function because the single-sided Laplace transform integral has a lower limit of zero. Therefore,

$$h(t)\,u(t) = 3t\,u(t)$$

Using (P-3), we determine $H(s)$.

$$t\,u(t) \Leftrightarrow \frac{1}{s^2}$$

$$H(s) = \frac{3}{s^2}$$

We then apply the right side of the operation to find the total Laplace transform

$$H(s + b) = \frac{3}{(s + 2)^2}$$

Therefore,

$$\mathcal{L}[3te^{-2t}\,u(t)] = \frac{3}{(s + 2)^2}$$

EXAMPLE 3-10

Find the Laplace transform.

$$i(t) = 5e^{-3t}\sin(7t)\,u(t)$$

Solution: This function requires the use of transform operation (O-2).

$$e^{-bt}\,h(t)\,u(t) \Leftrightarrow H(s + b)$$

Setting the time-domain side of this operation equal to the function, we can solve for $h(t)\,u(t)$.

$$e^{-bt}\,h(t)\,u(t) = 5e^{-3t}\sin(7t)\,u(t)$$

$$h(t)\,u(t) = 5\sin(7t)\,u(t)$$

$H(s)$ is found using transform pair (P-5).

$$\sin(\omega t)\, u(t) \iff \frac{\omega}{s^2 + \omega^2}$$

$$H(s) = \frac{(5)(7)}{s^2 + 7^2} = \frac{35}{s^2 + 49}$$

We then replace each s with $s + b$, as indicated in the transform operation. Therefore, we find

$$H(s + b) = \frac{35}{(s + 3)^2 + 49}$$

Therefore,

$$I(s) = \mathcal{L}[e^{-3t}5\,\sin(7t)\, u(t)] = \frac{35}{[(s + 3)^2 + 49]}$$

3-3-3 *t* Multiplier Operation

A function multiplied by t will appear occasionally. The time-domain operation of multiplying t times $h(t)\, u(t)$ can be performed in the Laplace domain by taking the negative derivative, with respect to s, of $H(s)$. This process is expressed in transform operation (O-3).

$$t\, h(t)\, u(t) \iff -\frac{d}{ds}\, H(s)$$

This involves taking the derivative of a function that, in most cases, is a fraction. The derivative of a fraction is given by

$$\frac{d}{ds}\left[\frac{F(s)}{G(s)}\right] = \frac{F'(s)\, G(s) - G'(s)\, F(s)}{G^2(s)} \qquad \text{(3-3)}$$

EXAMPLE 3-11

Find $\mathcal{L}[3te^{-2t}]$.

Solution: This is the same as Example 3-9, but in this example we will use transform operation (O-3).

$$t\, h(t)\, u(t) \iff -\frac{d}{ds}\, H(s)$$

Setting the time-domain side of this operation equal to the function, we solve for $h(t)\,u(t)$

$$t\,h(t)\,u(t) = 3te^{-2t}$$

$$h(t)\,u(t) = 3e^{-2t}$$

We notice that there is no $u(t)$ on the right side of this equation, but a $u(t)$ may be multiplied times the function because the single-sided Laplace transform integral has a lower limit of zero. Therefore,

$$h(t)\,u(t) = 3e^{-2t}\,u(t)$$

Using (P-4), we determine $H(s)$.

$$e^{-bt}\,u(t) \Leftrightarrow \frac{1}{(s+b)}$$

$$H(s) = \frac{3}{(s+2)}$$

We then apply the right side of the operation to find the total Laplace transform.

$$-\frac{d}{ds}H(s) = -\frac{d}{ds}\frac{3}{(s+2)}$$

$$= -\frac{(0)(s+2) - (1)(3)}{(s+2)^2}$$

$$= \frac{3}{(s+2)^2}$$

Therefore,

$$\mathcal{L}[3te^{-2t}\,u(t)] = \frac{3}{(s+2)^2}$$

The last example and Example 3-9 demonstrate how there is often more than one way to find the Laplace transform of a function.

EXAMPLE 3-12

Find the Laplace transform of the voltage

$$v(t) = 5t^2\,u(t)$$

Solution: There is no Laplace transform pair for t^2, but there is a transform pair for $t\,u(t)$ and a transform operation for a t multiplier. Therefore, we will consider $v(t)$ to be $5t$ multiplied by t and use the transform operation (O-3).

$$t\,h(t)\,u(t) \;\Leftrightarrow\; -\frac{d}{ds}\,H(s)$$

Setting the time-domain side of this operation equal to the function, we solve for $h(t)$ $u(t)$.

$$t\,h(t)\,u(t) = 5t^2\,u(t)$$

$$h(t)\,u(t) = 5t\,u(t)$$

Using (P-3), we determine $H(s)$.

$$t\,u(t) \;\Leftrightarrow\; \frac{1}{s^2}$$

$$H(s) = \frac{5}{s^2}$$

We then apply the right side of the operation to find the total Laplace transform.

$$V(s) = -\frac{d}{ds}\,H(s) = -\frac{d}{ds}\frac{5}{s^2}$$

$$= -\frac{(0)(s^2) - (2s)(5)}{s^4}$$

$$= \frac{10}{s^3}$$

Therefore,

$$\mathcal{L}[5t^2\,u(t)] = \frac{10}{s^3}$$

3-3-4 Derivative Operation

The Laplace transform can be used to find the derivative of functions that may be too complex to find in the time domain. There are actually two different operations that apply. The first Laplace operation (O-4a) applies to a differentiated continuous function that is switched on at $t = 0$. This would happen when a differentiating amplifier has a switch on its output to connect the signal to the rest of the circuit at $t = 0$. The second Laplace operation (O-4b) applies when the signal is switched on at $t = 0$ and then differentiated. This would happen when the differentiating amplifier has a switch on its input and the signal is applied to the differentiating circuit at $t = 0$.

The time-domain operation of differentiating a continuous function is expressed in transform operation (O-4a).

$$\left\{ \frac{d}{dt} [h(t)] \right\} u(t) \iff s\, H(s) - h(0)$$

and the time-domain operation of differentiating a function that is switched on at $t = 0$ is expressed in transform operation (O-4b).

$$\frac{d}{dt} [h(t)\, u(t)] \iff s\, H(s)$$

The difference in the Laplace side of these functions is the $h(0)$ term. This is the time-domain value of the function at $t = 0^+$. The reason for subtracting this term in (O-4a) stems from the fact that the Laplace integral's lower limit is 0, and, therefore, whenever we take the Laplace transform of any function, we in effect multiply it by $u(t)$. When we differentiate a function, $h(t)$, that contains a step at $t = 0$, it creates an impulse with an area equal to the value of the function at $t = 0$. Therefore, if our intent is to use a signal that is switched on at the output of a differentiating circuit, we must remove the impulse that is inherently created by taking the Laplace transform of the signal. The impulse is removed by subtracting $h(0)$ from the Laplace transform of the signal.

This issue can be completely avoided by using only functions that are equal to zero at $t = 0$. With these types of functions, both operation (O-4a) and (O-4b) will yield the same result. However, in a real circuit it is unlikely that the signal will always be zero at the time the signal is switched on.

EXAMPLE 3-13

Find the Laplace transform of the function $2\cos(3t)$ when it is switched on at the output of a differentiating circuit.

Solution: The signal is switched on at the output of the differentiating circuit, so we are finding the Laplace transform of

$$\left\{ \frac{d}{dt} [2\cos(3t)] \right\} u(t)$$

When the function is switched on at the output of the differentiating circuit, we must remove the impulse created by taking the Laplace transform of the signal. Therefore, we must use transform operation (O-4a).

$$\left\{ \frac{d}{dt} [h(t)] \right\} u(t) \iff s\, H(s) - h(0)$$

Setting the time-domain side of this operation equal to the function, we solve for $h(t)\,u(t)$

$$\left\{\frac{d}{dt}[h(t)]\right\}u(t) = \left\{\frac{d}{dt}[2\cos(3t)]\right\}u(t)$$

$$h(t)\,u(t) = 2\cos(3t)\,u(t)$$

In operation (O-4a) we will need the quantity $h(0)$; therefore,

$$h(t) = 2\cos(3t)$$

$$h(0) = 2\cos(0)$$

$$= 2$$

Using (P-6), we determine $H(s)$.

$$\cos(\omega t)\,u(t) \Leftrightarrow \frac{s}{s^2 + \omega^2}$$

$$H(s) = \frac{2s}{(s^2 + 9)}$$

We then apply the right side of the operation to find the total Laplace transform.

$$s\,H(s) - h(0) = s\left(\frac{2s}{(s^2 + 9)}\right) - 2$$

$$= \frac{-18}{s^2 + 9}$$

Therefore,

$$\mathcal{L}\left[\left\{\frac{d}{dt}[2\cos(3t)]\right\}u(t)\right] = \frac{-18}{s^2 + 9}$$

If we take the inverse of the answer in the last example, we see that the result is -6 $\sin(3t)\,u(t)$, which is the derivative of the function $2\cos(3t)$. The next example demonstrates the process for transform operation (O-4b).

EXAMPLE 3-14

Find the Laplace transform of the function $2\cos(3t)$ when it is switched on at the input of a differentiating circuit.

Solution: The signal is switched on at the input of the differentiating circuit, so we are finding the Laplace transform of

$$\frac{d}{dt}[2\cos(3t)\,u(t)]$$

When the function is switched on at the input of the differentiating circuit, it will cause the output of the differentiating circuit to have an impulse. This step function inherently occurs when we take the Laplace transform of a function, so we will not have to adjust our answer by removing the impulse. Therefore, we must use transform operation (O-4b).

$$\frac{d}{dt}[h(t)\,u(t)] \Leftrightarrow s\,H(s)$$

Setting the time-domain side of this operation equal to the function, we solve for $h(t)\,u(t)$.

$$\frac{d}{dt}[h(t)\,u(t)] = \frac{d}{dt}[2\cos(3t)\,u(t)]$$

$$h(t)\,u(t) = 2\cos(3t)\,u(t)$$

Note that in operation (O-4b) we do not need to find the quantity $h(0)$.

Using (P-6), we determine $H(s)$.

$$\cos(\omega t)\,u(t) \Leftrightarrow \frac{s}{s^2 + \omega^2}$$

$$H(s) = \frac{2s}{(s^2 + 9)}$$

We then apply the right side of the operation to find the total Laplace transform.

$$s\,H(s) = s\left(\frac{2s}{(s^2 + 9)}\right)$$

$$= \frac{2\,s^2}{s^2 + 9}$$

Therefore,

$$\mathcal{L}\left[\frac{d}{dt}[2\cos(3t)\,u(t)]\right] = \frac{2s^2}{s^2+9}$$

3-3-5 Integral Operation

This operation allows us to find the Laplace transform of the integral of a function. The time-domain operation of taking the integral from 0 to ∞ of the function $h(t)$ can be performed in the Laplace domain by dividing $H(s)$ by s. The integral sums the area under a function's curve and is not concerned with the slope of the function's curve, so we will not have the problem with a switched function's infinite slope as we did with the derivative operation. The integral process is expressed in transform operation (O-5).

$$\int_0^t h(t)\,u(t)\,dt \iff \frac{H(s)}{s}$$

This transform operation, along with the derivative operation, is significant in circuit analysis. To determine the values of voltage and current at any time in many ac circuits, we must use the integral and derivative. Therefore, the equations that we must solve in the time domain are simultaneous integral–differential equations. However, in Laplace transforms, the integral–differential equations are composed of algebraic expressions. This conversion from a calculus process to an algebraic process is the major reason for using Laplace transforms in circuit analysis.

EXAMPLE 3-15

Find the Laplace transform of $\int_0^t 7\,u(t)\,dt$ using operation (O-5) and transform pair (P-2).

Solution: From the Laplace transform table, operation (O-5) is

$$\int_0^t h(t)\,u(t)\,dt \iff \frac{H(s)}{s}$$

Setting the time-domain side of this operation equal to the function, we have

$$\int_0^t h(t)\,u(t)\,dt = \int_0^t 7\,u(t)\,dt$$

$$h(t)\,u(t) = 7\,u(t)$$

We can now determine $H(s)$ by using transform pair (P-2).

$$u(t) \Leftrightarrow \frac{1}{s}$$

$$H(s) = \frac{7}{s}$$

We then apply the right side of the operation to find the total Laplace transform.

$$\frac{H(s)}{s} = \frac{(7/s)}{s}$$

$$= \frac{7}{s^2}$$

Therefore, we get

$$\mathcal{L}\left[\int_0^t 7\, u(t)\, dt\right] = \frac{7}{s^2}$$

Notice that if we were not restricted in the problem to using (O-5) and (P-2), we could have found the integral in the time domain and found the Laplace transform using transform pair (P-3).

3-3-6 Combining Transform Operations

Often, a function requires the use of two or more transform operations. For example, a function may have an exponential multiplier and also be time shifted, requiring the use of both operation (O-2) and operation (O-1). Developing an organized approach when more than one operation is involved is critical. Once an organized approach is established, a function requiring two or more transform operations is no more difficult than working two or more separate problems, each requiring a single transform operation. The procedure for using multiple transform operations can be summarized as follows:

1. State the Laplace operations that may be required.
2. Set the time-domain side of one of the stated operations equal to the function that the Laplace transform is being found for, and solve for $h(t)\, u(t)$. If one of the operations is the time-shifted operation (O-1), then it must be used first. All the other operations may be used in any order.
3. If there is no transform pair to determine $H(s)$ from the resulting $h(t)\, u(t)$, then go back to step 2 and set this remaining function equal to another operation. Keep looping back until the inverse of an $H(s)$ can be found using a transform pair.

4. Once a transform pair is used, perform each Laplace operation applied in step 2 from the most recently used operation to the very first operations.

In the following examples, notice how subscripts are used to keep the operation in each loop separate. Also notice that when each new operation is introduced, the work shown is indented. The combination of these two techniques will help us keep organized in problems that require multiple operations.

EXAMPLE 3-16

Find the Laplace transform of

$$8e^{-3(t-7)} \sin[4(t-7)] u(t-7)$$

Solution: With a time-shifted function, we must first make sure that it is in a proper shifted form. Because all the occurrences of t are in the same form as the argument of the step function, $t-7$, this function is in the proper form. If it were not in the proper shifted form, we would use the techniques shown in Section 3-3-1 to reformat the function before proceeding.

The first step is to list all of the operations that may be required. In this case, we may need (O-1) and (O-2).

$$h(t-a)\,u(t-a) \Leftrightarrow e^{-as}\,H(s)$$

$$e^{-bt}\,h(t)\,u(t) \Leftrightarrow H(s+b)$$

Because the time shift operation is involved, we must use it first. If neither of the two operations were the time shift operation, then it would not matter which one we used first. We will subscript this function with 1 because it is the first operation used. For the second operation, we will indent it and subscript it with 2.

$$H(s) = \mathcal{L}[8e^{-3(t-7)} \sin[4(t-7)] u(t-7)]$$

$$h_1(t-a)\,u(t-a) \Leftrightarrow e^{-as}\,H_1(s)$$

$$h_1(t-a)\,u(t-a) = 8e^{-3(t-7)} \sin[4(t-7)] u(t-7)$$

$$h_1(t)\,u(t) \qquad = 8e^{-3t} \sin[4t]\,u(t)$$

$$H_1(s) \qquad = \mathcal{L}[8e^{-3t} \sin(4t)\,u(t)]$$

$$e^{-bt}\,h_2(t)\,u(t) \Leftrightarrow H_2(s+b)$$

$$e^{-bt}\,h_2(t)\,u(t) = 8e^{-3t} \sin[4t]\,u(t)$$

$$h_2(t)\,u(t) \qquad = 8 \sin[4t]\,u(t)$$

We can now use transform pair (P-5) to find the Laplace transform of $h_2(t)\, u(t)$. We then will perform each of the indicated operations in reverse order, unindenting as we go.

$$\sin(\omega t)\, u(t) \Leftrightarrow \frac{\omega}{s^2 + \omega^2}$$

$$H_2(s) \quad = \frac{32}{s^2 + 16}$$

$$H_1(s) = H_2(s + b) = \frac{32}{[(s + 3)^2 + 16]}$$

$$H(s) = e^{-as}\, H_1(s) = \frac{32e^{-7s}}{[(s + 3)^2 + 16]}$$

Therefore,

$$H(s) = \mathscr{L}[8e^{-3(t-7)}\sin[4(t - 7)]\, u(t - 7)] = \frac{32e^{-7s}}{[(s + 3)^2 + 16]}$$

EXAMPLE 3-17

Find the Laplace transform of

$$v(t) = \int_0^t 3t^2 e^{-5t}\, u(t)\, dt$$

Solution: The first step is to list all of the operations that may be required. In this case we may need (O-2), (O-3), and (O-5).

$$e^{-bt}\, h(t)\, u(t) \Leftrightarrow H(s + b)$$

$$t\, h(t)\, u(t) \Leftrightarrow -\frac{d}{ds}\, H(s)$$

$$\int_0^t h(t)\, u(t)\, dt \Leftrightarrow \frac{H(s)}{s}$$

We cannot algebraically factor out t^2 or e^{-5t} from the integral, so it only makes sense to use the integral operation first. We then can use the t multiplier or the exponential multiplier operation next. Let's arbitrarily choose to use the exponential multiplier operation and then the t multiplier operation after we use the integral operation.

$$V(s) = \mathcal{L}\left[\int_0^t 3t^2 e^{-5t}\, u(t)\, dt\right]$$

$$\int_0^t h_1(t)\, u(t)\, dt \iff \frac{H_1(s)}{s}$$

$$\int_0^t h_1(t)\, u(t)\, dt = \int_0^t 3t^2 e^{-5t}\, u(t)\, dt$$

$$h_1(t)\, u(t) \quad = 3t^2 e^{-5t}\, u(t)$$

$$H_1(s) \quad = \mathcal{L}[3t^2 e^{-5t}\, u(t)]$$

$$e^{-bt} h_2(t)\, u(t) \iff H_2(s+b)$$

$$e^{-bt} h_2(t)\, u(t) = 3t^2 e^{-5t}\, u(t)$$

$$h_2(t)\, u(t) \quad = 3t^2\, u(t)$$

$$H_2(s) \quad = \mathcal{L}[3t^2\, u(t)]$$

$$t\, h_3(t)\, u(t) \iff -\frac{d}{ds} H_3(s)$$

$$t\, h_3(t)\, u(t) = 3t^2\, u(t)$$

$$h_3(t)\, u(t) \quad = 3t\, u(t)$$

We can now use transform pair (P-3) to find the Laplace transform of $h_3(t)\, u(t)$. We then will perform each of the indicated operations in reverse order, unindenting as we go.

$$t\, u(t) \iff \frac{1}{s^2}$$

$$H_3(s) = \frac{3}{s^2}$$

$$H_2(s) = -\frac{d}{ds} H_3(s)$$

$$= -\frac{d}{ds} \frac{3}{s^2}$$

$$= -\frac{(0)(s^2) - (2s)(3)}{s^4}$$

$$= \frac{6}{s^3}$$

$$H_1(s) = H_2(s + b)$$

$$= \frac{6}{(s + 5)^3}$$

$$V(s) = \frac{H_1(s)}{s}$$

$$= \frac{6/(s + 5)^3}{s}$$

$$= \frac{6}{s(s + 5)^3}$$

Therefore,

$$V(s) = \mathscr{L}\left[\int_0^t 3t^2 e^{-5t} u(t) \, dt \right] = \frac{6}{s(s + 5)^3}$$

3-3-7 Extending the Table of Laplace Transform Pairs

When working Laplace problems in electronics or control systems, a particular type of function may be encountered over and over again. In a case like this, it will be expedient to have a Laplace transform pair that directly applies to this reoccurring function so that we can quickly find the Laplace and inverse Laplace transforms.

This section covers how to derive these additional transform pairs. In effect, we will be extending our table of transform pairs. Table A-4 in Appendix A is a table of extended transform pairs showing some of the more common pairs. To generate such a table is simply a matter of using the basic transform pairs and transform operations with variables instead of numerical values for the constants.

EXAMPLE 3-18

Find the Laplace transform of $t^n u(t)$.

Solution: To find the Laplace of t raised to any power, we will find t raised to the second power and then the third power. From these answers we can determine the pattern for t raised to any power.

First we find $\mathscr{L}[t^2 u(t)]$. This requires transform operation (O-3).

$$t\, h(t)\, u(t) \Leftrightarrow -\frac{d}{ds} H(s)$$

Setting the time-domain side of this operation equal to the function, we solve for $h(t)\, u(t)$.

$$t\, h(t)\, u(t) = t^2 u(t)$$

$$h(t)\, u(t) = t\, u(t)$$

Using (P-3), we determine $H(s)$.

$$t\, u(t) \Leftrightarrow \frac{1}{s^2}$$

$$H(s) = \frac{1}{s^2}$$

Applying the right side of the operation to find the Laplace transform, we have

$$-\frac{d}{ds} H(s) = -\frac{d}{ds}\frac{1}{s^2} = -\frac{(0)(s^2) - (2s)(1)}{(s^2)^2}$$

$$= \frac{2}{s^3}$$

Therefore,

$$\mathscr{L}[t^2 u(t)] = \frac{2}{s^3}$$

Now we will follow a similar procedure to find $\mathscr{L}[t^3 u(t)]$.

$$t\, h(t)\, u(t) \Leftrightarrow -\frac{d}{ds} H(s)$$

$$t\, h(t)\, u(t) = t^3 u(t)$$

$$h(t)\, u(t) = t^2 u(t)$$

Using the transform pair for t^2 that we just derived, we can find $H(s)$.

$$\mathcal{L}[t^2\, u(t)] = \frac{2}{s^3}$$

$$H(s) = \frac{2}{s^3}$$

Applying right side of the operation to find the Laplace transform, we have

$$-\frac{d}{ds}H(s) = -\frac{d}{ds}\frac{2}{s^3} = -\frac{(0)(s^3) - (3s^2)(2)}{(s^3)^2}$$

$$= \frac{6}{s^4}$$

Therefore,

$$\mathcal{L}[t^3\, u(t)] = \frac{6}{s^4}$$

From observing how the numbers work, we can predict that when t is raised to the fourth power, the Laplace transform will be 4 factorial over s raised to the fifth power. Therefore, we can conclude that the numerator is a factorial of the power of t and the power of s in the denominator is always the power of t plus 1. In equation form, we have

$$t^n\, u(t) \Leftrightarrow \frac{n!}{s^{(n+1)}}$$

This is the transform pair (P-7a) in Table A-4 in Appendix A.

The transform pair derived in the previous example is tailored toward finding the Laplace transform given the time-domain expression because we use values from the time-domain function to calculate the Laplace function values. When we need to find the inverse Laplace transform, it is more convenient to use quantities from the Laplace function to calculate the values for the time-domain function. In other words, we need a transform pair tailored toward finding the inverse Laplace transform from the time domain. There are two basic methods used to derive transform pairs that are tailored toward finding the inverse Laplace transform.

One method is to use the inverse Laplace transform equations described in later sections of this chapter. We will not do any examples of this technique because it is relatively simple and the inverse equations have not yet been presented. Briefly, to use this technique, apply the inverse Laplace transform equations to a time-domain function having variables rather than numbers for the constants.

A second method is to use an existing transform pair and modify both sides, similar to the way an equation is manipulated, using the transform operations. However, because it is not an equation, we can only perform an equivalent operation on each side. For example, we may let a non-time-dependent variable such as n be equal to $n + 3$ on both sides of the transform pair, but if we multiply the time-domain side by a time-dependent function such as e^{-3t}, then we must perform the equivalent operation on the Laplace domain side by setting $s = s + 3$.

EXAMPLE 3-19

Revise the following transform pair so that it is tailored toward finding the inverse Laplace transform.

$$t^n\, u(t) \Leftrightarrow \frac{n!}{s^{(n+1)}}$$

Solution: For this function we would like the Laplace side to have the form of

$$\frac{1}{s^n}$$

Therefore, the first step is to modify the denominator so that s is raised to the nth power. This can be accomplished by letting $n = n - 1$, and because n is not a time-dependent variable, we do exactly the same to both sides of the transform pair.

$$t^{(n-1)}\, u(t) \Leftrightarrow \frac{(n-1)!}{s^{(n-1+1)}}$$

$$t^{(n-1)}\, u(t) \Leftrightarrow \frac{(n-1)!}{s^n}$$

Next we remove the numerator on the Laplace side by dividing both sides by $(n - 1)!$

$$\frac{t^{(n-1)}}{(n-1)!}\, u(t) \Leftrightarrow \frac{(n-1)!}{(s^n)(n-1)!}$$

$$\frac{1}{(n-1)!}\, t^{(n-1)}\, u(t) \Leftrightarrow \frac{1}{s^n}$$

This is the transform pair (P-7b) in the table of extended transform pairs in Appendix A.

EXAMPLE 3-20

Modify the transform pair (P-7b) so that it has a denominator of $(s + b)^n$.

Solution: We first start with transform pair (P-7b).

$$\frac{1}{(n - 1)!} \, t^{(n-1)} \, u(t) \Leftrightarrow \frac{1}{s^n}$$

We then let $s = s + b$, but we must do the equivalent operation on the time-domain side. In this case, we apply transform operation (O-2), which is

$$e^{-bt} \, h(t) \, u(t) \Leftrightarrow H(s + b)$$

Therefore, we multiply the time-domain side by e^{-bt}.

$$\frac{1}{(n - 1)!} \, t^{(n-1)} e^{-bt} \, u(t) \Leftrightarrow \frac{1}{(s + b)^n}$$

This is the transform pair (P-8b) in the table of extended transform pairs in Appendix A.

These last few examples along with the other transform pairs in Table A-4 are a very small sampling of what is possible using Table A-1, Table A-2, and the inverse equations shown in a later section.

3-4 INVERSE LAPLACE TRANSFORMS USING TRADITIONAL TECHNIQUES

There are many ways to find the inverse Laplace transform of a function. A very traditional method involves the use of a technique called *partial fraction expansion*. In general, partial fraction expansion is a mathematical process of splitting up a complex fraction into a summation of simpler fractions. This technique is very useful when finding the inverse Laplace transform of a function.

In Laplace transforms we often have a complex function $F(s)$ that we do not have an entry for in our transform pair table. Using partial fraction expansion, we may split this function into a summation of simpler fractions that we do have entries for. We may then sum the inverse of each simpler fraction to find the inverse Laplace transform of the original complex function $F(s)$.

If you are familiar with the partial fraction expansion technique, you can skip this section and go directly to Section 3-5.

3-4-1 Inverse Transform of First-Order Real Poles

Before we can start finding inverses we must define a few terms and concepts. We will begin with *poles* and *zeros*. Poles are the values of s that when substituted into a function cause the function's value to become infinity. Therefore, poles are the roots of the denominator. Zeros are the values of s that when substituted into a function cause the function's value to become zero. Therefore, zeros are the roots of the numerator.

Once we are able to identify the poles and zeros, we can determine their order by observing the power to which the factor causing the pole or zero is raised. A first-order pole or a first-order zero means that the factor responsible for the pole or zero is raised to the first power. A multiple-order pole or multiple-order zero means that the factor responsible for the pole or zero is raised to second or higher power.

Three other terms we must be familiar with are $F(s)$, $N(s)$, and $D(s)$. $F(s)$ is defined as the Laplace function that we want to find the inverse of. It is composed of a numerator polynomial, $N(s)$, and a denominator polynomial, $D(s)$.

In order to use partial fraction expansion, the equation must be capable of being expressed as a numerator polynomial divided by a denominator polynomial; however, the function is usually expressed with a factored numerator and factored denominator.

$$F(s) = \frac{N(s)}{D(s)} = \frac{B_n s^n + B_{n-1} s^{n-1} + \cdots + B_1 s + B_0}{A_m s^m + A_{m-1} s^{m-1} + \cdots + A_1 s + A_0} \quad (3\text{-}4)$$

$$= \frac{(B_n/A_m) \text{ factored numerator}}{\text{factored denominator}}$$

In a practical system, the order of the numerator polynomial, n, will be equal to or less than the order of the denominator polynomial, m. If the numerator polynomial's order is equal to the denominator polynomial's order, then the inverse of this function will contain an impulse term. The magnitude of the impulse is found by dividing the numerator by the denominator of the function. This will allow us to express the function as a constant plus a fraction having the numerator polynomial one order less than the denominator polynomial's order. The inverse of the constant part will be an impulse, and the inverse of the fraction part will be found using the partial fraction expansion technique. The process of dividing out the constant is often referred to as *removing the impulse*. It is important to note that if the division is not performed and the partial fraction is performed, the inverse will be exactly the same as when the division is performed except that the impulse term will be missing from the answer.

EXAMPLE 3-21

Put the following function into a form suitable for using partial fraction expansion.

$$F(s) = \frac{9(s + 1)[(s + 5)^2 + 1]}{(s + 3)(s + 4)(s + 5)}$$

Solution: The numerator power is equal to the denominator power, so the fraction must be divided out. To do this, we first multiply out the numerator and denominator.

$$F(s) = \frac{9(s + 1)[(s + 5)^2 + 1]}{(s + 3)(s + 4)(s + 5)} = \frac{9(s^3 + 11s^2 + 36s + 26)}{s^3 + 12s^2 + 47s + 60}$$

Next, the numerator polynomial is divide by the denominator polynomial. In this example, the multiplying factor, 9, will be held out and multiplied back in after the division. Holding the constant out often makes the division process easier, but it works just as well if you multiply the constant through the numerator and include it in the division. The difficult part of holding out a multiplier is remembering to multiply it back in as the last step.

$$
\begin{array}{r}
1 \\
s^3 + 12s^2 + 47s + 60 \overline{\smash{\big)}\, s^3 + 11s^2 + 36s + 26} \\
\underline{s^3 + 12s^2 + 47s + 60} \\
-s^2 - 11s - 34
\end{array}
$$

$$
\therefore F(s) = \frac{9(s + 1)[(s + 5)^2 + 1]}{(s + 3)(s + 4)(s + 5)} = 9\left[1 + \frac{-s^2 - 11s - 34}{(s + 3)(s + 4)(s + 5)}\right]
$$

$$
= 9 + \frac{9(-s^2 - 11s - 34)}{(s + 3)(s + 4)(s + 5)}
$$

The resulting fraction part is now in a form on which it is suitable to perform partial fraction expansion. Notice that the numerator of the fraction part is not factored. This is commonly done, because this is an intermediate step in a series of steps used to find the inverse Laplace transform. The complete inverse will be equal to the inverse of 9, which is $9\,\delta(t)$, plus the inverse of the fraction.

After assuring that the numerator polynomial's power of $F(s)$ is at least one power less than the denominator polynomial's power, the denominator is expressed in a factored form. Next, $F(s)$ is set equal to a summation of fractions, where each fraction's denominator is one of the denominator factors of $F(s)$. The numerator of each fraction is called the *residue,* and it is denoted with the pole value as the subscript. This expansion is shown next.

$$
F(s) = \frac{N(s)}{(s + a_1)(s + a_2) \cdots (s + a_m)(\textit{other factors})} \tag{3-5}
$$

$$
= \frac{R_{-a_1}}{(s + a_1)} + \frac{R_{-a_2}}{(s + a_2)} + \cdots + \frac{R_{-a_m}}{(s + a_m)} + \textit{rest of expansion}
$$

where The first-order real pole values $= -a_1, -a_2, -a_3, \cdots -a_m$
R_{-a} is called the residue of $F(s)$ for the pole $s = -a$

In this expansion, only the first-order real poles are shown. All other factored forms of the denominator, which will be covered in later sections, are lumped together and expressed as "rest of expansion." The numerators on the expanded side are called the *residues,* and they will later found to be numerical values.

EXAMPLE 3-22

Show the expanded form of the equation

$$F(s) = \frac{14(s + 2)}{s(s + 7)}$$

Solution: This function is expanded as follows:

$$F(s) = \frac{14(s + 2)}{s(s + 7)} = \frac{R_0}{s} + \frac{R_{-7}}{(s + 7)}$$

EXAMPLE 3-23

Show the expanded form of the function shown.

$$F(s) = \frac{[(s + 5)^2 + 9]}{(s + 5)(s + 2)(s + 8)}$$

Solution: The power of the numerator polynomial is already less than the power of the denominator polynomial, so we do not need to divide out. Therefore,

$$F(s) = \frac{[(s + 5)^2 + 9]}{(s + 5)(s + 2)(s + 8)} = \frac{R_{-5}}{(s + 5)} + \frac{R_{-2}}{(s + 2)} + \frac{R_{-8}}{(s + 8)}$$

The next step is to isolate each residue and determine it's value. Let's find the value of the residue R_{-a_1} in Eq. (3-5). First we multiply through the equation by $(s + a_1)$. This factor will cancel with the denominator factor under R_{-a_1} and one of the factors in the denominator of $F(s)$.

$$F(s)(s + a_1) = \frac{N(s)\,(s + a_1)}{(s + a_1)(s + a_2) \cdots (s + a_m)(other\ factors)}(s + a_1)$$

$$= \frac{R_{-a_1}}{(s + a_1)}(s + a_1) + \frac{R_{-a_2}}{(s + a_2)}(s + a_1) + \cdots + \frac{R_{-a_m}}{(s + a_m)}(s + a_1)$$

$$+ \left[\begin{array}{c} rest\ of \\ expansion \end{array} \right](s + a_1)$$

Therefore,

$$\frac{N(s)}{(s + a_2) \cdots (s + a_m)(other\ factors)} = R_{-a_1} + \frac{R_{-a_2}}{(s + a_2)}(s + a_1) + \cdots$$

$$+ \frac{R_{-a_m}}{(s + a_m)}(s + a_1) + \left[\begin{array}{c} rest\ of \\ expansion \end{array} \right](s + a_1)$$

We then select an appropriate value of s that will allow us to determine the value of R_{-a_1}. If we select $s = -a_1$, then all of the terms on the right side of the equation will go to zero except R_{-a_1}, and because the left side of the equation contains all numbers except for s, the left side will become a numerical value. This numerical value will therefore be the value of R_{-a_1}.

Once the numerical value of R_{-a_1} is determined, the inverse can be found using the exponential transform pair (P-4).

$$\mathcal{L}^{-1}\left[\frac{R_{-a_1}}{(s + a_1)}\right] = R_{-a_1}e^{-a_1 t}\,u(t)$$

It is interesting to note that we could select any value of s, but we selected a value that made all the terms containing a residue other than R_{-a_1} go to zero. If we had selected a nonpole value of s, we could have generated an equation containing all the residues. In fact, by selecting several different nonpole values of s we could generate enough equations that all the residues could be found by solving these equations simultaneously. This is another method for finding the inverse Laplace transform that is more appropriate for computer software.

EXAMPLE 3-24

Find the inverse of

$$F(s) = \frac{10(s + 3)(s + 5)}{s(s + 1)(s + 6)}$$

Solution: The expansion is

$$F(s) = \frac{10(s + 3)(s + 5)}{s(s + 1)(s + 6)} = \frac{R_0}{s} + \frac{R_{-1}}{(s + 1)} + \frac{R_{-6}}{(s + 6)}$$

To determine the numerical value of R_0, we begin by multiplying both sides of the equation by s. This will cause the s under the R_0 term to be canceled out.

$$\left[\frac{10(s + 3)(s + 5)}{s(s + 1)(s + 6)}\right]s = \left[\frac{R_0}{s}\right]s + \left[\frac{R_{-1}}{(s + 1)}\right]s + \left[\frac{R_{-6}}{(s + 6)}\right]s$$

$$\left[\frac{10(s + 3)(s + 5)}{(s + 1)(s + 6)}\right] = [R_0] + \left[\frac{R_{-1}}{(s + 1)}\right]s + \left[\frac{R_{-6}}{(s + 6)}\right]s$$

Then we let $s = 0$. This causes all the terms on the right side of the equation except for R_0 to become zero.

$$\left[\frac{10(0 + 3)(0 + 5)}{(0 + 1)(0 + 6)}\right] = [R_0] + 0 + 0$$

Solving for the numerical value of R_0,

$$R_0 = \left[\frac{10(0 + 3)(0 + 5)}{(0 + 1)(0 + 6)}\right] = 25$$

Next we find the numerical value for R_{-1} using the same procedure.

$$\left[\frac{10(s + 3)(s + 5)}{s(s + 1)(s + 6)}\right](s + 1) = \left[\frac{R_0}{s}\right](s + 1) + \left[\frac{R_{-1}}{(s + 1)}\right](s + 1) + \left[\frac{R_{-6}}{(s + 6)}\right](s + 1)$$

$$\left[\frac{10(s + 3)(s + 5)}{s(s + 6)}\right] = \left[\frac{R_0}{s}\right](s + 1) + [R_{-1}] + \left[\frac{R_{-6}}{(s + 6)}\right](s + 1)$$

Setting $s = -1$,

$$\left[\frac{10(-1 + 3)(-1 + 5)}{-1(-1 + 6)}\right] = 0 + [R_{-1}] + 0$$

$$R_{-1} = \left[\frac{10(-1 + 3)(-1 + 5)}{-1(-1 + 6)}\right] = -16$$

Following a similar procedure, the numerical value for R_{-6} is found.

$$\left[\frac{10(s + 3)(s + 5)}{s(s + 1)(s + 6)}\right](s + 6) = \left[\frac{R_0}{s}\right](s + 6) + \left[\frac{R_{-1}}{(s + 1)}\right](s + 6) + \left[\frac{R_{-6}}{(s + 6)}\right](s + 6)$$

$$\left[\frac{10(s + 3)(s + 5)}{s(s + 1)}\right] = \left[\frac{R_0}{s}\right](s + 6) + \left[\frac{R_{-1}}{(s + 1)}\right](s + 6) + [R_{-6}]$$

Setting $s = -6$,

$$\left[\frac{10(-6 + 3)(-6 + 5)}{-6(-6 + 1)}\right] = 0 + 0 + [R_{-6}]$$

$$R_{-6} = \left[\frac{10(-6 + 3)(-6 + 5)}{-6(-6 + 1)}\right] = 1$$

The function now can be written using the numerical values of the residues, and from this the inverse Laplace transform can be found using transform pairs (P-2) and (P-4).

$$F(s) = \frac{10(s + 3)(s + 5)}{s(s + 1)(s + 6)} = \frac{R_0}{s} + \frac{R_{-1}}{(s + 1)} + \frac{R_{-6}}{(s + 6)}$$

$$= \frac{25}{s} + \frac{-16}{(s + 1)} + \frac{1}{(s + 6)}$$

$$f(t) = \mathcal{L}^{-1}\left[\frac{10(s+3)(s+5)}{s(s+1)(s+6)}\right] = \mathcal{L}^{-1}\left[\frac{25}{s}\right] + \mathcal{L}^{-1}\left[\frac{-16}{(s+1)}\right] + \mathcal{L}^{-1}\left[\frac{1}{(s+6)}\right]$$

$$f(t) = 25\,u(t) - 16e^{-t}\,u(t) + e^{-6t}\,u(t)$$

The procedure being shown in this section is often called the *cover-up method* because one covers up a factor in the denominator of $F(s)$ with a finger and substitutes in a value for each s in the remaining fraction. Once this is observed, the process can be streamlined. Watch for this in the next example.

EXAMPLE 3-25

Find the inverse of

$$F(s) = \frac{[(s+8)^2 + 36]}{(s+8)(s+20)}$$

Solution: Before we write the expansion, we will remove the impulse by dividing.

$$s^2 + 28s + 160\ \overline{\big)\ \begin{array}{l} 1 \\ s^2 + 16s + 100 \\ \underline{s^2 + 28s + 160} \\ \quad\ -12s -\ \ 60 \end{array}}$$

$$F(s) = \frac{[(s+8)^2 + 36]}{(s+8)(s+20)} = 1 + \frac{-12s - 60}{(s+8)(s+20)}$$

The 1 will inverse into $\delta(t)$. Now we will find the inverse of the leftover fraction.

$$\frac{-12s - 60}{(s+8)(s+20)} = \frac{R_{-8}}{(s+8)} + \frac{R_{-20}}{(s+20)}$$

First we find the value of R_{-8}.

$$\frac{-12s - 60}{(s+8)(s+20)}(s+8) = \frac{R_{-8}}{(s+8)}(s+8) + \frac{R_{-20}}{(s+20)}(s+8)$$

$$\frac{-12s - 60}{(s+20)} = R_{-8} + \frac{R_{-20}}{(s+20)}(s+8)$$

Setting $s = -8$ and solving for R_{-8},

$$\frac{-12(-8) - 60}{(-8 + 20)} = R_{-8} + 0$$

$$R_{-8} = \frac{-12(-8) - 60}{(-8 + 20)} = 3$$

Next we determine the value of R_{-20}.

$$\frac{-12s - 60}{(s + 8)(s + 20)}(s + 20) = \frac{R_{-8}}{(s + 8)}(s + 20) + \frac{R_{-20}}{(s + 20)}(s + 20)$$

$$\frac{-12s - 60}{(s + 8)} = \frac{R_{-8}}{(s + 8)}(s + 20) + R_{-20}$$

Setting $s = -20$ and solving for R_{-20},

$$\frac{-12(-20) - 60}{(-20 + 8)} = 0 + R_{-20}$$

$$R_{-20} = \frac{-12(-20) - 60}{(-20 + 8)} = -15$$

We now substitute in the numerical values for the residues and find the inverse using transform pair (P-4).

$$\frac{-12s - 60}{(s + 8)(s + 20)} = \frac{R_{-8}}{(s + 8)} + \frac{R_{-20}}{(s + 20)}$$

$$\frac{-12s - 60}{(s + 8)(s + 20)} = \frac{3}{(s + 8)} + \frac{-15}{(s + 20)}$$

$$\mathscr{L}^{-1}\left[\frac{-12s - 60}{(s + 8)(s + 20)}\right] = \mathscr{L}^{-1}\left[\frac{3}{(s + 8)}\right] + \mathscr{L}^{-1}\left[\frac{-15}{(s + 20)}\right]$$

$$3e^{-8t}u(t) - 15e^{-20t}u(t)$$

Finally, we add the impulse back in to find the total inverse Laplace transform.

$$f(t) = \delta(t) + 3e^{-8t}u(t) - 15e^{-20t}u(t)$$

3-4-2 Inverse Transform of Multiple-Order Real Poles

The procedure for finding the inverse Laplace transform for multiple-order real poles is similar to the procedure for finding the inverse of first-order poles. The difference is that there will be the same number of expanded fractions as the power of the factor in the denominator. For example, if $(s + a)^4$ is in the denominator of $F(s)$, we will have four fractions in the expansion due to this multiple-order pole. Most often in the traditional approach, the numerators of the fractions in the expansion will be simply called A, B, C, and so on. However, because our ultimate goal is to streamline the inverse Laplace transform process, the expansion shown here will use a form that includes a factorial. This form will prepare us for streamlining the procedure later on.

Multiple-order poles may be expanded as follows

$$F(s) = \frac{N(s)}{D(s)} = \frac{N(s)}{(s + a)^r (other\ factors)} = \sum_{n=1}^{r} \frac{R_{-a,n}/(n-1)!}{(s + a)^{r-n+1}} + \begin{matrix} rest \\ of \\ expansion \end{matrix} \qquad (3\text{-}6)$$

$$= \frac{R_{-a,1}/(1-1)!}{(s + a)^r} + \frac{R_{-a,2}/(2-1)!}{(s + a)^{r-1}} + \cdots + \frac{R_{-a,r}/(r-1)!}{(s + a)} + \begin{matrix} rest \\ of \\ expansion \end{matrix}$$

Note that $R_{-a,1}$ is a numerical value and $(1-1)!, (2-1)!, \ldots, (r-1)!$ are factorials also having numerical values. Therefore, the numerator of each term in the expansion will evaluate into a real number.

Because the terms in the numerator are numerical values divided by $(s + a)$ raised to a power, we will be able to use the transform pair (P-8b) to find the inverse Laplace transform.

$$\frac{1}{(n-1)!} t^{(n-1)} e^{-bt} u(t) \Leftrightarrow \frac{1}{(s + b)^n}$$

Finding the residues is a little more complicated because we can no longer simply multiply through by $(s + a)^r$ and let $s = -a$. This would only let us find $R_{-a,1}$. The procedure for isolating all the residues for the multiple-order real pole will involve taking derivatives. This will be demonstrated in the following example.

EXAMPLE 3-26

Find the inverse of

$$F(s) = \frac{5(s + 6)}{(s + 10)(s + 12)^2}$$

Solution: First the expansion is written.

$$F(s) = \frac{5(s + 6)}{(s + 10)(s + 12)^2} = \frac{R_{-10}}{(s + 10)} + \frac{R_{-12,1}/(1-1)!}{(s + 12)^2} + \frac{R_{-12,2}/(2-1)!}{(s + 12)}$$

Notice that the R_{-12} has two residues to find. This is because the $s = -12$ pole is a second-order pole.

We can now simplify this expression by evaluating the factorials. This will make the equation simpler to work with. Remember that 0! and 1! are both equal to 1.

$$F(s) = \frac{5(s + 6)}{(s + 10)(s + 12)^2} = \frac{R_{-10}}{(s + 10)} + \frac{R_{-12,1}}{(s + 12)^2} + \frac{R_{-12,2}}{(s + 12)}$$

Notice that if the pole had been raised to the fourth power, we would have had numerator terms of $(R_{-a,1})$, $(R_{-a,2})$, $(R_{-a,3}/2)$, and $(R_{-a,4}/6)$.

Finding the numerical value for the residue R_{-10} proceeds as usual.

$$\frac{5(s + 6)}{(s + 10)(s + 12)^2}(s + 10) = \frac{R_{-10}}{(s + 10)}(s + 10) + \frac{R_{-12,1}}{(s + 12)^2}(s + 10)$$

$$+ \frac{R_{-12,2}}{(s + 12)}(s + 10)$$

$$\frac{5(s + 6)}{(s + 12)^2} = R_{-10} + \frac{R_{-12,1}}{(s + 12)^2}(s + 10) + \frac{R_{-12,2}}{(s + 12)}(s + 10)$$

Setting $s = -10$ and solving for R_{-10}, we have

$$\frac{5(-10 + 6)}{(-10 + 12)^2} = R_{-10} + 0 + 0$$

$$R_{-10} = \frac{5(-10 + 6)}{(-10 + 12)^2} = -5$$

We can now substitute the value for R_{-10} into the expansion.

$$F(s) = \frac{5(s + 6)}{(s + 10)(s + 12)^2} = \frac{-5}{(s + 10)} + \frac{R_{-12,1}}{(s + 12)^2} + \frac{R_{-12,2}}{(s + 12)}$$

Then we find the residues for the second-order pole, $s = -12$. This begins by multiplying both sides of the expansion equation by $(s + 12)^2$.

$$\frac{5(s + 6)}{(s + 10)(s + 12)^2}(s + 12)^2 = \frac{-5}{(s + 10)}(s + 12)^2 + \frac{R_{-12,1}}{(s + 12)^2}(s + 12)^2$$

$$+ \frac{R_{-12,2}}{(s + 12)}(s + 12)^2$$

$$\frac{5(s + 6)}{(s + 10)} = \frac{-5}{(s + 10)}(s + 12)^2 + R_{-12,1} + R_{-12,2}(s + 12)$$

We then set $s = -12$ and solve for $R_{-12,1}$ as if we were finding the residue for a first-order pole.

$$\frac{5(-12 + 6)}{(-12 + 10)} = 0 + R_{-12,1} + 0$$

$$R_{-12,1} = \frac{5(-12 + 6)}{(-12 + 10)} = 15$$

Substituting the value for $R_{-12,1}$ into the expansion, we have

$$F(s) = \frac{5(s + 6)}{(s + 10)(s + 12)^2} = \frac{-5}{(s + 10)} + \frac{15}{(s + 12)^2} + \frac{R_{-12,2}}{(s + 12)}$$

Next, we need to isolate the $R_{-12,2}$. This process begins by multiplying both sides of the current equation by $(s + 12)^2$.

$$\frac{5(s + 6)}{(s + 10)} = \frac{-5}{(s + 10)}(s + 12)^2 + 15 + R_{-12,2}(s + 12)$$

Next we take the derivative with respect to s of both sides of the equation to isolate $R_{-12,2}$.

$$\frac{d}{ds}\left[\frac{5(s + 6)}{(s + 10)}\right] = \frac{d}{ds}\left[\frac{-5(s + 12)^2}{(s + 10)}\right] + \frac{d}{ds}[15] + \frac{d}{ds}[R_{-12,2}(s + 12)]$$

$$\left[\frac{20}{(s + 10)^2}\right] = \left[\frac{-5(s + 8)(s + 12)}{(s + 10)^2}\right] + [0] + [R_{-12,2}]$$

Notice how all the unknown values except for s and $R_{-12,2}$ are left. When finding the last residue of a multiple-order pole, we often find that we do not need to substitute a value in for s. Because $R_{-12,2}$ is the last residue for this multiple-order pole, we can substitute any value in for s, or no value at all, but traditionally we set $s = -12$ and solve for $R_{-12,2}$.

$$\left[\frac{20}{(-12 + 10)^2}\right] = \left[\frac{-5(-12 + 8)(-12 + 12)}{(-12 + 10)^2}\right] + [0] + [R_{-12,2}]$$

$$R_{-12,2} = \frac{20}{(-12 + 10)^2} = 5$$

Substituting all the values for the residues into the original expansion, we have

$$F(s) = \frac{5(s + 6)}{(s + 10)(s + 12)^2} = \frac{-5}{(s + 10)} + \frac{15/(1 - 1)!}{(s + 12)^2} + \frac{5/(2 - 1)!}{(s + 12)}$$

The inverse may now be found using transform pair (P-8b).

$$\frac{1}{(n-1)!}t^{(n-1)}e^{-bt}u(t) \Leftrightarrow \frac{1}{(s+b)^n}$$

$$f(t) = \mathcal{L}^{-1}\left[\frac{5(s+6)}{(s+10)(s+12)^2}\right] = \mathcal{L}^{-1}\left[\frac{-5}{(s+10)}\right] + \mathcal{L}^{-1}\left[\frac{15/(1-1)!}{(s+12)^2}\right]$$

$$+ \mathcal{L}^{-1}\left[\frac{5/(2-1)!}{(s+12)}\right]$$

$$= -5e^{-10t}u(t) + \frac{15/(1-1)!}{(2-1)!}t^{2-1}e^{-12t}u(t) + \frac{5/(2-1)!}{(1-1)!}t^{1-1}e^{-12t}u(t)$$

$$= -5e^{-10t}u(t) + 15te^{-12t}u(t) + 5e^{-12t}u(t)$$

The next example has only one fourth-order pole. With only one pole value, the problem will look very different from previous examples, but the procedure for the inverse of the multiple-order poles will be the same. This brings up a very important point that applies to working any problem. Because of the different ways the numbers look in each problem, it is imperative that we look for the procedure rather than looking for similarities in how the numbers flow in the examples. When the procedure is learned, it will be easy to see that all the problems are solved the same way every time no matter how different the numbers may look.

EXAMPLE 3-27

Find the inverse of

$$G(s) = \frac{3[(s+4.5)^2 + 9]}{(s+4.5)^4}$$

Solution: The order of the numerator polynomial is less than the order of the denominator polynomial, so we do not need to divide out the fraction, and we can write the expansion as

$$G(s) = \frac{3[(s+4.5)^2 + 9]}{(s+4.5)^4}$$

$$= \frac{R_{-4.5,1}/(1-1)!}{(s+4.5)^4} + \frac{R_{-4.5,2}/(2-1)!}{(s+4.5)^3} + \frac{R_{-4.5,3}/(3-1)!}{(s+4.5)^2} + \frac{R_{-4.5,4}/(4-1)!}{(s+4.5)}$$

Evaluating the factorials we have

$$\frac{3\,[(s+4.5)^2+9]}{(s+4.5)^4} = \frac{R_{-4.5,1}}{(s+4.5)^4} + \frac{R_{-4.5,2}}{(s+4.5)^3} + \frac{R_{-4.5,3}/2}{(s+4.5)^2} + \frac{R_{-4.5,4}/6}{(s+4.5)}$$

The first step in finding the numerical values for the residues is to multiply both sides of the equation by $(s+4.5)^4$.

$$\left[\frac{3[(s+4.5)^2+9]}{(s+4.5)^4}\right](s+4.5)^4 = \left[\frac{R_{-4.5,1}}{(s+4.5)^4} + \frac{R_{-4.5,2}}{(s+4.5)^3} + \frac{R_{-4.5,3}/2}{(s+4.5)^2}\right.$$
$$\left. + \frac{R_{-4.5,4}/6}{(s+4.5)}\right](s+4.5)^4$$

$$3[(s+4.5)^2+9] = R_{-4.5,1} + R_{-4.5,2}(s+4.5) + \frac{R_{-4.5,3}(s+4.5)^2}{2}$$
$$+ \frac{R_{-4.5,4}(s+4.5)^3}{6}$$

Notice how much different this step looks than it did in the previous examples even though we used the same procedure.

The first residue, $R_{-4.5,1}$, is found by setting $s = -4.5$.

$$3[(-4.5+4.5)^2+9] = R_{-4.5,1} + R_{-4.5,2}(-4.5+4.5) + \frac{R_{-4.5,3}(-4.5+4.5)^2}{2}$$
$$+ \frac{R_{-4.5,4}(-4.5+4.5)^3}{6}$$

$$3[(-4.5+4.5)^2+9] = R_{-4.5,1} + 0 + 0 + 0$$

$$R_{-4.5,1} = 3[(-4.5+4.5)^2+9] = 27$$

Next, $R_{-4.5,2}$ is isolated by taking the derivative of both sides of the equation that has been multiplied by $(s+4.5)^4$. Before we take the derivative we could substitute in the value of 27 just found for the $R_{-4.5,1}$ residue, but if we remember that it is simply a number and that the derivative of a number is zero, then substitution is not necessary.

$$\frac{d}{ds}3[(s+4.5)^2+9] = \frac{d}{ds}\left[R_{-4.5,1} + R_{-4.5,2}(s+4.5) + \frac{R_{-4.5,3}(s+4.5)^2}{2}\right.$$
$$\left. + \frac{R_{-4.5,4}(s+4.5)^3}{6}\right]$$

$$(6s+27) = 0 + R_{-4.5,2} + R_{-4.5,3}(s+4.5) + \frac{R_{-4.5,4}(s+4.5)^2}{2}$$

Then we let $s = -4.5$ and solve for $R_{-4.5,2}$.

$$(6s + 27) = 0 + R_{-4.5,2} + R_{-4.5,3}(s + 4.5) + \frac{R_{-4.5,4}(s + 4.5)^2}{2}$$

$$6(-4.5) + 27 = 0 + R_{-4.5,2} + R_{-4.5,3}(-4.5 + 4.5) + \frac{R_{-4.5,4}(-4.5 + 4.5)^2}{2}$$

$$6(-4.5) + 27 = 0 + R_{-4.5,2} + 0 + 0$$

$$R_{-4.5,2} = 6(-4.5) + 27 = 0$$

To isolate $R_{-4.5,3}$, we take the second derivative and set s equal to -4.5. Notice that when we found $R_{-4.5,2}$, we took the first derivative, so we are just simply continuing the derivative process.

$$\frac{d}{ds}[(6s + 27)] = \frac{d}{ds}\left[0 + R_{-4.5,2} + R_{-4.5,3}(s + 4.5) + \frac{R_{-4.5,4}(s + 4.5)^2}{2}\right]$$

$$(6) = 0 + 0 + R_{-4.5,3} + R_{-4.5,4}(s + 4.5)$$

$$(6) = 0 + 0 + R_{-4.5,3} + R_{-4.5,4}(-4.5 + 4.5)$$

$$R_{-4.5,3} = 6$$

To isolate $R_{-4.5,4}$, we take the third derivative and set $s = -4.5$. Notice that we have already found the first and second derivative previously in this example.

$$\frac{d}{ds}[(6)] = \frac{d}{ds}[0 + 0 + R_{-4.5,3} + R_{-4.5,4}(s + 4.5)]$$

$$(0) = 0 + 0 + 0 + R_{-4.5,4}$$

$$R_{-4.5,3} = 0$$

With all of the residues found, we can substitute back into the expansion and simplify.

$$G(s) = \frac{3[(s + 4.5)^2 + 9]}{(s + 4.5)^4} = \frac{27/(1 - 1)!}{(s + 4.5)^4} + \frac{0/(2 - 1)!}{(s + 4.5)^3} + \frac{6/(3 - 1)!}{(s + 4.5)^2} + \frac{0/(4 - 1)!}{(s + 4.5)}$$

$$= \frac{27/(1 - 1)!}{(s + 4.5)^4} + \frac{6/(3 - 1)!}{(s + 4.5)^2}$$

We can find the inverse using the transform pair (P-8b).

$$g(t) = \mathcal{L}^{-1}\left[\frac{3[(s + 4.5)^2 + 9]}{(s + 4.5)^4}\right] = \mathcal{L}^{-1}\left[\frac{27/(1-1)!}{(s + 4.5)^4}\right] + \mathcal{L}^{-1}\left[\frac{6/(3-1)!}{(s + 4.5)^2}\right]$$

$$= \frac{27/(1-1)!}{(4-1)!}\,t^{4-1}e^{-4.5t}\,u(t)$$

$$+ \frac{6/(3-1)!}{(2-1)!}\,t^{2-1}e^{-4.5t}\,u(t)$$

$$= 4.5t^3e^{-4.5t}\,u(t) + 3te^{-4.5t}\,u(t)$$

3-4-3 Inverse Transform of Complex-Poles—Complex-Numbers Method

In the traditional methods presented here, we will restrict ourselves to complex conjugate poles of the first order. Higher-order complex conjugate poles are almost impossible to find using traditional methods. However, using streamlined methods, which will be presented later, multiple-order complex conjugate poles are manageable.

There are two traditional methods to achieving the inverse. One inverse method expands the function into two separate fractions having complex conjugate poles. Then the same procedure as is used to find the inverse for real poles is followed. The second method uses only real numbers to generate simultaneous equations. This section will cover the first method, and the following section will cover the second method.

A function with complex poles may be expanded as follows:

$$F(s) = \frac{N(s)}{D(s)} = \frac{N(s)}{[(s + \alpha)^2 + \omega^2]\,(other\ factors)}$$

$$= \frac{k}{(s + \alpha - j\omega)} + \frac{k^*}{(s + \alpha + j\omega)} + rest\ of\ expansion$$

Each residue is found by using the same inverse procedure that is used for first-order real poles. The only difference is that the residues k and k^* are complex and therefore will have equal magnitudes but phase angles differing by 180°.

These complex residues will be converted into an exponential format using the relationship

$$Magnitude\ \angle\theta = Magnitude\ e^{j\theta} \tag{3-7}$$

Then these exponential factors will be converted into a sinusoidal function through the use of Euler's formula, which may be stated in two different forms:

$$e^{jx} = \cos(x) + j\sin(x) \tag{3-8}$$

and

$$e^{-jx} = \cos(x) - j\sin(x)$$

We will later find that the form of "x" in our case will always be $(\omega t + \theta)$, so we may rewrite the formula as

$$e^{j(\omega t + \theta)} = \cos(\omega t + \theta) + j\sin(\omega t + \theta) \tag{3-9}$$

and

$$e^{-j(\omega t + \theta)} = \cos(\omega t + \theta) - j\sin(\omega t + \theta)$$

Let's apply this in an example.

EXAMPLE 3-28

Using the complex number method, find the inverse of

$$V(s) = \frac{30s(s + 2)}{(s + 5)[(s + 1)^2 + (3)^2]}$$

Solution: The function can be expanded into

$$V(s) = \frac{30s(s + 2)}{(s + 5)[(s + 1)^2 + (3)^2]} = \frac{R_{-5}}{(s + 5)} + \frac{k}{(s + 1 - j3)} + \frac{k^*}{(s + 1 + j3)}$$

We will first find the magnitude of the real pole R_{-5} by multiplying both sides of the equation by $(s + 5)$ and then setting $s = -5$.

$$\frac{30s(s + 2)}{(s + 5)[(s + 1)^2 + (3)^2]}(s + 5) = \frac{R_{-5}}{(s + 5)}(s + 5) + \frac{k}{(s + 1 - j3)}(s + 5)$$

$$+ \frac{k^*}{(s + 1 + j3)}(s + 5)$$

$$\frac{30s(s + 2)}{[(s + 1)^2 + (3)^2]} = R_{-5} + \frac{k}{(s + 1 - j3)}(s + 5)$$

$$+ \frac{k^*}{(s + 1 + j3)}(s + 5)$$

$$\frac{30(-5)(-5 + 2)}{[(-5 + 1)^2 + (3)^2]} = R_{-5} + 0 + 0$$

$$R_{-5} = \frac{30(-5)(-5 + 2)}{[(-5 + 1)^2 + (3)^2]} = 18$$

The procedure for finding k is exactly the same. We begin by multiplying both sides by $(s + 1 - j3)$.

$$\frac{30s(s + 2)}{(s + 5)[(s + 1)^2 + (3)^2]}(s + 1 - j3) = \frac{R_{-5}}{(s + 5)}(s + 1 - j3)$$

$$+ \frac{k^*}{(s + 1 - j3)}(s + 1 - j3)$$

$$+ \frac{k^*}{(s + 1 + j3)}(s + 1 - j3)$$

$$\frac{30s(s + 2)}{(s + 5)[s + 1 + j3]} = \frac{R_{-5}}{(s + 5)}(s + 1 - j3)$$

$$+ k + \frac{k}{(s + 1 + j3)}(s + 1 - j3)$$

Then we let $s = -1 + j3$.

$$\frac{30(-1 + j3)[(-1 + j3) + 2]}{[(-1 + j3) + 5][(-1 + j3) + 1 + j3]} = 0 + k + 0$$

$$k = \frac{30(-1 + j3)((-1 + j3) + 2)}{[(-1 + j3) + 5][(-1 + j3) + 1 + j3]} = 10\,\angle 53.130°$$

To find k^* we could go through the same procedure, but this is not necessary. The value will be the conjugate of k. Therefore, we may simply state

$$k^* = 10\,\angle -53.1301°$$

Substituting these values into the expansion, we have

$$V(s) = \frac{30s(s + 2)}{(s + 5)[(s + 1)^2 + (3)^2]} = \frac{R_{-5}}{(s + 5)} + \frac{k}{(s + 1 - j3)} + \frac{k^*}{(s + 1 + j3)}$$

$$= \frac{18}{(s + 5)} + \frac{10\,\angle 53.131°}{(s + 1 - j3)} + \frac{10\,\angle -53.131°}{(s + 1 + j3)}$$

Next we convert the complex residues to an exponential form using the identity

Magnitude $\angle\theta$ = *Magnitude* $e^{j\theta}$

$$V(s) = \frac{30s(s + 2)}{(s + 5)[(s + 1)^2 + (3)^2]} = \frac{18}{(s + 5)} + \frac{10e^{j53.131°}}{(s + 1 - j3)} + \frac{10e^{-j53.131°}}{(s + 1 + j3)}$$

and then we take the inverse of both sides of the equation using transform pair (P-4).

$$v(t) = \mathcal{L}^{-1}\left[\frac{30s(s+2)}{(s+5)[(s+1)^2+(3)^2]}\right] = \mathcal{L}^{-1}\left[\frac{18}{(s+5)}\right] + \mathcal{L}^{-1}\left[\frac{10e^{j53.131°}}{(s+1-j3)}\right]$$

$$+ \mathcal{L}^{-1}\left[\frac{10e^{-j53.131°}}{(s+1+j3)}\right]$$

$$= 18e^{-5t}u(t) + 10e^{j53.131°}e^{(-1+j3)t}u(t) + 10e^{-j53.131°}e^{(-1-j3)t}u(t)$$

Let's pull out the two terms containing the imaginary exponents and convert them into a sinusoidal function.

$$10e^{j53.131°}e^{(-1+j3)t}u(t) + 10e^{-j53.131°}e^{(-1-j3)t}u(t)$$

First we separate out the exponents.

$$10e^{j53.131°}e^{-t}e^{j3t}u(t) + 10e^{-j53.131°}e^{-t}e^{-j3t}u(t)$$

Combining the exponents that have a "j" multiplier, we may rewrite the expression as

$$10e^{-t}e^{j(3t+53.131°)}u(t) + 10e^{-t}e^{-j(3t+53.131°)}u(t)$$

Next we factor out the $10e^{-t}$ and the $u(t)$.

$$10e^{-t}[e^{j(3t+53.131°)} + e^{-j(3t+53.131°)}]u(t)$$

Then using Euler's formula, we substitute in the sinusoidal expressions for the exponential functions.

$$10e^{-t}[\cos(3t+53.131°) + j\sin(3t+53.131°) + \cos(3t+53.131°)$$
$$- j\sin(3t+53.131°)]u(t)$$

Simplifying the expression, we get

$$20e^{-t}\cos(3t+53.131°)u(t)$$

Then we convert the cosine expression into a sine expression.

$$20e^{-t}\sin(3t+143.13°)u(t)$$

We then add back in the $18e^{-5t}u(t)$ for the final answer.

$$v(t) = 18e^{-5t}u(t) + 20e^{-t}\sin(3t+143.13°)u(t)$$

The last example demonstrates how the procedure for finding the residues for the complex poles is the same as the procedure for finding the residues for the real poles. The big difference is in converting the exponential expressions into a sinusoidal function. The initial conversion from exponential to sinusoidal results in a cosine, but it is customary to change this into a sine function. Because the steps are the same, an equation could be developed to go directly to the sine function.

3-4-4 Inverse Transform of Complex-Poles—Real-Numbers Method

In this section, we will go through another traditional method that uses only real numbers and does not require Euler's formula. The main disadvantage is that this method requires simultaneous equations. For every set of complex conjugate poles, two simultaneous equations are required. Thus this method becomes a little awkward when more than one set of complex conjugate poles are involved.

Complex–conjugate poles may be expanded into the following form:

$$F(s) = \frac{N(s)}{D(s)} = \frac{N(s)}{[(s + \alpha)^2 + \omega^2]\,(other\,factors)} = \frac{As + B}{[(s + \alpha)^2 + \omega^2]} + \begin{matrix} rest \\ of \\ expansion \end{matrix} \qquad \textbf{(3-10)}$$

Notice that the complex roots are combined into a single term and that the numerator has two unknowns, A and B. When the magnitudes for the other expanded fractions are known, two equations can be generated and used to solve for A and B.

The two equations required to solve for the values of A and B are generated by substituting two different real values of s into the equation of $F(s)$ and its expanded form. Any real value of s other than a pole value of $F(s)$ will work because the pole values will cause $F(s)$ to go to infinity. Once the numerical values of A and B are determined, the inverse of this term is found using the transform pair (P-12b).

$$\sqrt{A^2 + K^2}\, e^{-\alpha t} \sin(\omega t + \theta)\, u(t)$$

where $K = \dfrac{(B - A\alpha)}{\omega}$

$\theta = \tan^{-1}\left(\dfrac{A}{K}\right)$

The signs of K (x-direction) and A (y-direction) determine the quadrant of the angle.

In this transform pair we must observe the sign of the numerator and denominator for the arctangent when determining θ. Some calculators take into account the signs of K and A and return the value of the angle in the correct quadrant, but most do not. When the calculator does not take into account the quadrant, we must add 180° to the calculated value of θ when K is negative and A is positive, and we must subtract 180° from the calculated value of θ when K is negative and A is negative. In the subsequent examples, we will assume that the calculator does not take the signs of K and A into account.

EXAMPLE 3-29

Using real numbers only, find the inverse Laplace transform of

$$\frac{8[(s + 2.4)^2 + 9]}{s[(s + 1.5)^2 + 1.44]}$$

Solution: The function will expand into

$$F(s) = \frac{8[(s + 2.4)^2 + 9]}{s[(s + 1.5)^2 + 1.44]} = \frac{R_0}{s} + \frac{As + B}{[(s + 1.5)^2 + (1.2)^2]}$$

The magnitude of $R(s)_0$ is found as usual by multiplying both sides of the equation by s and setting $s = 0$.

$$\frac{8[(s + 2.4)^2 + 9]}{s[(s + 1.5)^2 + 1.44]} s = \frac{R_0}{s} s + \frac{As + B}{[(s + 1.5)^2 + (1.2)^2]} s$$

$$\frac{8[(s + 2.4)^2 + 9]}{[(s + 1.5)^2 + 1.44]} = R_0 + \frac{As + B}{[(s + 1.5)^2 + (1.2)^2]} s$$

$$\frac{8[(s + 2.4)^2 + 9]}{[(s + 1.5)^2 + 1.44]} = R_0 + 0$$

$$R_0 = \frac{8[(0 + 2.4)^2 + 9]}{[(0 + 1.5)^2 + 1.44]} = 32$$

This value is then substituted back into the expansion.

$$\frac{8[(s + 2.4)^2 + 9]}{s[(s + 1.5)^2 + 1.44]} = \frac{32}{s} + \frac{As + B}{[(s + 1.5)^2 + (1.2)^2]}$$

To solve for A and B, we will substitute in a couple of values of s. We can not use $s = 0, -1.5 + j1.2$, or $-1.5 - j1.2$ because these are the pole values of the function, and when they are substituted in, the value of the function will go to infinity. We must use values of s that will give us equations containing only finite numbers, and the simpler the numbers the better. Therefore, we select values such as $0, 1, -1, 2, -2$, or even the values of the zeros of the function. The main idea is to generate two different equations. In this example, let's start with $s = 1$.

$$\frac{8[(1 + 2.4)^2 + 9]}{(1)[(1 + 1.5)^2 + 1.44]} = \frac{32}{1} + \frac{A(1) + B}{[(1 + 1.5)^2 + (1.2)^2]}$$

Putting A and B on the left side and the constant on the right side, we have

$$A + B = -81.6$$

Next we choose another value for s. Let's use $s = -1$.

$$\frac{8[(-1 + 2.4)^2 + 9]}{(-1)[(-1 + 1.5)^2 + 1.44]} = \frac{32}{-1} + \frac{A(-1) + B}{[(-1 + 1.5)^2 + (1.2)^2]}$$

Putting A and B on the left side and the constant on the right side, we have

$$-A + B = -33.6$$

Therefore, the two equations to solve simultaneously are

$$\begin{cases} A + B = -81.6 \\ -A + B = -33.6 \end{cases}$$

Any method such as Cramer's rule, substitution, or matrix algebra may be used to solve the set of equations. Solving the set of equations for A and B, we have

$$A = -24 \quad \text{and} \quad B = -57.6$$

We can now substitute these values back into the expansion and find the inverse.

$$\frac{8[(s + 2.4)^2 + 9]}{s[(s + 1.5)^2 + 1.44]} = \frac{32}{s} + \frac{-24s - 57.6}{[(s + 1.5)^2 + (1.2)^2]}$$

The inverse of the $32/s$ term simply becomes $32\, u(t)$ by using transform pair (P-2), but the inverse of the second term requires the transform pair (P-12b) from the extended transform pairs in Appendix A.

$$\sqrt{A^2 + K^2}\, e^{-\alpha t} \sin(\omega t + \theta)\, u(t)$$

where $K = \dfrac{(B - A\alpha)}{\omega}$

$$\theta = \tan^{-1}\left(\frac{A}{K}\right)$$

The signs of $K(x\text{-direction})$ and $A(y\text{-direction})$ determine the quadrant of the angle.

We first determine the K value.

$$K = \frac{(B - A\alpha)}{\omega}$$

$$= \frac{(-57.6 - (-24)(1.5))}{1.2}$$

$$= -18$$

Next we determine θ. Since both K and A are negative, the angle is in the third quadrant, and, therefore, we must subtract $180°$ from the inverse tangent.

$$\theta = \tan^{-1}\left(\frac{A}{K}\right) - 180°$$

$$= \tan^{-1}\left(\frac{-24}{-18}\right) - 180°$$

$$= -126.87°$$

With the values of K and θ determined, we can find the inverse due to the complex pole.

$$\mathcal{L}^{-1}\left[\frac{-24s - 57.6}{[(s + 1.5)^2 + (1.2)^2]}\right] = \sqrt{A^2 + K^2}\,e^{-\alpha t}\sin(\omega t + \theta)\,u(t)$$

$$= \sqrt{(-24)^2 + (-18)^2}\,e^{-1.5t}\sin(1.2t - 126.87)\,u(t)$$

$$= 30e^{-1.5t}\sin(1.2t - 126.87°)\,u(t)$$

Therefore, the total inverse is

$$f(t) = \mathcal{L}^{-1}\left[\frac{8[(s + 2.4)^2 + 9]}{s[(s + 1.5)^2 + 1.44]}\right] = \mathcal{L}^{-1}\left[\frac{32}{s}\right] + \mathcal{L}^{-1}\left[\frac{-24s - 57.6}{[(s + 1.5)^2 + (1.2)^2]}\right]$$

$$f(t) = 32\,u(t) + 30e^{-1.5t}\sin(1.2t - 126.87°)\,u(t)$$

3-5 STREAMLINED INVERSE LAPLACE TRANSFORMS

The traditional techniques of the previous section allow us to see how the mathematics isolate each value required to find the inverse Laplace transform of a function. Some of the steps shown are informative, giving the details of how the numbers are manipulated, but these informative steps are not really necessary to justify the inverse. In this section, the inverse Laplace transform will be approached in a form more suitable for use in electronics than the mathematical approach of the previous section. The techniques in this section are, for the most part, the same, but they have been streamlined so that only the steps necessary to justify the inverse Laplace transform are used. An understanding of the traditional mathematical approach of the previous section is not necessary to be able to use the techniques shown here.

3-5-1 Inverse Equations Definitions

$F(s)$, the function for which the inverse Laplace transform is to be found, can be represented as a numerator polynomial, $N(s)$, divided by a denominator polynomial, $D(s)$.

$$F(s) = \frac{N(s)}{D(S)}$$

(3-11)

where the degree of the $N(s)$ polynomial is less than the $D(s)$ degree. The roots of $D(s)$ are called *poles* because the function $F(s)$ will approach infinity when s is equal to any of the poles. (When graphed, this looks like a flagpole.) The roots of $N(s)$ are called *zeros* since the function $F(s)$ is zero when s is equal to one of these roots.

Before beginning the inverse Laplace process, it is very important to have the function $F(s)$ in a standard form, as described in Section 3-1-2. This form guarantees that common factors in the numerator and denominator have been canceled, and it allows us to quickly identify the type of each pole in the function. Identifying the pole type is important because the type of pole will determine which inverse equation to use.

The first step in finding the inverse Laplace transform is to remove the impulse function, if there is one, from the original $F(s)$ function. A function that has an impulse will have the numerator polynomial equal to the denominator polynomial. When this is the case, we divide the numerator polynomial by the denominator polynomial. This will result in a constant plus a new fraction, $N(s)/D(s)$. The constant will invert into an impulse, and then we will find the inverse Laplace transform of the resulting $N(s)/D(s)$. It should be noted that it is unnecessary to factor the new $N(s)$ in this resulting function, because we will not need to identify the zeros in this $N(s)$.

EXAMPLE 3-30

Put the following function in the proper form of $N(s)/D(s)$.

$$F(s) = \frac{6(s + 2)(s + 3)(s + 5)}{(s + 1)(3s^2 + 18s + 24)}$$

Solution: The numerator is already in a properly factored form; however, the denominator needs to be factored. We also see that this function needs to be divided out because both numerator and denominator polynomials are of the same degree. First let's factor out the 3 in the denominator's quadratic.

$$F(s) = \frac{6(s + 2)(s + 3)(s + 5)}{3(s + 1)(s^2 + 6s + 8)} = \frac{2(s + 2)(s + 3)(s + 5)}{(s + 1)(s^2 + 6s + 8)}$$

Next we could factor the rest of the denominator and look for terms to cancel, or we could multiply out and do the division first. If we take the first choice and if there is a cancellation, the division will be easier. Let's factor first.

$$F(s) = \frac{2(s + 2)(s + 3)(s + 5)}{(s + 1)(s + 2)(s + 4)} = \frac{2(s + 3)(s + 5)}{(s + 1)(s + 4)}$$

Next we multiply out the numerator and the denominator to get the constant plus the new $N(s)/D(s)$.

$$s^2 + 5s + 4 \overline{\smash{\big)}\, \begin{array}{c} 1 \\ s^2 + 8s + 15 \end{array}}$$
$$\underline{s^2 + 5s + 4}$$
$$3s + 11$$

The result is

$$F(s) = 2\left[1 + \frac{3(s + 3.6667)}{(s + 1)(s + 4)}\right]$$

$$= 2 + \frac{6s + 22}{(s + 1)(s + 4)}$$

The mathematical process of finding the inverse Laplace transform of a function is explained in detail in Section 3-4. Briefly, we expand a function into a summation of simpler fractions that can be inverted using tabulated Laplace transform pairs. This summation of simpler functions is based on the poles. In the expansion, each first-order real pole will have only one fraction, and multiple-order real poles will have multiple fractions. The numerator for each of the simpler fractions will be a constant, which is called a *residue*. The following example shows the general form that real poles expand into. Note in the example how the subscripts are used for the residue constants.

EXAMPLE 3-31

What will the summation of fractions look like for the following function?

$$F(s) = \frac{N(s)}{(s + 2)(s + 7)^3}$$

Solution: The numerator of each function will be a constant.

$$F(s) = \frac{N(s)}{(s + 2)(s + 7)^3} = \frac{R_{-2}}{(s + 2)} + \frac{R_{-7,1}}{(s + 7)^3} + \frac{R_{-7,2}}{(s + 7)^2} + \frac{R_{-7,3}}{(s + 7)}$$

The value of the numerator of each expanded fraction shown in Example 3-31 will have a numerical value. From this point on we will not be interested in the mathematical intricacies of how to isolate the residues. Instead, we will use presolved equations to determine the values directly.

Before we see the full inverse equations, it would be a good idea to get used to the residue equation for first-order real poles and first-order complex–conjugate poles. The residue for a first-order real pole can be found by the equation

$$R_{-a} = F(s)\,(s + a)\,\Big|_{s=-a} \qquad\qquad (3\text{-}12)$$

and for first-order complex–conjugate poles by

$$R_{-\alpha+j\omega} = F(s)\,[(s + a)^2 + \omega^2]\,\Big|_{s=-\alpha+j\omega} \qquad\qquad (3\text{-}13)$$

In both equations, the term that is multiplied times the function $F(s)$ must be a factor of the denominator of $F(s)$. The following example demonstrates how these equations are applied.

EXAMPLE 3-32

For the function shown, find R_{-15}, R_{-6}, and R_{-9+j3}.

$$F(s) = \frac{s(s + 6)}{(s + 15)[(s + 9)^2 + 9]}$$

Solution: R_{-15} is found by applying Eq. (3-13), which is for a real pole.

$$R_{-a} = F(s)\,(s + a)\,\Big|_{s=-a}$$

$$R_{-15} = \frac{s(s + 6)}{(s + 15)[(s + 9)^2 + 9]}\,(s + 15)\,\Big|_{s=-15}$$

$$= \frac{s(s + 6)}{[(s + 9)^2 + 9]}\,\Big|_{s=-15}$$

$$= \frac{-15(-15 + 6)}{[(-15 + 9)^2 + 9]}$$

$$= 3$$

A value for R_{-6} does not exist because this is a zero. The equations for finding residues only apply to the poles.

R_{-9+j3} is found by applying Eq. (3-14) because this is for a complex–conjugate pole.

$$R_{-\alpha+j\omega} = F(s) \, [(s+a)^2 + \omega^2] \,\Big|_{s=-\alpha+j\omega}$$

$$R_{-9+j3} = \frac{s(s+6)}{(s+15)[(s+9)^2+9]} \, [(s+9)^2 + 9] \,\Big|_{s=-9+j3}$$

$$= \frac{s(s+6)}{(s+15)} \,\Big|_{s=-9+j3}$$

$$= \frac{(-9+j3)((-9+j3)+6)}{[(-9+j3)+15]}$$

$$= 6 \angle -90°$$

3-5-2 Inverse Transform of Real Poles

The process of finding the inverse Laplace transform of real poles can be summarized by the following three equations:

$$R_{-a,n} = \frac{d^{n-1}}{ds^{n-1}} F(s) \, (s+a)^r \,\Big|_{s=-a} \tag{3-14}$$

$$F(s) = \sum_{n=1}^{r} \frac{R_{-a,n}/(n-1)!}{(s+a)^{r-n+1}} + \textit{rest of function} \tag{3-15}$$

$$f(t) = \left[\sum_{n=1}^{r} \frac{R_{-a}}{(n-1)! \, (r-n)!} \, t^{r-n} \right] e^{-at} u(t) + \textit{rest of inverse} \tag{3-16}$$

where $n = 1, 2, 3, \ldots, r$ for Eqs. $(3 - 14)$ through $(3 - 16)$.

In Eq. (3-14) the value calculated for $R_{-a,n}$ will always be a real number. This real number is substituted into Eq. (3-15) to find the terms that $F(s)$ will expand into. However, when we are working with electronic circuits, we are looking for the voltages and currents in the time domain and are relatively uninterested in how the Laplace function can be expanded. Therefore, to streamline the inverse Laplace transform process even more, we will use only Eq. (3-14) and Eq. (3-16).

We will begin using these equations on first-order poles. In a first-order pole, $r = 1$ and, therefore, n will only go to 1. When we substitute $r = 1$ and $n = 1$ into these equations, they reduce to

$$R_{-a} = F(s) \, (s+a) \,\Big|_{s=-a} \tag{3-17}$$

$$f(t) = R_{-a} e^{-at} u(t) + \textit{rest of inverse} \tag{3-18}$$

Most functions contain real poles raised to the first power, so these equations are used most of the time. Equation (3-15) was not listed in this group because it is of little significance to us in determining the inverse Laplace transform. This group of equations is often referred to as the *cover-up method*: R_{-a} is the value of $F(s)$ with one of the pole factors covered up while substituting in a value for s. Watch for this in the following examples.

Most of the following examples are the same ones used in Section 3-4 so that a comparison of the two approaches may be easily made. The streamlined approach is easier and faster to use, and because it has fewer and more concise steps, this approach minimizes the chance for errors.

EXAMPLE 3-33

Find the inverse of

$$F(s) = \frac{10(s + 3)(s + 5)}{s(s + 1)(s + 6)}$$

Solution: In this function the numerator polynomial's power is less than the denominator polynomial's power so we do not have an impulse function to remove. Equation (3-17) is applied to find the value of the residue R_{-a} for the pole $s = 0$.

$$R_{-a} = F(s)(s + a)\Big|_{s=-a}$$

$$R_{-0} = \frac{10(s + 3)(s + 5)}{s(s + 1)(s + 6)}(s)\Big|_{s=0} = \frac{10(s + 3)(s + 5)}{(s + 1)(s + 6)}\Big|_{s=0}$$

$$= \frac{10(0 + 3)(0 + 5)}{(0 + 1)(0 + 6)}$$

$$R_{-0} = 25$$

We then find the inverse Laplace due to the pole $s = 0$ using Eq. (3-18).

$$f(t) = R_{-a}e^{-at}u(t)$$

$$f_0(t) = 25e^0 u(t)$$

$$= 25 u(t)$$

Notice how a subscript has been added to the $f(t)$ to help us keep track of what pole the $25 u(t)$ applies to.

Next, the residue for the pole $s = -1$ is found.

$$R_{-a} = F(s)(s + a) \Big|_{s=-a}$$

$$R_{-1} = \frac{10(s + 3)(s + 5)}{s(s + 1)(s + 6)}(s + 1)\Big|_{s=-1} = \frac{10(s + 3)(s + 5)}{s(s + 6)}\Big|_{s=-1}$$

$$= \frac{10(-1 + 3)(-1 + 5)}{(-1)(-1 + 6)}$$

$$R_{-1} = -16$$

The inverse Laplace due to the pole $s = -1$ is therefore

$$f(t) = R_{-a}e^{-at}u(t)$$

$$f_{-1}(t) = -16e^{-t}u(t)$$

Finally, the inverse due to the pole at $s = -6$ is

$$R_{-a} = F(s)(s + a)\Big|_{s=-a}$$

$$R_{-6} = \frac{10(s + 3)(s + 5)}{s(s + 1)(s + 6)}(s + 6)\Big|_{s=-6} = \frac{10(s + 3)(s + 5)}{s(s + 1)}\Big|_{s=-6}$$

$$= \frac{10(-6 + 3)(-6 + 5)}{(-6)(-6 + 1)}$$

$$R_{-6} = 1$$

$$f(t) = R_{-a}e^{-at}u(t)$$

$$f_{-6}(t) = e^{-6t}u(t)$$

We then sum all of the inverses due to each pole to find the inverse of the function.

$$f(t) = \mathscr{L}^{-1}\left[\frac{10(s + 3)(s + 5)}{s(s + 1)(s + 6)}\right]$$

$$= 25\,u(t) - 16\,e^{-t}\,u(t) + e^{-6t}\,u(t)$$

EXAMPLE 3-34

Find the inverse of

$$F(s) = \frac{[(s + 8)^2 + 36]}{(s + 8)(s + 20)}$$

Solution: Before we write the expansion we will remove the impulse by dividing

$$
\begin{array}{r}
1 \\
s^2 + 28s + 160 \overline{\smash{\big)}\, s^2 + 16s + 100} \\
\underline{s^2 + 28s + 160} \\
-12s - 60
\end{array}
$$

$$F(s) = \frac{[(s + 8)^2 + 36]}{(s + 8)(s + 20)} = 1 + \frac{-12s - 60}{(s + 8)(s + 20)}$$

The 1 will inverse into $\delta(t)$. Now we will find the inverse of the leftover fraction. The inverse due to the pole at $s = -8$ is

$$R_{-a} = F(s)(s + a)\Big|_{s = -a}$$

$$R_{-8} = \frac{-12s - 60}{(s + 8)(s + 20)}(s + 8)\Big|_{s = -8} = \frac{-12s - 60}{(s + 20)}\Big|_{s = -8}$$

$$= \frac{-12(-8) - 60}{(-8 + 20)}$$

$$R_{-8} = 3$$

$$f(t) = R_{-a}e^{-at}\,u(t)$$

$$f_{-8}(t) = 3e^{-8t}\,u(t)$$

Next, the inverse due to $s = -20$ is found.

$$R_{-a} = F(s)\,(s + a)\Big|_{s = -a}$$

$$R_{-20} = \frac{-12s - 60}{(s + 8)(s + 20)}(s + 20)\Big|_{s = -20} = \frac{-12s - 60}{(s + 8)}\Big|_{s = -20}$$

$$= \frac{-12(-20) - 60}{(-20 + 8)}$$

$$R_{-20} = -15$$

$$f(t) = R_{-a} e^{-at} u(t)$$

$$f_{-20}(t) = -15 e^{-20t} u(t)$$

The inverse due to the impulse and the poles are added together to find the total inverse Laplace transform.

$$f(t) = \mathcal{L}^{-1}\left[\frac{[(s+8)^2 + 36]}{(s+8)(s+20)} \right]$$

$$= \delta(t) + 3 e^{-8t} u(t) - 15 e^{-20t} u(t)$$

With multiple-order poles, $r > 1$, the variable n goes from 1 to the value of r, and we must use the full equations in Eqs. (3-14) and (3-16).

EXAMPLE 3-35

Find the inverse of

$$F(s) = \frac{5(s+6)}{(s+10)(s+12)^2}$$

Solution: There is no impulse to remove so we begin by finding the inverse Laplace due to the pole $s = -10$.

$$R_{-a} = F(s)\,(s+a) \bigg|_{s=-a}$$

$$R_{-10} = \frac{5(s+6)}{(s+10)(s+12)^2}(s+10) \bigg|_{s=-10} = \frac{5(s+6)}{(s+12)^2} \bigg|_{s=-10}$$

$$= \frac{5(-10+6)}{(-10+12)^2}$$

$$R_{-10} = -5$$

$$f(t) = R_{-a} e^{-at} u(t)$$

$$f_{-10}(t) = -5 e^{-10t} u(t)$$

Next we find the inverse due to the second-order pole $s = -12$ using Eq. (3-14). In this case, $r = 2$ so we will use $n = 1, 2$.

$$R_{-a,n} = \frac{d^{n-1}}{ds^{n-1}} F(s) (s + a)^r \Big|_{s = -a}$$

$$R_{-12,1} = \frac{d^0}{ds^0} \frac{5(s + 6)}{(s + 10)(s + 12)^2} (s + 12)^2 \Big|_{s = -12} = \frac{5(s + 6)}{(s + 10)} \Big|_{s = -12}$$

$$= \frac{5(-12 + 6)}{(-12 + 10)}$$

$$R_{-12,1} = 15$$

$$R_{-12,2} = \frac{d}{ds} \frac{5(s + 6)}{(s + 10)(s + 12)^2} (s + 12)^2 \Big|_{s = -12} = \frac{d}{ds} \frac{5(s + 6)}{(s + 10)} \Big|_{s = -12}$$

$$= \frac{20}{(s + 10)^2} \Big|_{s = -12}$$

$$= \frac{20}{(-12 + 10)^2}$$

$$R_{-12,2} = 5$$

The inverse Laplace transform due to the poles at $s = -12$ is found using Eq. (3-16).

$$f(t) = \left[\sum_{n=1}^{r} \frac{R_{-a,n}}{(n - 1)! (r - n)!} t^{r-n} \right] e^{-at} u(t)$$

$$f_{-12}(t) = \left[\frac{15}{(1 - 1)! (2 - 1)!} t^{2-1} + \frac{5}{(2 - 1)! (2 - 2)!} t^{2-2} \right] e^{-12t} u(t)$$

$$= [15t + 5] e^{-12t} u(t)$$

Combining the inverse due to each pole, we have

$$f(t) = \mathcal{L}^{-1} \left[\frac{5(s + 6)}{(s + 10)(s + 12)^2} \right]$$

$$= -5e^{-10t} u(t) + 15te^{-12t} u(t) + 5e^{-12t} u(t)$$

When the pole is equal to zero, the inverse of multiple-order poles appears to be a special case; however, it is not.

EXAMPLE 3-36

Find

$$v(t) = \mathcal{L}^{-1}\left[\frac{3(s + 2)^2}{s^4}\right]$$

Solution: Using Eq. (3-14), where $n = 1, 2, 3, 4$ since $r = 4$,

$$R_{-a,n} = \frac{d^{n-1}}{ds^{n-1}} F(s)\,(s + a)^r \bigg|_{s = -a}$$

$$R_{0,1} = \frac{d^0}{ds^0} \frac{3(s + 2)^2}{s^4} s^4 \bigg|_{s = 0} = 3(s + 2)^2 \bigg|_{s = 0}$$

$$= 3(0 + 2)^2$$

$$R_{0,1} = 12$$

$$R_{0,2} = \frac{d}{ds} 3(s + 2)^2 \bigg|_{s = 0} = 6s + 12 \bigg|_{s = 0}$$

$$= 6(0) + 12$$

$$R_{0,2} = 12$$

Equation (3-14) now requires the second derivative. The function we have just used is the first derivative, so we will simply take the derivative of this function.

$$R_{0,3} = \frac{d}{ds}(6s + 12) \bigg|_{s = 0} = 6 \bigg|_{s = 0}$$

$$= 6$$

$$R_{0,3} = 6$$

For the third derivative, we will take the derivative of the last function used.

$$R_{0,4} = \frac{d}{ds} 6 \bigg|_{s = 0} = 0 \bigg|_{s = 0}$$

$$= 0$$

$$R_{0,4} = 0$$

The inverse Laplace transform can now be found using Eq. (3-16).

$$f(t) = \left[\sum_{n=1}^{r} \frac{R_{-a,n}}{(n-1)! \, (r-n)!} \, t^{r-n} \right] e^{-at} \, u(t)$$

$$f_0(t) = \left[\frac{12}{(1-1)! \, (4-1)!} \, t^{4-1} + \frac{12}{(2-1)! \, (4-2)!} \, t^{4-2} \right.$$

$$\left. + \frac{6}{(3-1)! \, (4-3)!} \, t^{4-3} + \frac{0}{(4-1)! \, (4-4)!} \, t^{4-4} \right] e^{0} \, u(t)$$

$$= [2t^3 + 6t^2 + 3t] \, u(t)$$

Therefore, the inverse Laplace transform is

$$v(t) = \mathcal{L}^{-1} \left[\frac{3(s+2)^2}{s^4} \right]$$

$$= [2t^3 + 6t^2 + 3t] \, u(t)$$

3-5-3 Inverse Transform of Complex Poles

Each complex–conjugate pole in $F(s)$ will generate a complex exponential time-domain value. When conjugate time-domain values are combined, they can always be converted into a sinusoidal function by using Euler's identities. In electronic circuits, complex poles will always occur in conjugates. Therefore, we are able to solve directly for the sinusoidal function. The equations for finding the inverse Laplace transform for complex–conjugate poles are

$$R_{-\alpha+j\omega,\,n} = \frac{d^{n-1}}{ds^{n-1}} F(s) \, [(s+\alpha)^2 + \omega^2]^r \Big|_{s=-\alpha+j\omega} \qquad (3\text{-}19)$$

$$M_n \angle\theta_n \text{ is defined as} \qquad (3\text{-}20)$$

$$M_1 \angle\theta_1 = R_{-\alpha+j\omega,1}$$

$$M_2 \angle\theta_2 = R_{-\alpha+j\omega,2} - \frac{r}{2j\omega} R_{-\alpha+j\omega,1}$$

$$M_3 \angle\theta_3 = R_{-\alpha+j\omega,3} - \frac{2r}{2j\omega} R_{-\alpha+j\omega,2} + \frac{r(r+1)}{(2j\omega)^2} R_{-\alpha+j\omega,1}$$

$$M_4 \angle\theta_4 = R_{-\alpha+j\omega,4} - \frac{3r}{2j\omega} R_{-\alpha+j\omega,3} + \frac{3r(r+1)}{(2j\omega)^2} R_{-\alpha+j\omega,2} - \frac{r(r+1)(r+2)}{(2j\omega)^3} R_{-\alpha+j\omega,1}$$

$$f(t) = \left[\sum_{n=1}^{r} \frac{2\,(-1)^{int(r/2)}M_n}{(2\omega)^r(n-1)!\,(r-n)!}\, t^{r-n} \sin\left(\omega t + \theta_n + \frac{1+(-1)^r}{2}(90°)\right)\right]e^{-\alpha t}u(t)$$

$$+ \; rest\;of\;inverse \tag{3-21}$$

where in Eqs. (3-20) and (3-21),

  1. $n = 1, 2, 3, \ldots, r$
  2. $int(r/2)$ means the integer portion. Therefore, $int(5/2) = int(4/2) = 2$.

Not all of the $M_n \angle \theta_n$ equations are shown, but only enough to determine the pattern. Notice how the numerical constants in these equations can be found by using Pascal's triangle.

Fortunately, it is rare to see r greater than 1 in complex–conjugate poles. Because first-order complex–conjugate poles are the most common, we typically use a simplified version of the preceding equations. When $r = 1$, it causes n to have only the value of 1, and we therefore only require M_1. Because M_1 is the magnitude, or absolute value, of $R_{-\alpha+j\omega}$, we may use $|R_{-\alpha+j\omega}|$ instead of M_1, and we will use θ to represent the phase angle of $R_{-\alpha+j\omega}$. With these simplifications, the equations for finding the inverse Laplace transform of first-order complex–conjugate poles are

$$R_{-\alpha+j\omega} = F(s)\left[(s+\alpha)^2 + \omega^2\right]\Big|_{s=-\alpha+j\omega} \tag{3-22}$$

$$f(t) = \frac{|R_{-\alpha+j\omega}|}{\omega}\, e^{-\alpha t} \sin(\omega t + \theta)\, u(t) + rest\;of\;inverse \tag{3-23}$$

EXAMPLE 3-37

Find the inverse transform of

$$V(s) = \frac{30s(s+2)}{(s+5)[(s+1)^2 + (3)^2]}$$

Solution: The numerator polynomial's power is less than the denominator polynomial's power so we do not have an impulse to remove. The inverse due to the first-order real pole at $s = -5$ proceeds as usual.

$$R_{-a} = F(s)\,(s+a)\,\Big|_{s=-a}$$

$$R_{-5} = \frac{30s(s+2)}{(s+5)[(s+1)^2 + (3)^2]}\,(s+5)\,\Big|_{s=-5} = \frac{30s(s+2)}{[(s+1)^2 + (3)^2]}\,\Big|_{s=-5}$$

$$= \frac{30(-5)(-5+2)}{[(-5+1)^2 + (3)^2]}$$

$$R_{-5} = 18$$

$$f(t) = R_{-a}e^{-at}\,u(t)$$

$$f_{-5}(t) = 18e^{-5t}\,u(t)$$

Next we find the inverse due to the complex–conjugate poles using Eqs. (3-22) and (3-23)

$$R_{-\alpha+j\omega} = F(s)\,[(s+\alpha)^2 + \omega^2]\Big|_{s=-\alpha+j\omega}$$

$$R_{-1+j3} = \frac{30s(s+2)}{(s+5)[(s+1)^2+(3)^2]}\,[(s+1)^2+(3)^2]\Big|_{s=-1+j3}$$

$$= \frac{30s(s+2)}{(s+5)}\Big|_{s=-1+j3}$$

$$= \frac{30(-1+j3)\,((-1+j3)+2)}{[(-1+j3)+5]}$$

$$R_{-1+j3} = 60 \angle 143.13°$$

$$f(t) = \frac{|R_{-\alpha+j\omega}|}{\omega}\,e^{-\alpha t}\sin(\omega t + \theta)\,u(t)$$

$$f_{-1+j3}(t) = \frac{60}{3}e^{-t}\sin(3t + 143.13°)\,u(t)$$

$$= 20e^{-t}\sin(3t + 143.13°)\,u(t)$$

Combining the inverse due to each pole, we have

$$v(t) = \mathcal{L}^{-1}\left[\frac{30s(s+2)}{(s+5)[(s+1)^2+(3)^2]}\right]$$

$$= 18e^{-5t}\,u(t) + 20e^{-t}\sin(3t + 143.13°)$$

EXAMPLE 3-38

Find the inverse Laplace transform of

$$\frac{8[(s+2.4)^2 + 9]}{s[(s+1.5)^2 + 1.44]}$$

Solution: We find the inverse due to the pole $s = 0$ first.

$$R_{-a} = F(s)(s + a) \Big|_{s = -a}$$

$$R_{-0} = \frac{8[(s + 2.4)^2 + 9]}{s[(s + 1.5)^2 + 1.44]}(s) \Big|_{s=0} = \frac{8[(s + 2.4)^2 + 9]}{[(s + 1.5)^2 + 1.44]} \Big|_{s=0}$$

$$= \frac{8[(0 + 2.4)^2 + 9]}{[(0 + 1.5)^2 + 1.44]}$$

$$R_{-0} = 32$$

$$f(t) = R_{-a}e^{-at}u(t)$$

$$f_0(t) = 32e^0\,u(t)$$

$$= 32\,u(t)$$

Next we find the inverse due to the complex–conjugate poles using Eqs. (3-22) and (3-23).

$$R_{-\alpha+j\omega} = F(s)\,[(s + \alpha)^2 + \omega^2] \Big|_{s = -\alpha + j\omega}$$

$$R_{-1.5+j1.2} = \frac{8[(s + 2.4)^2 + 9]}{s[(s + 1.5)^2 + 1.44]}[(s + 1.5)^2 + 1.44] \Big|_{s=-1.5+j1.2}$$

$$= \frac{8[(s + 2.4)^2 + 9]}{s} \Big|_{s=-1.5+j1.2}$$

$$= \frac{8[((-1.5 + j1.2) + 2.4)^2 + 9]}{(-1.5 + j1.2)}$$

$$R_{-1.5+j1.2} = 36 \angle -126.87°$$

$$f(t) = \frac{|R_{-\alpha+j\omega}|}{\omega}e^{-\alpha t}\sin(\omega t + \theta)\,u(t)$$

$$f_{-1.5+j1.2}(t) = \frac{36}{1.2}e^{-1.5t}\sin(1.2t - 126.87°)\,u(t)$$

$$= 30e^{-1.5t}\sin(1.2t - 126.87°)\,u(t)$$

Combining the inverse due to each pole, we have

$$f(t) = \mathcal{L}^{-1}\left[\frac{8[(s + 2.4)^2 + 9]}{s[(s + 1.5)^2 + 1.44]}\right]$$

$$= 32\,u(t) + 30e^{-1.5t}\sin(1.2t - 126.87°)\,u(t)$$

Example 3-37 and Example 3-38 had complex poles that contained both a real part and an imaginary part. It is common to find complex poles containing only an imaginary part: For example, $s^2 + 25$. This type of complex pole looks completely different; however, if we put in the 0 for the real part, it will look the same $[(s + 0)^2 + 25]$. This is similar to the case with real poles, as shown in Example 3-36.

EXAMPLE 3-39

Find the inverse Laplace transform of

$$\frac{50[(s + 8)^2 + 1]}{(s + 10)[s^2 + 25]}$$

Solution: We find the inverse due to the pole $s = -10$ first.

$$R_{-a} = F(s)(s + a)\Big|_{s=-a}$$

$$R_{-10} = \frac{50[(s + 8)^2 + 1]}{(s + 10)[s^2 + 25]}(s + 10)\Big|_{s=-10} = \frac{50[(s + 8)^2 + 1]}{[s^2 + 25]}\Big|_{s=-10}$$

$$= \frac{50[((-10) + 8)^2 + 1]}{[(-10)^2 + 25]}$$

$$R_{-10} = 2$$

$$f(t) = R_{-a}e^{-at}u(t)$$

$$f_{-10}(t) = 2e^{-10t}u(t)$$

Next we find the inverse due to the complex conjugate poles using Eqs. (3-22) and (3-23).

$$R_{-\alpha + j\omega} = F(s)[(s + \alpha)^2 + \omega^2]\Big|_{s=-\alpha+j\omega}$$

$$R_{j5} = \frac{50[(s + 8)^2 + 1]}{(s + 10)[s^2 + 25]}[s^2 + 25]\Big|_{s=j5} = \frac{50[(s + 8)^2 + 1]}{(s + 10)}\Big|_{s=j5}$$

$$= \frac{50[(j5 + 8)^2 + 1]}{(j5 + 10)}$$

$$R_{j5} = 400 \angle 36.870°$$

$$f(t) = \frac{|R_{-\alpha+j\omega}|}{\omega} e^{-\alpha t} \sin(\omega t + \theta)\, u(t)$$

$$f_{j5}(t) = \frac{400}{5} e^{0t} \sin(5t + 36.870°)\, u(t)$$

$$= 80 \sin(5t + 36.870°)\, u(t)$$

Combining the inverse due to each pole, we have

$$f(t) = \mathcal{L}^{-1}\left[\frac{50[(s+8)^2 + 1]}{(s+10)[s^2+25]} \right]$$

$$= 2e^{-10t}\, u(t) + 80 \sin(5t + 36.870°)\, u(t)$$

The next example shows the inverse of a multiple-order complex pole. This is a simple example so that the order of what to do can be easily seen.

EXAMPLE 3-40

Find

$$f(t) = \mathcal{L}^{-1}\left[\frac{(s+2)(s+3)}{[(s+1)^2 + 25]^2} \right]$$

Solution: Because r is raised to the second power, we will need to find $R_{-1+j5,1}$ and $R_{-1+j5,2}$ using Eq. (3-19).

$$R_{-\alpha+j\omega,n} = \frac{d^{n-1}}{ds^{n-1}} F(s)\, [(s+\alpha)^2 + \omega^2]^r \,\bigg|_{s=-\alpha+j\omega}$$

$$R_{-1+j5,1} = \frac{d^0}{ds^0} \frac{(s+2)(s+3)}{[(s+1)^2 + 25]^2} [(s+1)^2 + 25]^2 \,\bigg|_{s=-1+j5}$$

$$= (s+2)(s+3)\,\bigg|_{s=-1+j5}$$

$$= [(-1+j5) + 2][(-1+j5) + 3]$$

$$R_{-1+j5,1} = 27.459 \angle 146.89°$$

$$R_{-1+j5,2} = \frac{d}{ds} \frac{(s+2)(s+3)}{[(s+1)^2+25]^2} [(s+1)^2+25]^2 \Big|_{s=-1+j5}$$

$$= 2s + 5 \Big|_{s=-1+j5}$$

$$= 2(-1+j5) + 5$$

$$R_{-1+j5,2} = 10.440 \angle 73.300°$$

Next we calculate the first two $M_n \angle \theta_n$ values using Eq. (3-20).

$$M_1 \angle \theta_1 = R_{-\alpha+j\omega,1}$$

$$= 27.459 \angle 146.89°$$

$$M_2 \angle \theta_2 = R_{-\alpha+j\omega,2} - \frac{r}{2j\omega} R_{-\alpha+j\omega,1}$$

$$= 10.440 \angle 73.300° - \frac{2}{2j(5)} 27.459 \angle 146.89°$$

$$= 5.4 \angle 90°$$

Finally, we substitute into Eq. (3-21).

$$f(t) = \left[\sum_{n=1}^{r} \frac{2(-1)^{int(r/2)} M_n}{(2\omega)^r (n-1)!\, (r-n)!} t^{r-n} \sin\left(\omega t + \theta_n + \frac{1+(-1)^r}{2}(90°)\right) \right] e^{-\alpha t} u(t)$$

$$f_{-1+j5} = \left[\frac{2(-1)^{int(2/2)} 27.459}{[2(5)]^2 (1-1)!\, (2-1)!} t^{2-1} \sin\left(5t + 146.89° + \frac{1+(-1)^2}{2}(90°)\right) \right.$$

$$\left. + \frac{2(-1)^{int(2/2)} 5.4}{[2(5)]^2 (2-1)!\, (2-2)!} t^{2-2} \sin\left(5t + 90° + \frac{1+(-1)^2}{2}(90°)\right) \right] e^{-t} u(t)$$

$$= [-0.54918t \sin(5t + 236.89°) - 0.108 \sin(5t + 180°)] e^{-t} u(t)$$

The sine functions are then put in a standard form for the final result.

$$f(t) = [0.54918t \sin(5t + 56.889°) + 0.108 \sin(5t)] e^{-t} u(t)$$

3-6 USING LAPLACE TRANSFORMS TO SOLVE DIFFERENTIAL EQUATIONS

As mentioned previously, the point of using Laplace transforms is to convert the process of solving an integral–differential equation from a calculus procedure to an algebraic procedure. There are so many fields of study that rely heavily on solving integral–differential and differential equations that a short discussion on solving these equations is in order.

In control systems and in physics, a linear integral–differential equation form that often occurs is

$$C_1 \frac{d\,y(t)}{dt} + C_2\, y(t) + C_3 \int_0^t y(t)\, dt = f(t) \tag{3-24}$$

where C_1, C_2, and C_3 are constants.

To solve this equation or differential equations discussed later, initial conditions must be known. Because we are interested in the solution of these equations for $t > 0$, the initial conditions are given at $t = 0^+$. However, if the initial conditions are not given, then the system that the equation applies to must be available so that the initial conditions can be determined. In this section, the initial conditions will always be given.

To solve Eq. (3-24), we start by taking the Laplace transform of both sides.

$$\mathcal{L}\left[C_1 \frac{d\,y(t)}{dt} + C_2\, y(t) + C_3 \int_0^t y(t)\, dt \right] = \mathcal{L}[f(t)]$$

$$C_1\, \mathcal{L}\left[\frac{d\,y(t)}{dt}\right] + C_2\, \mathcal{L}[y(t)] + C_3\, \mathcal{L}\left[\int_0^t y(t)\, dt\right] = \mathcal{L}[f(t)]$$

The integral–differential equation may be for either a continuous system or a system that is switched on at $t = 0$. This difference is determined by the nature of $f(t)$. If the function $f(t)$ contains a $u(t)$, then this indicates that the system is switched on at $t = 0$ and is not a continuous system. If the $f(t)$ function does not contain a $u(t)$ and the system is not noted as a switched system, then the system is considered to be a continuous system. This makes a difference as to whether we use the operation (O-4a) or operation (O-4b) when finding the Laplace transform of the first term in the preceding expression. Most differential equations are considered continuous, and operation (O-4a) is used. Whether the system is continuous or switched is not an issue for the integral term, and operation (O-5) is always applied.

Assuming a continuous system and applying operation (O-4a) this can then be written as

$$C_1\,[sY(s) - y(0)] + C_2\, Y(s) + C_3 \left[\frac{Y(s)}{s}\right] = F(s) \tag{3-25a}$$

Remember that $y(0)$ is the value of the time-domain function $y(t)$ at $t = 0^+$. If the integral–differential equation is for a system that is switched on at $t = 0$, then we do not subtract the value of $y(0)$ from the transform of the derivative part, and the Laplace transform of the integral–differential in this case becomes

$$C_1 [sY(s)] + C_2 Y(s) + C_3 \left[\frac{Y(s)}{s} \right] = F(s) \tag{3-25b}$$

The equation is then solved for $Y(s)$ and the inverse Laplace transform is found to determine the time-domain value of $y(t)$. This is best demonstrated in an example.

EXAMPLE 3-41

Solve the following integral–differential equation assuming $w(0) = 5$.

$$2 \frac{d\,w(t)}{dt} + 2 \int_0^t w(t)\,dt = 20$$

Solution: Notice that in this equation there is no $w(t)$ term between the derivative term and the integral term, but this does not change the procedure. Also notice that on the right side of the equal sign, the $f(t)$, as defined in Eq. (3-24), does not have a $u(t)$ appearing in it; therefore, the system is continuous.

   Our first step is to take the Laplace transform of both sides.

$$\mathcal{L} \left[\frac{2\,d\,w(t)}{dt} + 2 \int_0^t w(t)\,dt \right] = \mathcal{L}[20]$$

$$\mathcal{L} \left[2 \frac{d\,w(t)}{dt} \right] + \mathcal{L} \left[2 \int_0^t w(t)\,dt \right] = \mathcal{L}[20]$$

$$2\,[\,sW(s) - 5] + 2 \left[\frac{W(s)}{s} \right] = \frac{20}{s}$$

Solving for $W(s)$, we have

$$W(s) = \frac{5(s + 2)}{(s^2 + 1)}$$

Then using the inverse techniques shown previously in this chapter, we can solve for $w(t)$.

$$w(t) = 11.180 \sin(t + 26.566°)$$

Notice that because we were solving a continuous integral–differential equation, there is no $u(t)$ in the answer.

Often, we have an equation involving only derivatives that is referred to as a differential equation. In fact, we could take the derivative of the integral–differential equation in Eq. (3-24) to convert it into a second-order differential equation. The equation is second order because the highest derivative in the equation is second order. To solve a differential equation of multiple degree, we must develop a transform operation for second and higher derivatives. This is done by using techniques similar to the way we extended the transform pairs in Section 3-3-7.

In deriving this operation, we will find it easier to use the "prime notation." This is where a superscript of "'" is used to denote the derivative order of a function. In symbolic form this may be stated as

$$h'(t) = \frac{d\,h(t)}{dt}$$

$$h''(t) = \frac{d^2\,h(t)}{dt^2}$$

$$h'''(t) = \frac{d^3\,h(t)}{dt^3}$$

$$\cdot$$
$$\cdot$$
$$\cdot$$

$$h^n(t) = \frac{d^n\,h(t)}{dt^n} \tag{3-26}$$

The multiple-order derivative operation may be developed for a continuous or switched system. Continuous is the most often used, so we will develop this operation. We will find the Laplace operation for a first-order derivative, then a second-order derivative in terms of the first-order derivation, and so on until a pattern can be established that will lead us to a general operation.

Using operation O-4a, we may write

$$H'(s) = \mathcal{L}[h'(t)] = \mathcal{L}\left[\frac{d\,h(t)}{dt}\right] = s\,H(s) - y(0) \tag{3-27a}$$

The second derivative is, therefore,

$$H''(s) = \mathcal{L}[h''(t)] = \mathcal{L}\left[\frac{d^2\,h(t)}{dt^2}\right] = \mathcal{L}\left[\frac{d\,h'(t)}{dt}\right] = s\,H'(s) - y'(0)$$

Substituting in Eq. (2-27a) for the $H'(s)$, we find

$$H''(s) = s\,H'(s) + y'(0) = s\,(s\,H(s) - y(0)) - y'(0)$$

Simplifying this expression, we find

$$H''(s) = s^2\,H(s) - s\,y(0) - y'(0) \tag{3-27b}$$

The third derivative is found using the same technique.

$$H'''(s) = \mathcal{L}[h'''(t)] = \mathcal{L}\left[\frac{d^3 h(t)}{dt^3}\right] = \mathcal{L}\left[\frac{d\,h''(t)}{dt}\right] = s\,H''(s) - y''(0)$$

Substituting in Eq. (2-27b) for $H''(s)$, we find

$$H'''(s) = s\,H''(s) - y''(0) = s\,(s^2\,H(s) - s\,y(0) - y'(0)) - y''(0)$$

Simplifying this we have

$$H'''(s) = s^3\,H(s) - s^2\,y(0) - s\,y'(0) - y''(0) \qquad \textbf{(3-27b)}$$

We see a pattern emerging and may write the general form as

$$H^n(s) = s^n\,H(s) - s^{(n-1)}\,y(0) - s^{(n-2)}\,y'(0) - s^{(n-3)}\,y''(0) - \cdots - y^n(0) \quad \textbf{(3-28)}$$

We may now apply this general form to solving a higher-order differential equation.

EXAMPLE 3-42

Solve the following differential equation. Assume $q(0) = 3$ and $q'(0) = 0$.

$$2\frac{d^2\,q(t)}{dt^2} + 10\frac{d\,q(t)}{dt} + 12\,q(t) = 36t$$

Solution: Taking the Laplace transform using the transform pair (P-3) and Eq. (3-28),

$$2\,[s^2\,Q(s) - s\,q'(0) - q(0)] + 10\,[s\,Q(s) - q(0)] + 12\,Q(s) = \frac{36}{s^2}$$

Substituting in the values of $q(0)$ and $q'(0)$, we have

$$2\,[s^2\,Q(s) - s\,(0) - 3] + 10\,[s\,Q(s) - 3] + 12\,Q(s) = \frac{36}{s^2}$$

Next we solve for $Q(s)$, which results in

$$Q(s) = \frac{18(s^2 + 1)}{s^2(s + 2)(s + 3)}$$

Finally, the inverse Laplace transform is found.

$$q(t) = -2.5 + 3t + 22.5e^{-2t} - 20e^{-3t}$$

3-7 USING MATLAB

MATLAB is capable of finding the Laplace transform and inverse Laplace transform of a function in a single command line, but this will not give us any insight into how the parameters affect the transform or inverse transform. On the other hand, when we already have an insight, we just want to be able to find the answer quickly. The next two examples demonstrate the range of detail that MATLAB is capable of. The first example will show how to find the Laplace transform with a single command line, and the second example will show how MATLAB can be used to show a great level of detail in finding the inverse Laplace transform.

EXAMPLE 3-43

Find the Laplace transform of the following function using the Laplace integral shown in Eq. (3-1):

$$f(t) = 3t \sin(7t) \, u(t)$$

Solution: This example requires that the Symbolic toolbox be installed so that we can use symbolic integration in MATLAB. First the variables "t" and "s" are defined as symbolic variables by entering the following command line in the Command Window:

```
t = sum('t');s = sym('s');
```

Next the function is integrated. Because $u(t)$ is 1 for values of t greater than 0 and because the integral we are using is only for values of t greater than 0, we will replace the $u(t)$ with the value of 1. Using this substitution for the expression, the command for integrating this function is then entered.

```
ft = int(3*t*sin(7*t)*exp(-s*t),t);
```

This expression has two variables, so the variable to integrate needs to be explicitly stated. This is done by the ",t" at the end of the expression. Also note that this expression uses the "*" operator instead of the ".*" because this is a symbolic expression. The result of the integration is stored in the variable "ft".

Next we find the expression with the upper limit substituted in for t, minus the expression with the lower limit substituted in for t. The upper limit is infinity, and MATLAB, as with any other calculating program, has trouble evaluating a finite value for an expression with an infinite value substituted in for a variable. Fortunately, the Laplace integral always results in an expression that has a multiplying factor of e^{-st}. When the upper limit of infinity is substituted in for the value of t, this factor goes to zero, causing the lower limit value to always be equal to zero. Therefore, the final value of the integral will be equal to the negative of the expression with t set equal to zero. Understanding this, the Laplace transform is found by entering the following command line:

```
-1*subs(ft,t,0)
```

```
>> t = sym('t');s = sym('s');
>> pretty(-1*subs(int(3*t*sin(7*t)*exp(-s*t),t), t, 0))
```

$$42 \, \frac{s}{(s^2 + 49)^2}$$

Figure 3-2 *MATLAB Command Window for Example 3-43*

To have the output print in a nicer format, we can use the follow command line instead of the previous command:

```
Pretty(-1*subs(ft,t,0))
```

All of this can be done on a single command line as follows:

```
pretty(-1*subs(int(3*t*sin(7*t)*exp(-s*t),t),t,0))
```

The Command Window printout is shown in Fig. 3-2. From this printout we can conclude that

$$\mathcal{L}[3t \sin(7t)] = \frac{42s}{(s^2 + 49)^2}$$

EXAMPLE 3-44

Find the residue values required to find the inverse Laplace transform of the following function:

$$F(s) = \frac{2(s + 1)}{(s + 3)[(s + 2)^2 + 1]}$$

Solution: Although the Symbolic toolbox has an inverse Laplace command, we are going to use MATLAB to help us go through the inverse of this function step-by-step. First the variables *t* and *s* are defined as symbolic variables, and then the expression for *F(s)* is entered as a symbolic expression. Remember that in symbolic expressions the multiplication and division operations are written without a period in front of them. We then enter the following command lines:

```
t = sym('t');s = sym('s');
Fs = sym('2*(s+1)/((s+3)*((s+2)^2+1))');
```

We will first find the residue for the real pole $s = -3$. To do this we use the symbolic multiplication operation.

```
temp = symmul (Fs, '(s+3)');
```

This will store in the variable "temp" the $F(s)$ function with the $(s + 3)$ denominator factor removed. The residue is then found by setting $s = -3$ by using the substitution operation as follows:

```
R1 = subs(temp,-3,'s')
```

In this operation, "temp" is the expression to be evaluated, "-3" is the value to substitute in for the specified variable, and the "'s'" is the specified variable.

As we have seen before, commands can be combined. The last two commands for finding the residue for the pole at $s = -3$ can be written as

```
R1 = subs (symmul (Fs, '(s+3)'),-3,'s')
```

The next value we need to calculate is the residue for the complex poles. This can be found by using individual commands, or we can combine the commands into a single command as

```
R2 = subs (symmul (Fs, '((s+2)^2+1)'),-2+1i,'s')
```

This will return the value in rectangular form. Therefore, we need to find the value of the magnitude and the phase angle in degrees. This is accomplished by the following two commands:

```
M = abs (R2)

Theta = angle (R2) .* (180/pi)
```

Unless you have assigned the variable "pi" to some other value, MATLAB has the value for π built into the "pi" variable.

Figure 3-3 shows the Command Window printout for finding the residues. Therefore, we may state that $R_{-3} = -2$ and $R_{-2+j} = 2 \angle 90°$.

Since MATLAB can take the derivative of a function with the Symbolic toolbox, it is possible to find the inverse Laplace transform for functions that have multiple-order poles. This could then be programmed into an M-file. However, MATLAB already has the Laplace and inverse Laplace operations ready to use in the Symbolic toolbox if you just want the answer.

```
>> t = sym('t');s = sym('s');
>> Fs = sym('2*(s+1)/((s+3)*((s+2)^2+1))');
>> R1 = subs(symmul(Fs, '(s+3)'),-3,'s')

R1 =

     -2

>>R2 = subs(symmul(Fs, '((s+2)^2+1)'),-2+1i,'s');
>>M = abs(R2)
M =
     2

>> Theta = angle (R2) .* (180/pi)

Theta =

     90
```

Figure 3-3   *MATLAB Command Window for Example 3-44*

3-8 USING THE TI-89

The TI-89 is capable of finding the Laplace transform in one step, and, with some programming, it is capable of finding the inverse Laplace transform in one step. When the detailed steps involved in finding the Laplace transform and the inverse Laplace transform are not shown, little insight into the solution is gained. However, if we already possess the insight and just require the answer, the one-step approach is often the best. The next two examples show the range of detail that can be used in solving Laplace problems with this calculator.

EXAMPLE 3-45

Find the Laplace transform of the following function using the Laplace integral shown in Eq. (3-1):

$$f(t) = 3t \sin(7t) \, u(t)$$

Solution: To find the Laplace transform, the integral is found first. The upper limit is then substituted into the integral for t, and then the lower limit is substituted into the integral for t. Finally, the quantity for the lower limit is subtracted from the quantity for the upper limit. When one of these limits of integration is infinity, however, the calculator is unable to simplify the expression, but there is a way around this problem.

The integral for which the limits are substituted will always have an e^{-st} multiplier. When we substitute $t = \infty$ into the expression for the upper limit, the result will always be zero because $e^{-s\infty} = 0$. This means that all we need to do is substitute the lower limit of $t = 0$ into the integral result and multiply the expression by -1.

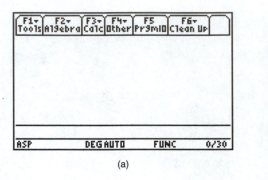

(a)

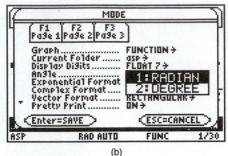

(b)

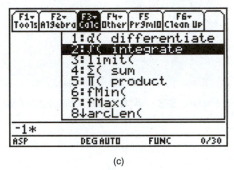

(c)

Figure 3-4

First set the calculator to the normal toolbar, as shown in Fig. 3-4a. If the custom toolbar is shown at the top instead of the normal toolbar, press the following keys:

2nd [CUSTOM]

Because we will be finding the integral of a sinusoidal function, we must be in the correct mode for the argument of the sine function. In this case, the sine function has a radian argument. We must therefore enter the following keystrokes to put the calculator in radian mode:

MODE (scroll to Angle) ▶ 1:RADIAN ENTER ENTER

Figure 3-4b shows how the screen should appear just before the Enter key is pressed twice.

Next enter "−1 *" on the Entry line, and then press the following keystrokes to access the integral operation:

F3 (scroll to 2: ∫ (integrate) ENTER

Figure 3-4c shows how the screen should appear just before pressing the Enter key.

Figure 3-5

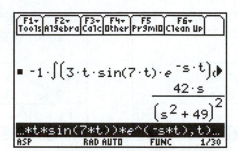

The time-domain function multiplied by e^{-st} is then entered as the function to integrate, but the $u(t)$ is dropped from the expression because it is always 1 for $t > 0$. The rest of the function is then entered onto the Entry line. The completed line should be

$$-1*\int ((3*t*\sin(7*t))*e\wedge(-s*t),t)|t = 0$$

The ",t" at the end of the integral process tells the calculator to integrate using t as the variable, and at the end of this entry line the variable t is set to equal 0.

Figure 3-5 shows that when the Enter key is pressed the Laplace transform is

$$\frac{42s}{(s^2 + 49)^2}$$

Note that since the TI-89 is capable of symbolic integration, the Laplace transform has been found without showing any intermediate steps.

EXAMPLE 3-46

Find the residue values required to find the inverse Laplace transform of the following function:

$$F(s) = \frac{2(s + 1)}{(s + 3)[(s + 2)^2 + 1]}$$

Solution: We will start by finding R_{-3}. There are many ways to approach this. One way is to enter the function without the $(s + 3)$ in the denominator and a -3 value for all the occurrences of s. A better way is to take advantage of the symbolic nature of this calculator. We start by entering the entire function on the Entry line as

$$2*(s + 1)/((s + 3)((s + 2)\wedge 2 + 1))$$

When the Enter key is pressed, the display will look like Fig 3-6a. With this in the History Area, we can recall it later to use for other calculations without reentering the function.

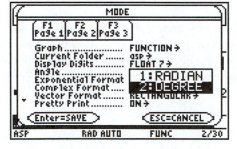

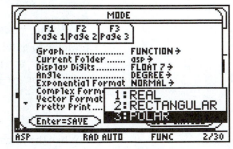

Figure 3-6

After having pressed the Entry key, press the ⊙ key once. Then delete the $(s + 3)$ term from the Entry line. At the end of the entry line add

$$|s = -3$$

and then press the Enter key. Figure 3-6b shows the display after pressing the Enter key, and, as shown in the figure, we have calculated the value of R_{-3} to be -2.

Next we find the residue for the complex–conjugate poles. Before we calculate this, the calculator should be in a mode that displays the magnitude with the phase angle in degrees for complex number results. This is accomplished by pressing the following keystrokes:

MODE (scroll to Angle) ⊙ 2:DEGREE ENTER

Figure 3-7a shows the display just before the Enter key is pressed once. Next the following keystrokes are entered:

(scroll to Complex Format) ⊙ 3:POLAR ENTER ENTER

Figure 3-7b shows the display just before the Enter key is pressed twice.

Figure 3-7

Figure 3-8

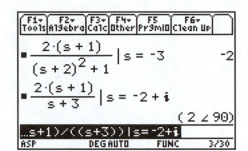

Press the Clear key once to remove the last entry from the Entry line. Next press 2nd [ENTRY] enough times in a row until the full function that was entered at the beginning of the examples is displayed. Remove the $*((s + 2)^2 + 1)$ expression from the Entry line. Then go to the end of the Entry line and add

$$|s = -2 + i$$

Press the Enter key once. Figure 3-8 shows the result of this calculation, which is $R_{-2+j} = 2 \angle 90°$.

Because the TI-89 is capable of taking the derivative of a function, the technique shown in the last example could be extended to multiple-order poles. The whole process could be made into a program for finding the inverse Laplace transform. It is also not unreasonable to construct a program for finding the Laplace transform of a time-domain function.

PROBLEMS

Section 3-1

1. Find the Laplace transform of $t\, u(t)$ using the Laplace integral Eq. (3-1).

2. Find $V(s)$.

    (a) $v(t) = 10\, u(t)$
    (b) $v(t) = 4e^{-7t}\, u(t)$
    (c) $v(t) = \delta(t) + 5t\, u(t)$

3. Find $I(s)$.

    (a) $i(t) = 3\cos(12t)\, u(t)$
    (b) $i(t) = 5\sin(2\pi t)\, u(t)$
    (c) $i(t) = 4\cos(3t + 30°)\, u(t)$

Section 3-2

4. Find $i(t)$.

    (a) $I(s) = \dfrac{4}{s^2}$

    (b) $I(s) = 20$

5. Find $v(t)$.

$$V(s) = \frac{14\,s}{(s^2 + 16)}$$

6. Find $f(t)$.

$$F(s) = \frac{s^2 + 2s}{s^3}$$

7. Find

$$\mathcal{L}^{-1}\left[\frac{4(s + 2)}{s^2 + 4}\right]$$

Section 3-3

8. Find the Laplace transform of the following functions using the shifted function operation presented in Section 3-3-1.

    (a) $3u(t - 2) - 2\delta(t - 4)$

    (b) $e^{-3(t-1)}u(t - 1) + \sin[3(t - 2)]\,u(t - 2)$

9. Find the Laplace transform of the following functions using the methods presented in Section 3-3-1.

    (a) $4(t - 3)\,u(t - 2)$

    (b) $\cos(2t - 60°)\,u(t - 1)$

10. Find the Laplace transform of the function using the exponential multiplier operation (O-2) presented in Section 3-3-2.

$$v(t) = 2te^{-7t}\,u(t)$$

11. Find the Laplace transform of the function using the exponential multiplier operation (O-2) presented in Section 3-3-2.

$$i(t) = 12e^{-2t}\cos(3t - 30°)\,u(t)$$

12. Find the Laplace transform of the following function using the t multiplier operation (O-3) presented in Section 3-3-3.

$$e^{-4t}t\,u(t)$$

13. Find the Laplace transform of the following function using the t multiplier operation (O-3) presented in Section 3-3-3.

$$t \cos(3t) \, u(t)$$

14. Find the Laplace transform of the following functions using one of the derivative operations (O-4a) or (O-4b).

    (a) $\left\{ \dfrac{d}{dt} [5] \right\} u(t)$

    (b) $\dfrac{d}{dt} [5 \, u(t)]$

15. Find the Laplace transform of the following functions using one of the derivative operations (O-4a) or (O-4b).

    (a) $\left\{ \dfrac{d}{dt} [4e^{-6t}] \right\} u(t)$

    (b) $\dfrac{d}{dt} [4e^{-6t} \, u(t)]$

16. Find the Laplace transform of the following functions using one of the derivative operations (O-4a) or (O-4b).

    (a) $\left\{ \dfrac{d}{dt} [3t] \right\} u(t)$

    (b) $\dfrac{d}{dt} [3t \, u(t)]$

17. Using the integral operation (O-5), find the Laplace transform of

    (a) $\displaystyle\int_0^t 12e^{-4t} \, u(t) \, dt$

    (b) $\displaystyle\int_0^t 10\delta(t) \, dt$

18. Using the integral operation (O-5), find the Laplace transform of

    (a) $\displaystyle\int_0^t 5 \cos(2t) \, u(t) \, dt$

    (b) $\displaystyle\int_0^t [2 \, u(t) + 3t \, u(t)] \, dt$

19. Using the techniques presented in Section 3-3-6 on multiple operations, find the Laplace transform of

$$v(t) = 4te^{-10t} \sin(30t) \, u(t)$$

20. Using the techniques presented in Section 3-3-6 on multiple operations, find the Laplace transform of

$$i(t) = \int_0^t 7te^{-4t} u(t) \, dt$$

21. Using the techniques presented in Section 3-3-6 on multiple operations, find the Laplace transform of

$$v(t) = \frac{d}{dt} [e^{-4(t-9)}\cos(3\pi(t-9)) \, u(t-9)]$$

22. Using the techniques presented in Section 3-3-6 on multiple operations, find the Laplace transform of

$$f(t) = 6(t-8)e^{-7(t-8)}\sin(4(t-8)) \, u(t-8)$$

23. Derive the Laplace transform of the function shown using the techniques shown in Section 3-3-7.

  (a) $f(t) = e^{-\alpha t} \sin(\omega t) \, u(t)$
  (b) $f(t) = e^{-\alpha t} \cos(\omega t) \, u(t)$

24. Derive the Laplace transform of the function below using the techniques shown in Section 3-3-7.

  (a) $f(t) = \sin(\omega t + \theta) \, u(t)$
  (b) $f(t) = e^{-\alpha t} \sin(\omega t + \theta) \, u(t)$

Section 3-4

25. Find the inverse Laplace transform using the traditional techniques shown in Section 3-4-1.

$$F(s) = \frac{10}{(s+8)(s+10)}$$

26. Find the inverse Laplace transform using the traditional techniques shown in Section 3-4-1.

$$F(s) = \frac{(s+4.5)}{(s+9)(s+12)}$$

27. Find the inverse Laplace transform using the traditional techniques shown in Section 3-4-1.

$$I(s) = \frac{5(s+4)}{s(s+10)}$$

28. Find the inverse Laplace transform using the traditional techniques shown in Section 3-4-1.

$$I(s) = \frac{s(s+13)}{(s+5)(s+9)}$$

29. Find the inverse Laplace transform using the traditional techniques shown in Section 3-4-1.

$$F(s) = \frac{3[(s + 8)^2 + 4]}{(s + 4)(s + 6)}$$

30. Find the inverse Laplace transform using the traditional techniques shown in Section 3-4-1.

$$V(s) = \frac{(s + 8)(s + 10)}{(s + 2.5)(s + 1)(s + 15)}$$

31. Find the inverse Laplace transform using the traditional techniques shown in Section 3-4-2.

$$F(s) = \frac{7[(s + 10)^2 + 4]}{(s + 6)^2(s + 8)}$$

32. Find the inverse Laplace transform using the traditional techniques shown in Section 3-4-2.

$$F(s) = \frac{4(s + 12)}{s(s + 2)^3}$$

33. Find the inverse Laplace transform using the traditional techniques shown in Section 3-4-2.

$$I(s) = \frac{(s + 2)^2(s + 3)}{s^4}$$

34. Find the inverse Laplace transform using the traditional techniques shown in Section 3-4-2.

$$V(s) = \frac{[s^2 + 64]}{(s + 4)^2(s + 8)^2}$$

35. Find the inverse Laplace transform using the complex-number method shown in Section 3-4-3.

$$F(s) = \frac{10s(s + 8)}{(s + 6.5)[(s + 6)^2 + (4)^2]}$$

36. Find the inverse Laplace transform using the complex-number method shown in Section 3-4-3.

$$V(s) = \frac{5(s + 4)}{(s + 1)[s^2 + 4]}$$

37. Find the inverse Laplace transform using the real-number method shown in Section 3-4-4.

$$F(s) = \frac{20[(s + 4)^2 + 25]}{s[(s + 2.5)^2 + 4]}$$

38. Find the inverse Laplace transform using the real-number method shown in Section 3-4-4.

$$I(s) = \frac{6s(s + 10)}{(s + 2.5)[(s + 4)^2 + 9]}$$

Section 3-5

39. Put the function in the proper form of $N(s)/D(s)$.

$$\frac{(s^2 + 8s)(4s + 8)}{(2s^2 + 17s + 26)(2s^2 + 24s + 104)}$$

40. Put the function in the proper form of $N(s)/D(s)$.

$$\frac{(2s^2 + 16s + 30)}{(s^2 + 14s + 48)}$$

41. For the function, find R_{-5}, R_{-6}, and R_{-4+j2}.

$$\frac{4(s + 6)(s + 10)}{(s + 5)[(s + 4)^2 + 2^2]}$$

42. For the function, find R_{j2}, R_{-4}, and R_{-3+j1}.

$$\frac{[(s + 3)^2 + 1]}{[s^2 + 4](s + 4)}$$

43. Find the inverse Laplace transform of the function using the techniques shown in Section 3-5-2.

$$\frac{3}{(s + 7)(s + 10)}$$

44. Find the inverse Laplace transform of the function using the techniques shown in Section 3-5-2.

$$\frac{(s + 4.5)}{(s + 9)(s + 12)}$$

45. Find the inverse Laplace transform of the function using the techniques shown in Section 3-5-2.

$$\frac{(s + 5)}{s(s + 1)}$$

46. Find the inverse Laplace transform of the function using the techniques shown in Section 3-5-2.

$$\frac{2s(s + 11)}{(s + 6)(s + 8)}$$

47. Find the inverse Laplace transform of the function using the techniques shown in Section 3-5-2.

$$\frac{[(s + 18)^2 + 9]}{(s + 9)(s + 6)}$$

48. Find the inverse Laplace transform of the function using the techniques shown in Section 3-5-2.

$$\frac{5(s + 7)(s + 8)}{(s + 3.5)(s + 6)(s + 14)}$$

49. Find the inverse Laplace transform of the function using the techniques shown in Section 3-5-2.

$$\frac{[(s + 9)^2 + 1]}{(s + 6)(s + 4)^2}$$

50. Find the inverse Laplace transform of the function using the techniques shown in Section 3-5-2.

$$\frac{(s + 15)}{s(s + 1)^3}$$

51. Find the inverse Laplace transform of the function using the techniques shown in Section 3-5-2.

$$\frac{2(s + 14)(s + 9)}{s^4}$$

52. Find the inverse Laplace transform of the function using the techniques shown in Section 3-5-2.

$$\frac{[s^2 + 4]}{(s + 5)^2(s + 6)^2}$$

53. Find the inverse Laplace transform of the function using the techniques shown in Section 3-5-3.

$$\frac{s(s + 9)}{(s + 5)[(s + 6)^2 + (3)^2]}$$

54. Find the inverse Laplace transform of the function using the techniques shown in Section 3-5-3.

$$\frac{4[(s + 2)^2 + 4]}{[(s + 2)^2 + 16]}$$

55. Find the inverse Laplace transform of the function using the techniques shown in Section 3-5-3.

$$\frac{7(s + 8)}{(s + 2)[s^2 + 16]}$$

56. Find the inverse Laplace transform of the function using the techniques shown in Section 3-5-3.

$$\frac{s}{[(s + 7.2)^2 + 9]}$$

57. Find the inverse Laplace transform of the function using the techniques shown in Section 3-5-3.

$$\frac{40(s + 1)[(s + 7)^2 + 1]}{(s + 4)^2[(s + 2)^2 + (4)^2]}$$

58. Find the inverse Laplace transform of the function

$$i(t) = \frac{s^2(s + 4)}{[(s + 1)^2 + 4]^2}$$

59. Find the inverse Laplace transform of the function

$$v(t) = \frac{s^2}{[(s + 2)^2 + 1]^3}$$

60. Find the inverse Laplace transform of the function

$$F(s) = \frac{s^2 + 9s + 25}{(s + 1)[(s + 1)^2 + 1]^2}$$

Section 3-6

61. Solve the following integral–differential equation assuming $i(0) = 0$.

$$\frac{d\,i(t)}{dt} + 6\,i(t) + 5\int_0^t i(t)\,dt = 10$$

62. Solve the following integral–differential equation assuming $v(0) = 1$.

$$2\frac{d\,v(t)}{dt} + 24\,v(t) + 90\int_0^t v(t)\,dt = 8e^{-5t}$$

63. Solve the following integral–differential equation. Assume $p(0) = 4$ and $p'(0) = 24$.

$$\frac{d^2\,p(t)}{dt^2} + 10\frac{d\,p(t)}{dt} + 41\,p(t) = 656$$

64. Solve the following integral–differential equation. Assume $w(0) = 0$ and $w'(0) = 2$.

$$3\frac{d^2\,w(t)}{dt^2} + 9\frac{d\,w(t)}{dt} + 6\,w(t) = 96t$$

Circuit Analysis Using Laplace Transforms

OBJECTIVES

Upon successful completion of this chapter, you should be able to:

- Calculate the initial voltages and currents for capacitors and inductors, respectively.
- Convert the basic R, L, C components having or not having an initial condition into their equivalent Laplace schematic representation.
- Discuss and identify steady-state terms and transient terms, and identify which terms make up the steady-state and transient responses.
- Calculate the Laplace voltage or Laplace current in single-source circuits using the basic circuit analysis techniques of Ohm's law, the voltage-divider rule, the current-divider rule, Kirchhoff's voltage law, and Kirchhoff's current law.
- Calculate the Laplace voltage or Laplace current in a multiple-source circuit using the analysis methods of superposition, mesh current, and node voltage.
- Determine and draw Thévenin and Norton equivalent circuits.
- Identify and define the four types of dependent sources—VCVS, CCVS, VCCS, and CCCS—and be able to convert from one dependent source to any one of the other three types.
- Define and identify circuit order from either a schematic diagram or a Laplace transform function.
- Identify when a second-order circuit's impulse or step response is underdamped, critically damped, or overdamped.
- Identify the natural response, ω_n, and the damping ratio, ζ, of a quadratic.
- Discuss how the damping ratio, ζ, can be used to determine if a circuit is underdamped, critically damped, or overdamped.
- Calculate the values of an R, L, or C element in a circuit that will force the output to be either underdamped, critically damped, or overdamped.

4-0 INTRODUCTION

Knowing Laplace transforms, we can solve circuits for the complete solution of voltages and currents. First we convert the time-domain circuit to the Laplace-domain circuit. Then the analysis proceeds the same as dc analysis would. Once the Laplace voltage or current is found, we find the inverse Laplace for the time-domain value.

Because we have not yet solved circuits in the time domain, it may seem that Laplace is the long method. This is not true. When Laplace is not used, simultaneous differential equations may have to be solved.

4-1 INITIAL VOLTAGES AND CURRENTS

A circuit may contain switches that may be mechanical switches or more typically electronic switches. These will cause reactive components to have initial conditions starting at the time we want to analyze the circuit.

We can analyze a circuit starting at any arbitrary time, which we will consider to be $t = 0$. This is done by drawing an equivalent circuit that is mathematically identical to the physical circuit. We must know the initial conditions of the circuit elements at our selected $t = 0$ time in order to draw the equivalent circuit. Laplace transforms could be used to find the initial conditions, but it is more practical to find the initial conditions in the time domain because frequently the circuit is in a dc steady-state condition.

Reactive components store energy and later release the energy into the circuit, so they may have an initial condition associated with them. Resistors do not store energy, and therefore will never cause an initial condition.

To determine the initial condition of a reactive component, we consider its stored energy at our selected time $t = 0$. The condition of the energy contained in the component will be the same at $t = 0^-$ and 0^+ because this is the same instant in time. In this time period, the current through the inductor stays the same, so we will use the current to describe its initial condition. Because the voltage across a capacitor also stays the same in this time period, we will use this voltage to describe the capacitor's initial condition.

In addition to determining the initial conditions of the reactive components, we will also replace switched power supplies with nonswitched power supplies that are multiplied by the $u(t)$. Figure 4-1 shows how a dc supply of 6 V is replaced by a supply of $6\,u(t)$ without a switch. This converts the physical circuit into a more mathematical representation.

In this text, unless otherwise noted, we will assume that all switches are drawn in their $t = 0^-$ position and that they have been in this position since $t = -\infty$. There will be an arrow drawn across the switch connector indicating the position that the switch will be in for the $t = 0^+$ time.

Figure 4-2 shows a two contact switch that we will also consider to be electronically switched. This drawing shows that the switch will be connected to position 1 from $t = -\infty$ to 0^- and connected to position 2 from $t = 0^+$ on. It will always be connected to one position or the other and will never be disconnected from both at the same time. If we need to simulate a mechanical switch, we would put a switch position between position 1 and position 2 and analyze the circuit in each of the three positions.

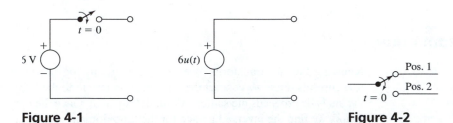

Figure 4-1

Figure 4-2

Removing switches from a circuit to make an equivalent circuit takes several steps. First, we determine the initial conditions of the reactive components by finding the current through the inductors and the voltage across the capacitors at $t = 0^-$. These values will be the initial conditions at $t = 0^+$. Next, we replace the switches with connections that connect the circuit to where the switches will be at their $t = 0^+$ position. Finally, we multiply all the sources by $u(t)$.

EXAMPLE **4-1**

At $t = 0$ the switch in Fig. 4-3 is thrown. Find the initial condition on the inductor. Draw the equivalent circuit.

Solution: When we look at the circuit from $t = -\infty$ to 0^-, we have the circuit shown in Fig. 4-4a. The circuit is at steady state at $t = 0^-$, and the inductor therefore looks like a short, as shown in Fig. 4-4b. The current through the inductor at $t = 0^-$ (just before the switch is thrown) will be the same at $t = 0^+$ because current will not change instantaneously through an inductor.

$$i(0^-) = i(0^+) = \frac{5}{2} = 2.5 \text{ A}$$

The circuit after $t = 0$ looks like a 5-V source switched on at $t = 0$ to a circuit containing an initially fluxed inductor and series resistor. Figure 4-5 shows this circuit.

The circuits of Figs. 4-3 and 4-5 are mathematically identical. Note that $u(t)$ is associated with the initial condition so that everything becomes active at $t = 0$.

EXAMPLE **4-2**

Draw an equivalent circuit for Fig. 4-6 without switches.

Solution: Figure 4-7a shows the circuit just before the switch is opened. Notice that the capacitor has been replaced with an open circuit because this is what the capacitor looks like in dc steady state. The circuit is a series circuit with the voltage across the R_2 being the same voltage as across the capacitor. For convenience, we measure this voltage (Fig. 4-7a), so we get a positive voltage. The voltage measured at $t = 0^-$ is

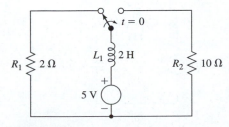

Figure 4-3

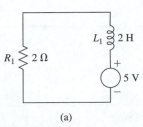

(a)

(b)

Figure 4-4

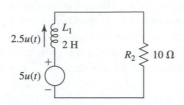

Figure 4-5

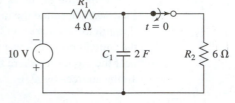

Figure 4-6

$$V_{R_2} = \frac{ER_2}{R_1 + R_2} = \frac{10(6)}{4 + 6} = 6 \text{ V}$$

Because the voltage across the capacitor cannot change instantaneously, it will have the same value at $t = 0^+$.

$$v(0^-) = v(0^+) = V_{R_2}$$

Figure 4-7b shows the circuit without switches and with everything starting at $t = 0$.

EXAMPLE 4-3

Draw an equivalent circuit for Fig. 4-8 without switches.

Solution: Figure 4-9a shows the circuit at $t = 0^-$, which is just before the switch is changed. The capacitors have been replaced with an open and the inductors have been replaced with a short because the circuit is in dc steady state. Notice that the inductor L_2 shorts any voltage across C_2. This circuit is essentially a voltage source across a 10 Ω resistor, and the current through the series is

$$I = \frac{E}{R} = \frac{5}{10} = 0.5 \text{ A}$$

This current must go through both inductors, which makes both their currents 0.5 A at $t = 0^-$. The voltage across capacitor C_1 at this time is the voltage across R_1, as seen by using Kirchhoff's voltage law.

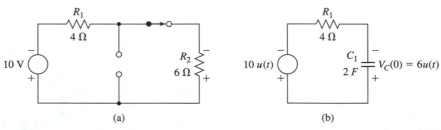

(a)   (b)

Figure 4-7

Figure 4-8

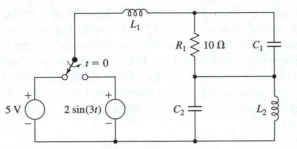

Because the current through an inductor and the voltage across a capacitor cannot change instantaneously, the values at $t = 0^-$ will be the same at $t = 0^+$. Figure 4-9b shows the equivalent circuit. Notice that because the switch has been removed, the sinusoidal source must have $u(t)$ added to it.

In this book we will limit ourselves to circuits that are in dc steady state with dc sources when initial conditions are to be found. This will not be the case in some applications. In a case where the circuit is not in steady state at your chosen $t = 0$, find the currents through the inductors and the voltages across the capacitors using Laplace transform methods instead of using dc techniques. Then put these values in the equivalent circuit for initial conditions and proceed from there as usual.

4-2 LAPLACE IMPEDANCE

Each circuit element, R, L, and C, has a value in the Laplace domain. Their value in the Laplace domain will be called the *Laplace impedance*. The Laplace impedance is the ratio of the Laplace voltage to the Laplace current.

$$Z(s) = \frac{E(s)}{I(s)} \tag{4-1}$$

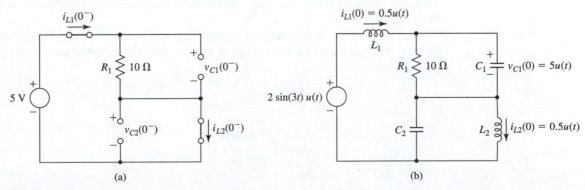

(a)   (b)

Figure 4-9

The mathematically correct units for the voltage, current, and impedance are unimportant to us, but we may call them Laplace volts, Laplace amperes, and Laplace ohms.

From the preceding section we found that inductors and capacitors may have an initial condition associated with them. When there is an initial value, it may be expressed as either a voltage or a current. We found that the natural way of expressing an initial condition for an inductor was with a current, and the natural way of expressing an initial condition for a capacitor was with a voltage. This form will be called the *natural equivalent circuit*.

4-2-1 Laplace Resistance

Figure 4-10 shows a resistor with an arbitrary voltage applied to it. By using Ohm's law, the current through it is

$$i(t) = \frac{e(t)}{R}$$

Because these are functions of time, we may take the Laplace transform of both sides of the equation. R is a constant multiplier so it will not be affected by this process. Taking the Laplace transform gives

$$I(s) = \frac{E(s)}{R}$$

or

$$R = \frac{E(s)}{I(s)}$$

Therefore, by Eq. (4-1) the Laplace impedance for resistance is

$$Z_R(s) = R \tag{4-2}$$

For convenience, the function notation may be dropped, and the notation of Z_R may be used. The subscript is not necessary, other than to indicate to which component in a circuit the impedance refers.

Resistors cannot store energy, so there will never be an initial condition caused by a resistor. A resistor may have a current at $t = 0^-$, but it does not store this energy to be released later.

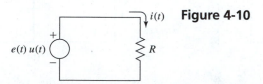

Figure 4-10

4-2-2 Laplace Inductance

Figure 4-11 shows an inductor with an arbitrary voltage applied to it. The current through the inductor is

$$i(t) = \frac{1}{L} \int_0^t e(t)\, u(t)\, dt$$

Taking the Laplace transform of both sides using (0–5) yields

$$I(s) = \frac{1}{L}\left[\frac{E(s)}{s}\right] = \frac{E(s)}{sL}$$

and solving for $E(s)/I(s)$ gives

$$sL = \frac{E(s)}{I(s)}$$

Therefore, by Eq. (4-1) the Laplace impedance for inductance is

$$Z_L(s) = sL \qquad\qquad\qquad (4\text{-}3)$$

Here again this may be expressed as Z_L.

Inductors store energy and therefore may have an initial condition. The initial condition can be represented as a current, $i(0)$, which will always be a constant. In the time domain this adds a current to the integral.

$$i(t) = \frac{1}{L} \int_0^t e(t)\, dt + i(0)\, u(t)$$

Taking the Laplace of the equation reveals the same Laplace equation for an unfluxed inductor with another current caused by the initial condition. The current due to the initial condition will be added if it has been measured in the same direction as $i(t)$ and subtracted if measured in the opposite direction.

$$I(s) = \frac{E(s)}{sL} + \frac{i(0)}{s}$$

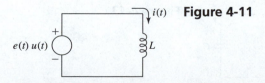

$i(t)$ **Figure 4-11**

$e(t)\, u(t)$ L

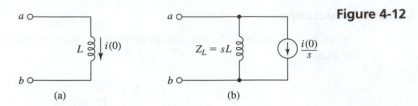

Figure 4-12

This equation is a sum of currents, and can be represented by an initially unfluxed inductor in parallel with a dc current source. Figure 4-12a shows an inductor with an initial current. Figure 4-12b shows the equivalent Laplace circuit, which is called the *natural equivalent circuit for the inductor.*

We must remember that Fig. 4-12b is an equivalent circuit. The two circuits, Figs. 4-12a and 4-12b, are equivalent only as seen from the terminals a and b. When we want to find the current through or voltage across the inductor, it must be done at the terminals, which will include the initial-condition current source. It is essential to understand this in order to use equivalent circuits.

4-2-3 Laplace Capacitance

Figure 4-13 shows a capacitor with an arbitrary voltage applied to it. The current through the capacitor is

$$i(t) = C\frac{d}{dt}\left[e(t)\, u(t)\right]$$

Taking the Laplace transform using (0-4b) gives

$$I(s) = Cs\, E(s)$$

and solving for $E(s)/I(s)$, we obtain

$$\frac{1}{sC} = \frac{E(s)}{I(s)}$$

Therefore, by Eq. (4-1) the Laplace impedance for capacitance is

$$Z_C(s) = \frac{1}{sC} \tag{4-4}$$

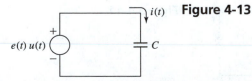

$i(t)$ **Figure 4-13**

As with the resistor and inductor notation, we may drop the reference to s and simply write Z_C.

Capacitors store energy and therefore may have an initial condition. The initial condition can be represented as a voltage, $v(0)$, which will always be a constant. In the time domain, this initial voltage affects the total amount of voltage that the capacitor reacts to. If the applied voltage and the initial-condition voltage measured in a loop are both rises or both drops, then the capacitance reacts to the sum of these two voltages, but if they are not, then the capacitance reacts to applied voltage minus the initial-condition voltage. Assuming the last case we can write,

$$i(t) = C\frac{d}{dt}\left[e(t)\,u(t) - v(0)\,u(t)\right]$$

$$i(t) = C\frac{d}{dt}\left[e(t)\,u(t)\right] - Cv(0)\,\delta(t)$$

Taking the Laplace of both sides and solving for $E(s)$ reveals the same Laplace equation for the voltage across an uncharged capacitor with another voltage caused by the initial condition.

$$I(s) = Cs\,E(s) - Cv(0)$$

$$E(s) = \frac{I(s)}{Cs} + \frac{v(0)}{s}$$

This equation is a sum of voltages and can be represented by an initially uncharged capacitor in series with a dc voltage source. Figure 4-14a shows a capacitor with an initial voltage. Figure 4-14b shows the equivalent Laplace circuit, which is called the *natural equivalent circuit for the capacitor.*

This is an equivalent circuit and is equivalent only at terminals a and b. When the voltage across or current through the capacitor is found, it must be found at the terminals, which will include the initial-condition voltage source.

4-2-4 Laplace Impedance and Source Conversion

In some cases, we need the fluxed inductor to be represented as an impedance with a series voltage source. We may also require a charged capacitor to be represented as an impedance with a parallel current source. This is accomplished by using source conversion.

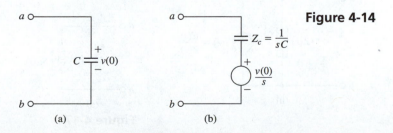

Figure 4-14

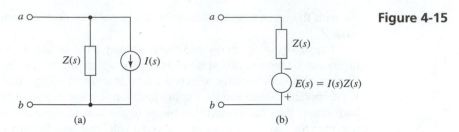

Figure 4-15

Source conversion is done exactly as it is done in dc circuit analysis. Keep in mind that the circuit, as with all equivalent circuits, is equivalent only at the terminals where the conversion was done.

Figure 4-15 shows conversion of a parallel current source and impedance (Norton equivalent) to a series voltage source and impedance (Thévenin equivalent). Figure 4-16 shows a series voltage source and impedance conversion to a parallel current source and impedance. The value of the impedance is the same in both equivalent circuits; however, the voltage and current for this impedance may be different. Only the voltage and current at terminals a and b are the same.

Notice the similarity between source conversion and Ohm's law. To get a voltage, you multiply a current times an impedance, and to get a current, you divide a voltage by an impedance.

To determine the plus/minus signs or current direction for a source in the conversion, begin by placing a pure resistance across the terminals a and b of the circuit to be converted. Note the direction that the current flows through this resistor. Then attach a resistor across the terminals of the converted circuit, and assign the plus/minus or current direction of the source so that current flows through this resistor in the same directions as it does through the resistor in the original circuit.

EXAMPLE 4-4

Draw the Laplace natural equivalent circuit for the circuits in Fig. 4-17. Use source conversion, and draw the resulting Laplace equivalent circuit.

Solution: Figure 4-18 shows the Laplace natural equivalent circuits.

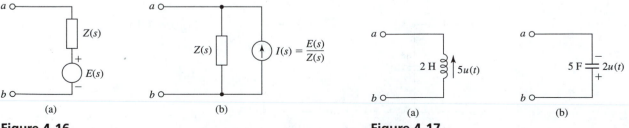

Figure 4-16

Figure 4-17

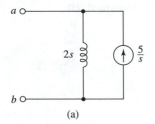

(a)

Figure 4-18

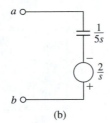

(b)

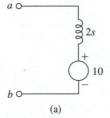

(a)

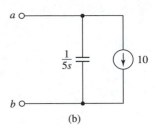

(b)

Figure 4-19

Converting the parallel current source to a series voltage source, we have

$$E(s) = Z(s)\,I(s) = (2s)\left(\frac{5}{s}\right) = 10$$

The equivalent circuit is shown in Fig. 4-19a.

Converting the series voltage source in Fig. 4-18b to a parallel current source, we have

$$I(s) = \frac{E(s)}{Z(s)} = \frac{\left(\dfrac{2}{s}\right)}{\left(\dfrac{1}{5s}\right)} = 10$$

The resulting circuit is shown in Fig. 4-19b.

From Example 4-4, we notice in the converted circuits of Fig. 4-19 that we have a constant for the source values. This corresponds to an impulse in the time domain. Therefore, in the time domain we could represent a fluxed inductor as an unfluxed inductor in series with an impulse voltage source and a charged capacitor as an uncharged capacitor in parallel with an impulse current source.

4-3 LAPLACE CIRCUIT

To find the Laplace circuit, we find the initial conditions and convert each element to its natural equivalent circuit. Depending on the circuit analysis method used, we may have to convert to all voltage sources or all current sources. A few examples will demonstrate this.

EXAMPLE 4-5

Convert the time-domain circuit shown in Fig. 4-20 to the Laplace domain circuit. Convert all sources to voltage sources.

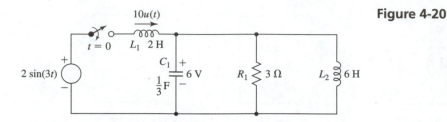

Figure 4-20

Solution: Using the methods described in Section 4-2, we replace the components that have an initial condition with their natural equivalent circuit, as shown in Fig. 4-21a. Z_{L1} is the only component with a current source, and the only component on which we need to use source conversion. The resulting circuit is shown in Fig. 4-21b.

EXAMPLE 4-6

For the circuit shown in Fig. 4-22, draw the Laplace circuit with the reactive components in their natural equivalent circuit.

Solution: At $t = 0^-$ the resistor R_3 and capacitor C_1 are shorted. The circuit, just before the switch is thrown, is shown in Fig. 4-23a. The only reactive component with an initial condition is L_1, which has an initial current of

$$I(0^-) = I(0^+) = \frac{E}{R_1 \parallel R_2} = \frac{10}{2} = 5 \text{ A}$$

We can now remove the switch of Fig. 4-22, replace the voltage source with $10\, u(t)$, and indicate the initial current through L_1. This is shown in Fig. 4-23b. From Fig. 4-23b, we can convert the circuit to the Laplace domain as shown in Fig. 4-23c.

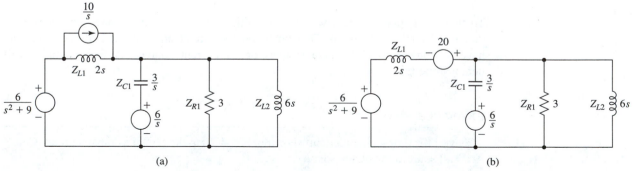

(a)   (b)

Figure 4-21

Figure 4-22

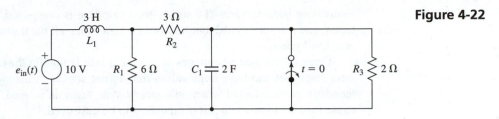

4-4 SOLVING SIMPLE CIRCUITS

Once we have the Laplace circuit, we solve for the Laplace voltages, $V(s)$, and/or Laplace currents, $I(s)$, as if the circuit was a dc circuit involving only resistors. The difference is that the values in the Laplace domain will be functions of s instead of numerical values. We then find the inverse of $V(s)$ and/or $I(s)$, which will give us the complete response from $t = 0$ to ∞.

The inverse Laplace transform will contain two types of terms—transient terms and steady-state terms. Transient terms will always contain an e^{-at} multiplier, and the steady-state will not have an exponential multiplier. The transient terms will be considered to be zero when e^{-at} is past five time constants. This may be as short as 1 nsec or as long as several days. The steady-state terms will be active from $t = 0$ until $t = \infty$.

The response of the circuit's output voltage or current can be similarly divided into two types of responses—the transient response and the steady-state response. The transient response is composed of all the transient terms and all the steady-state terms, and it is the output of the circuit's voltage or current from $t = 0$ until the time when all the transient

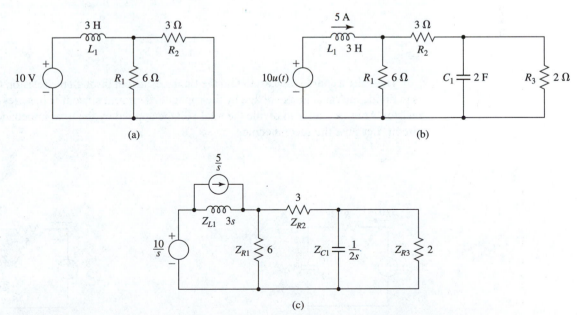

(a) (b)

(c)

Figure 4-23

terms have gone to zero. The steady-state response is composed of only the steady-state terms, and it is the circuit's output from the time when all the transients terms have gone to zero until infinity.

One of the major advantages of using Laplace is finding the transient response. Transient responses can have large values that do not appear in the steady-state response and therefore go undetected when only steady-state analysis is used. These large values can cause fuses to blow or circuit components to be destroyed.

A few examples will demonstrate how to use Laplace to analyze a circuit. Once the basic concept is understood, it is just a matter of reviewing dc circuit analysis. A review of basic dc circuit analysis equations is given in Appendix C.

EXAMPLE 4-7

Find the Laplace voltage $V_C(s)$ and the Laplace current $I_R(s)$ in Fig. 4-24.

Solution: First we convert the circuit into a Laplace-domain circuit. This is shown in Fig. 4-25.

The voltage across the capacitor can be found by using the voltage divider rule.

$$V_C(s) = \frac{E(s)\, Z_C}{Z_R + Z_C}$$

$$= \frac{\left(\dfrac{20}{s}\right)\left(\dfrac{1}{2s}\right)}{5 + \left(\dfrac{1}{2s}\right)}$$

$$= \frac{2}{s(s + 0.1)}$$

There are a couple of ways to find the Laplace current through the resistor. One way is to divide the capacitor's voltage by the capacitor's current since it is in series with the resistor. Another way is to divide the total voltage applied by the total impedance of the circuit. Let's use the second choice.

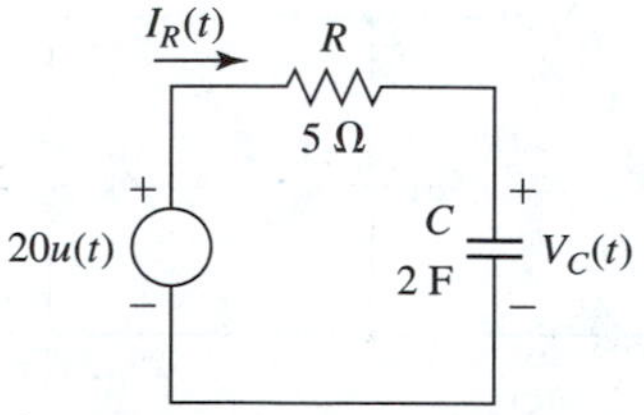

Figure 4-24

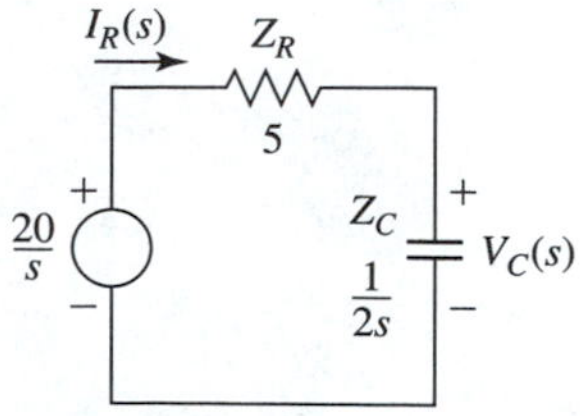

Figure 4-25

$$I_R(s) = \frac{E(s)}{Z_R + Z_C}$$

$$= \frac{\left(\dfrac{20}{s}\right)}{5 + \left(\dfrac{1}{2s}\right)}$$

$$= \frac{4}{(s + 0.1)}$$

EXAMPLE 4-8

Find the Laplace current, $I_R(s)$, through R_1 in Fig. 4-26, and then find $i_R(t)$.

Solution: We must first convert the time-domain circuit to a Laplace-domain circuit. This is shown in Fig. 4-27.

Because this is a current split between parallel branches, we should use the current divider rule.

$$I_R(s) = \frac{I_{in}(s)\,(Z_{C1} + Z_{C2})}{Z_{R1} + (Z_{C1} + Z_{C2})}$$

$$= \frac{\dfrac{3}{s}\left(\dfrac{1}{2s} + \dfrac{1}{4s}\right)}{5 + \left(\dfrac{1}{2s} + \dfrac{1}{4s}\right)}$$

$$= \frac{0.45}{s(s + 0.15)}$$

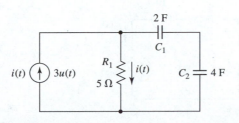

Figure 4-26

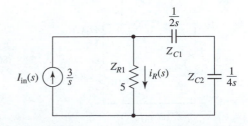

Figure 4-27

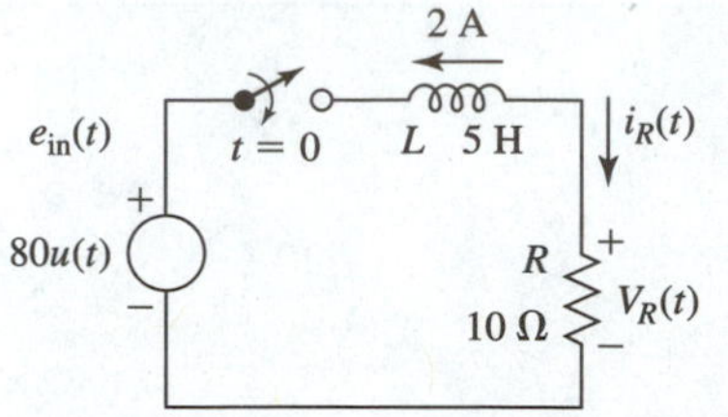

Figure 4-28

Notice that the series capacitors are added rather than being the product over the sum as in time-domain circuit analysis. When we use Laplace methods, all impedances and sources are used as if they were resistors or dc sources.

Using the inverse Laplace transform techniques for first-order real poles shown in Chapter 3, we find the time-domain current to be

$$i_R(t) = 3\,u(t) - 3e^{-0.15t}\,u(t)$$

EXAMPLE 4-9

The circuit in Fig. 4-28 shows an inductor that was switched from one circuit path to the one shown. Previous to the switching, 2 A of current was established through the inductor. Find $i_R(t)$ and $v_R(t)$. Find the steady-state value of the current.

Solution First we convert the time-domain circuit to the Laplace circuit, as shown in Fig. 4-29a. The initial condition current source is then converted to a series voltage source in Fig. 4-29b. Because this initial voltage source is in series with the supply voltage source, we can combine them as shown in Fig. 4-29c.

$$\text{Voltages combined} = \left(\frac{80}{s}\right) - (10)$$

$$= \frac{-10(s-8)}{s}$$

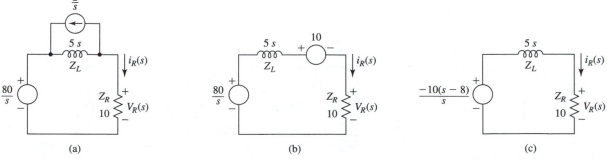

(a) (b) (c)

Figure 4-29

We have two approaches that could easily be applied. We could divide the resulting voltage source by the total circuit impedance, because it is a series circuit, and find the voltage across the impedance, $Z_R(s)$, using Ohm's law. The other approach would be to use the voltage divider rule to find the voltage across the impedance, $Z_R(s)$, and Ohm's law to find the current. Let's use the second approach.

$$V_R(s) = \frac{E(s)\, Z_R}{Z_R + Z_L}$$

$$= \frac{\left(\dfrac{-10(s-8)}{s}\right)(10)}{10 + 5s}$$

$$= \frac{-20(s-8)}{s(s+2)}$$

The inverse Laplace is then

$$v_R(t) = 80\, u(t) - 100e^{-2t}\, u(t)$$

Because the component to find the current across is a resistor, we could divide by the resistance value in the time domain. If the component were a capacitor, we could take the derivative of the voltage and multiply by the capacitance value. An inductor would involve the integral. In the Laplace domain, however, all cases are done by using $I(s) = E(s)/Z(s)$.

In this case, it would be simpler to divide the time-domain voltage by the resistance, but we will use the Laplace approach to demonstrate the Laplace method.

$$I_R(s) = \frac{V_R}{Z_R}$$

$$= \frac{\left(\dfrac{-20(s-8)}{s(s+2)}\right)}{10}$$

$$= \frac{-2(s-8)}{s(s+2)}$$

Notice that the only quantity changed is the constant multiplier. The inverse Laplace is therefore

$$i_R(t) = 8\, u(t) - 10e^{-2t}\, u(t)$$

Because the transient response includes both steady-state terms and transient terms, the current shown above is the transient current. To find the steady-state response, we remove the transient terms. The transient terms can be easily identified because they will always contain an exponential factor that causes them to go to zero, or "die out," in five time constants. In this, we remove the second term to give us the steady-state response

$$i_{ss}(t) = 8 \text{ A}$$

This is the answer that we would find using steady-state circuit analysis. Notice how the $u(t)$ factor is removed. This is because in steady state $u(t)$ is 1.

EXAMPLE 4-10

For the Laplace circuit shown in Fig. 4-30, find the voltage $v_{ab}(t)$. Notice that the Laplace current through R_2 has already been solved and given in the drawing.

Solution: The only way to solve for a voltage across an open circuit is to use Kirchhoff's voltage law. We must first find all the other voltages around a loop. Because we know the current through Z_{R2}, we can find the voltage across Z_{R2} using Ohm's law. By using Kirchhoff's current law at the top node and using Ohm's law, we can find the current and voltage for Z_{R1}.

Using Fig. 4-31, we can apply Kirchhoff's current law at the top node.

$$I_{in} - I_1 - I_2 = 0$$

$$I_1 = I_{in} - I_2$$

$$= 4 - \frac{s + 1}{s + 0.25}$$

$$= \frac{3s}{s + 0.25}$$

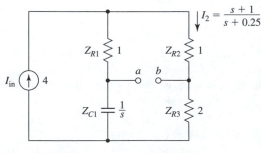

Figure 4-30

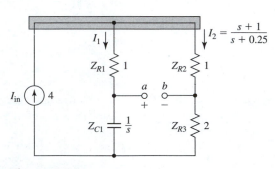

Figure 4-31

Using Ohm's law, we can find the voltage across Z_{R1} and Z_{R2}.

$$V_1 = (1)\frac{3s}{s + 0.25} = \frac{3s}{s + 0.25}$$

$$V_2 = (1)\frac{s + 1}{s + 0.25} = \frac{s + 1}{s + 0.25}$$

We are looking for V_{ab}, so we measure a with respect to b as shown in Fig. 4-31. (The first subscript is always the point measured, and the second subscript is always used as reference.) Writing Kirchhoff's voltage law gives

$$V_1 - V_2 + V_{ab} = 0$$

$$V_{ab} = V_2 - V_1$$

$$= \frac{s + 1}{s + 0.25} - \frac{3s}{s + 0.25}$$

$$= \frac{-2(s - 0.5)}{s + 0.25}$$

The inverse Laplace transform is

$$v_{ab}(t) = -2\,\delta(t) + 1.5e^{-0.25t}\,u(t)$$

4-5 SOLVING MULTIPLE-SOURCE CIRCUITS

A network with multiple sources is nearly impossible to solve in the time domain when inductors and capacitors are present. Using the Laplace transform method will make the process much easier. Because Laplace transforms make the analysis a dc process, this section is written as a review of some dc network techniques. There are a number of methods in dc circuit analysis used to solve multiple-source circuits that are all valid to use in Laplace. We will concentrate on three methods.

4-5-1 Superposition

This method is best suited for finding one or a few voltages and currents when a multiple-source network is being solved. It also allows us to see the effect that each independent source has on a network or component because we work with each source separately using single-source techniques.

Superposition Method

1. Indicate the direction to measure the desired current or the voltage.
2. Reduce all the independent sources to zero except for one source. Voltage sources become shorts, and current sources become opens.

3. Find the voltage or current as indicated in step 1.
4. Do steps 2 and 3, leaving in a different source each time, until the voltage or current due to each source is found.
5. Sum the voltages or currents due to each source to find the total voltage or current.

EXAMPLE 4-11

Find the Laplace current through the inductor of Fig. 4-32 using the superposition method.

Solution: We first convert the circuit to a Laplace circuit, as shown in Fig. 4-33. An arbitrary current direction is chosen because a direction was not indicated in the problem.

We will find the current due to the initial voltage from the capacitor first. We remove all sources except the source we are working with, as shown in Fig. 4-34a. The circuit can be simplified to the circuit shown in Fig. 4-34b. We will call the current I'. Using Ohm's law, the current is

$$I' = \frac{\left(\dfrac{4}{s}\right)}{\left(4s + 8 + \dfrac{4}{s}\right)} = \frac{4}{4s^2 + 8s + 4} = \frac{1}{(s + 1)^2}$$

Next we will find the current due to the current source. We will call this current I''. Figure 4-35a shows all the sources removed except the current source, and Fig. 4-35b shows the circuit simplified. Notice that the current I'' is opposite to the direction of current flow we choose. Therefore, I'' will be a negative quantity. Using the current divider rule yields

$$I'' = \frac{-5\left(8 + \dfrac{4}{s}\right)}{\left(8 + \dfrac{4}{s}\right) + 4s} = \frac{-5(8s + 4)}{4s^2 + 8s + 4} = \frac{-10(s + 0.5)}{(s + 1)^2}$$

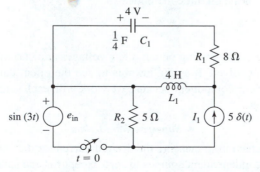

Figure 4-32

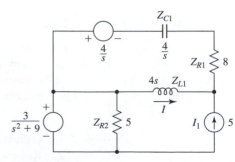

Figure 4-33

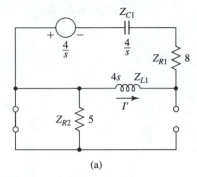

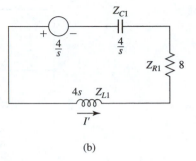

(a)

(b)

Figure 4-34

Finally, we will find the current due to the sinusoidal voltage source. Figure 4-36 shows this current, I'''. Notice that there is no complete path for current from the source to flow through the inductor. The current, therefore, is

$$I''' = 0$$

We measured the currents due to each source in the same direction, so we sum the results.

$$I_T = I' + I'' + I'''$$

$$= \frac{1}{(s + 1)^2} + \frac{-10(s + 0.5)}{(s + 1)^2} + 0$$

$$= \frac{-10(s + 0.4)}{(s + 1)^2}$$

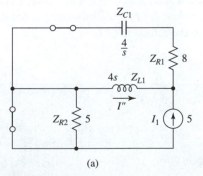

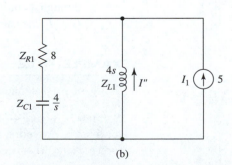

(a)

(b)

Figure 4-35

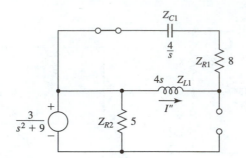

Figure 4-36

4-5-2 Mesh Current

When there are several quantities to find in a network, it is often easier to solve for all of them at the same time. The mesh current method solves for the loop currents. Once these are known, any current or voltage in the circuit can be solved. There are a few variations of this method. We will use the following form:

Mesh Current Method

1. Draw the circuit so that none of the wires cross over each other.
2. Convert all independent and dependent current sources to voltage sources.
3. Draw a clockwise loop current inside each enclosed area (called a *window*) of the network to indicate the direction the loop currents are measured.
4. For each loop write an equation inside each window as follows:

    (a) Sum all the components, except sources, in the window and multiply by the loop current.
    (b) Locate any component, other than sources, that has another loop current passing through it. For each component, subtract the component value times this other loop current.
    (c) Set the terms found above equal to the clockwise algebraic summation of the rises (added) and drops (subtracted) of the voltage sources.

5. Solve the simultaneous equations for the currents. This is usually done with determinants.
6. The current through a component showing two loop currents is equal to the difference of the two loop currents. This value of the current is measured in the direction of the current subtracted from. Therefore, a component with I_3 and I_4 through it could be either: (a) $I_3 - I_4$, measured in the direction of I_3, or (b) $I_4 - I_3$, measured in the direction of I_4.

EXAMPLE 4-12

Write the mesh equations for Fig. 4-37.

Solution: Figure 4-38a shows the Laplace circuit. The initial condition for the inductor must be converted to a voltage source. Figure 4-38b shows this conversion, and the loop currents have been indicated.

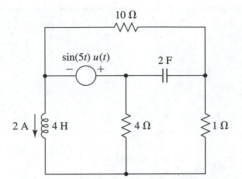

Figure 4-37

The first equation is found in the top, or I_1, window. There are two components and one voltage source in this window. The capacitor is the only component that has another current going through it. Voltage sources are treated separately and not considered as components in this context. The first equation is

$$\left(10 + \frac{1}{2s}\right)I_1 - 0I_2 - \frac{1}{2s}I_3 = \frac{-5}{s^2 + 25}$$

For the second window, or I_2 loop, the 4-Ω resistor is the only component with two currents through it. The equation is

$$-0I_1 + (4s + 4)I_2 - 4I_3 = \frac{5}{s^2 + 25} - 8$$

Notice how the voltage sources are added or subtracted depending on the way they are measured and using a clockwise summation.

The last equation from the third loop is

$$-\frac{1}{2s}I_1 - 4I_2 + \left(1 + 4 + \frac{1}{2s}\right)I_3 = 0$$

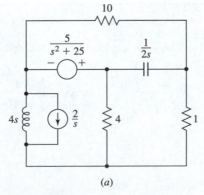

(a)

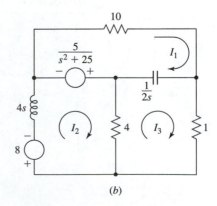

(b)

Figure 4-38

The complete set of equations from this circuit is therefore,

$$\left(10 + \frac{1}{2s}\right)I_1 \quad - (0)\ I_2 - \left(\frac{1}{2s}\right) \quad I_3 = \frac{-5}{s^2 + 25}$$

$$-(0) \quad I_1 + (4s + 4)I_2 - (4) \quad I_3 = \frac{5}{s^2 + 25} - 8$$

$$-\left(\frac{1}{2s}\right) I_1 \quad - (4)\ I_2 + \left(1 + 4 + \frac{1}{2s}\right)I_3 = \quad 0$$

In this text, simultaneous equations will be shown in this format so that their relationship to a matrix format, can be easily seen. Experience will show that it is best to remove fractions from these equations before solving them. To remove the fractions from an equation, we multiply through both sides by an expression that would be the common denominator if all the terms were added. In the first equation, the expression is $2s(s^2 + 25)$; in the second equation, the expression is $(s^2 + 25)$; and in the third equation, the expression is $(2s)$. Multiplying through by these factors, we have the simultaneous equations in a form suitable for solving.

$$(s^2 + 25)(20s + 1)I_1 - \quad (0) \quad I_2 - (s^2 + 25)\ I_3 = \quad -10s$$

$$-(0) \quad I_1 + (s^2 + 25)(4s + 4)\ I_2 - (4)(s^2 + 25)\ I_3 = -8s^2 - 195$$

$$-(1) \quad I_1 - \quad (8s) \quad I_2 + (10s + 1)\ I_3 = \quad 0$$

EXAMPLE 4-13

Using mesh analysis, find the voltage across the capacitor in Fig. 4-39.

Solution: Figure 4-40 shows the Laplace circuit with all sources in a voltage source form. The loop equations are

$$\left(2s + \frac{2}{s}\right)I_1 - \left(\frac{2}{s}\right) I_2 = 4 - \frac{3}{s}$$

$$-\left(\frac{2}{s}\right) I_1 + \left(\frac{2}{s} + 3\right)I_2 = \frac{3}{s} - \frac{2}{s}$$

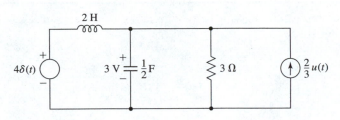

Figure 4-39

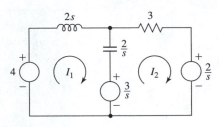

Figure 4-40

Removing the fractions gives

$$(2s^2 + 2)I_1 \quad - (2) \; I_2 = 4s - 3$$

$$-(2) \quad I_1 + (3s + 2)I_2 = \quad 1$$

Solving the denominator determinant, we obtain

$$D = \begin{vmatrix} 2s^2 + 2 & -2 \\ -2 & 3s + 2 \end{vmatrix} = (2s^2 + 2)(3s + 2) - (-2)(-2)$$

$$= 6s^3 + 4s^2 + 6s$$

$$= 6s[(s + 0.33333)^2 + 0.88889]$$

Solving for the currents I_1 and I_2 yields

$$I_1 = \frac{\begin{vmatrix} 4s - 3 & -2 \\ 1 & 3s + 2 \end{vmatrix}}{D} = \frac{12s^2 - s - 4}{D}$$

$$I_2 = \frac{\begin{vmatrix} 2s^2 + 2 & 4s - 3 \\ -2 & 1 \end{vmatrix}}{D} = \frac{2s^2 + 8s - 4}{D}$$

Assuming that we want to find the current through the capacitor measured in the downward, or I_1, direction,

$$I_C = I_1 - I_2 = \frac{[12s^2 - s - 4] - [2s^2 + 8s - 4]}{6s[(s + 0.33333)^2 + 0.88889]}$$

$$= \frac{5(s - 0.9)}{3[(s + 0.33333)^2 + 0.88889]}$$

When finding the voltage across the capacitor, we must include the entire equivalent circuit of the capacitor. The voltage across the capacitor part of the equivalent circuit will be the current I_C times $(2/s)$. This voltage is measured at the top of the capacitor with respect to the bottom of the capacitor. The voltage across the voltage source part of the equivalent circuit of the capacitor is also measured at the top with respect to the bottom. Therefore, the voltage across the capacitor will be the current times the capacitor part of the equivalent circuit plus the voltage part of the equivalent circuit.

$$V_C(s) = \frac{1}{sC} I_C + \frac{3}{s}$$

$$= \left(\frac{2}{s}\right) \left(\frac{5(s - 0.9)}{3[(s + 0.33333)^2 + 0.88889]}\right) + \frac{3}{s}$$

$$= \frac{3(s + 1.7778)}{[(s + 0.33333)^2 + 0.88889]}$$

The complete voltage across the capacitor is

$$v_c(t) = 5.4886e^{-t/3} \sin(0.94281t + 33.133°) u(t)$$

A good test for this result is to let $t = 0$ and see if we get $v(0) = 3$.

4-5-3 Node Voltage

This method is similar to mesh except that we solve for the node voltages with respect to a chosen reference node. This method is frequently used for solving op-amp filters.

Node Voltage Method

1. Draw the circuit so that none of the wires cross over each other.
2. Convert all independent and dependent voltage sources to current sources.
3. Choose a node to reference voltages to. Number the remaining nodes. Each node should include as many components as possible.
4. For each node except the reference node, write an equation as follows:

    (a) Sum all the admittances (reciprocal of impedances) attached to the node and multiply by the node voltage.

(b) Locate any component, other than sources, that has its other electrode attached to a node other than the reference node. Subtract this component admittance times the other node voltage.

(c) Set the terms found above equal to the algebraic summation of the current sources entering (added) and leaving (subtracted) the node.

5. Solve the simultaneous equations for the node voltages. This is usually done with determinants.

6. The voltage across a component is equal to the difference of the two voltages where its electrodes are attached. This value of the voltage is measured at the voltage node that is subtracted from, using the voltage location of the subtracting node as reference. Therefore, a component between V_3 and V_4 could be either: (a) $V_3 - V_4$, measured at V_3 with respect to V_4, or (b) $V_4 - V_3$, measured at V_4 with respect to V_3.

EXAMPLE 4-14

Write the node equation for the circuit in Fig. 4-41.

Solution: Figure 4-42 shows the Laplace circuit with all current sources. From the circuit we write the equations

$$\left(\frac{1}{2s} + \frac{1}{4} + 1\right)V_1 - \quad (1) \quad V_2 - \quad (0) \quad V_3 = \frac{4}{s(s^2 + 16)}$$

$$-(1) \quad V_1 + \left(1 + 2s + \frac{1}{3}\right)V_2 - \left(\frac{1}{3}\right) V_3 = \quad 20$$

$$-(0) \quad V_1 - \quad \left(\frac{1}{3}\right) \quad V_2 + \left(\frac{1}{3} + \frac{1}{3}\right)V_3 = \quad \frac{-2}{s}$$

Multiplying the three equations by $4s(s^2 + 16)$, 3, and $3s$, respectively, and combining terms, we have

$$(s^2 + 16)(5s + 2) V_1 - 4s(s^2 + 16) V_2 - (0) \quad V_3 = 16$$

$$-(3) \quad V_1 + (6s + 4) \quad V_2 - (1) \quad V_3 = 60$$

$$-(0) \quad V_1 - (s) \quad V_2 + (s + 3) V_3 = -6$$

These equations are ready to be solved by determinants. Notice that if we replaced the Vs with Is, the equations would look just like a set of mesh equations.

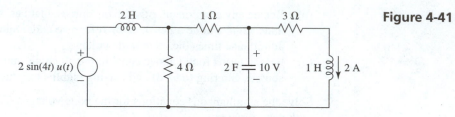

Figure 4-41

With one modification, the node voltage method is typically used to analyze op-amp filters. We can modify this method so that we may leave voltage sources in the network instead of converting to current sources. If a voltage source has one terminal connected to the reference node, we may treat it as another voltage node. Because we know the value of this voltage, we do not write an equation at this known voltage node.

EXAMPLE 4-15

For the Laplace circuit shown in Fig. 4-43, write the node equations using the modified node voltage method.

Solution: This circuit is written as a four-node circuit. The equations generated from the circuit are

$$\left(\frac{1}{3} + \frac{1}{2s} + \frac{1}{2}\right)V_1 \quad - \left(\frac{1}{2}\right) \quad V_2 - \left(\frac{1}{3}\right)V_x \quad - (0)\,(-V_y) = 0$$

$$-\left(\frac{1}{2}\right) \quad V_1 + \left(\frac{1}{2} + s + \frac{1}{4}\right)V_2 - (0)\quad V_x - \left(\frac{1}{4}\right)(-V_y) = 0$$

In the second equation, V_y is entered as negative because the node voltages are always measured with respect to the reference.

The values of V_x and V_y are then substituted into the equations.

$$\left(\frac{1}{3} + \frac{1}{2s} + \frac{1}{2}\right)V_1 \quad - \left(\frac{1}{2}\right) \quad V_2 - \left(\frac{1}{3}\right)\left(\frac{2}{s}\right) - (0)\quad (-5) = 0$$

$$-\left(\frac{1}{2}\right) \quad V_1 + \left(\frac{1}{2} + s + \frac{1}{4}\right)V_2 - (0)\quad \left(\frac{2}{s}\right) - \left(\frac{1}{4}\right)(-5) = 0$$

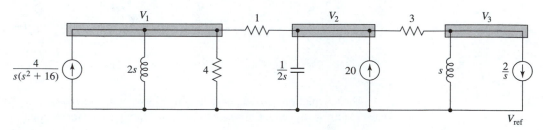

Figure 4-42

Figure 4-43

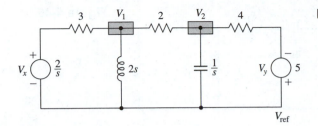

These terms will now become constants, so we move them to the right side of the equation.

$$\left(\frac{1}{3} + \frac{1}{2s} + \frac{1}{2}\right)V_1 - \left(\frac{1}{2}\right)V_2 = \frac{2}{3s}$$

$$-\left(\frac{1}{2}\right)V_1 + \left(\frac{1}{2} + s + \frac{1}{4}\right)V_2 = \frac{-5}{4}$$

These equations are the same equations as we would find if we did source conversion and then wrote the equations. With practice, we can go directly to these equations by doing the algebra mentally.

4-6 SOLVING DEPENDENT SOURCE CIRCUITS

Whenever we work with an active circuit, we typically have a dependent source. A dependent source has a value dependent on a voltage or current from another location, usually in the same circuit.

A dependent source may be either a voltage source or a current source whose output value is equal to a multiplier, K, times a voltage or current from another location. There are four different types of dependent sources shown in Fig. 4-44: voltage-controlled voltage

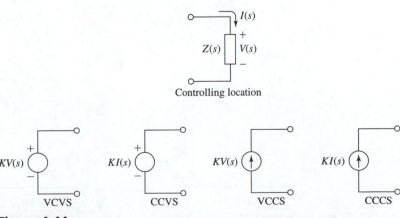

Figure 4-44

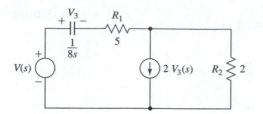

Figure 4-45

source (VCVS), current-controlled voltage source (CCVS), voltage-controlled current source (VCCS), and current-controlled current source (CCCS). By using Ohm's law and source conversion, we can change one type of source to any other type of source.

EXAMPLE 4-16

Convert the VCCS in Fig. 4-45 to a CCVS.

Solution: First we convert the current source to a voltage source using source conversion, as shown in Fig. 4-46a. This makes it a VCVS.
    We then convert from a voltage-controlled voltage source to a current-controlled voltage source. This is done by developing an equation for $V_3(s)$ in terms of a current. In this case, Ohm's law will work.

$$V_3(s) = \frac{1}{8s} I_3(s)$$

We then substitute this value of $V_3(s)$ into the expression for the VCVS of Fig. 4-46a

$$4V_3(s)$$

$$4\left(\frac{1}{8s}\right)I_3(s)$$

$$\frac{1}{2s} I_3(s)$$

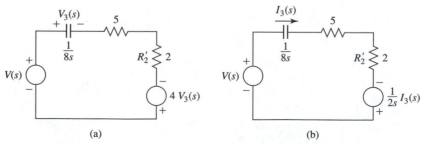

(a)   (b)

Figure 4-46

Figure 4-46b shows the resulting circuit using a CCVS.

In this example we could have converted first to a CCCS and then used source conversion, which would result in the same final circuit. In any case, because source conversion was used, we must remember that the voltage across R_2 is now the voltage across R'_2 plus the voltage across the CCVS.

EXAMPLE 4-17

Find $I(s)$ in terms of $V_{in}(s)$ for Fig. 4-47.

Solution: We must first find $V_2(s)$ using the voltage divider rule.

$$V_2(s) = \frac{V_{in}(10/s)}{2 + (10/s)} = \frac{5V_{in}}{s + 5}$$

Now we can find the value of the VCVS by substitution.

$$10V_2(s) = \frac{50V_{in}}{s + 5}$$

By using Ohm's law, we find the current through the inductor.

$$I(s) = \frac{\dfrac{50V_{in}}{s + 5}}{5s} = \frac{10V_{in}}{s(s + 5)}$$

When we have multiple dependent sources, we use the multiple-source techniques, as seen in previous sections. Because the dependent sources have a variable associated with them, we will have to move their values algebraically to the correct location for that variable.

EXAMPLE 4-18

Find the current through the inductor in Fig. 4-48 using the mesh current method.

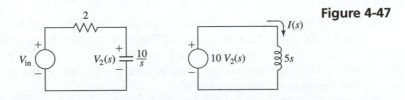

Figure 4-47

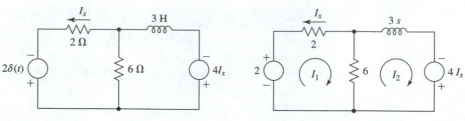

Figure 4-48   **Figure 4-49**

Solution: Figure 4-49 shows the Laplace circuit set up for mesh analysis. We must get the dependent sources in terms of variables in the mesh equations. Notice that I_x is the same but the opposite of I_1.

$$4I_x = -4I_1$$

Writing the loop equations, we have

$$(2 + 6)I_1 + \quad -6 \quad I_2 = 2$$

$$-(6)I_1 + (3s + 6)I_2 = -4I_1$$

In the second equation, we need to move $-4I_1$ to the left side. Doing this and combining terms, the two equations become

$$8I_1 \qquad - 6 \quad I_2 = 2$$

$$-2I_1 + (3s + 6)I_2 = 0$$

At this point, we work the problem as any other mesh problem. The Laplace current and resulting time-domain current are

$$I_2(s) = \frac{1}{6(s + 1.5)}$$

$$i(t) = 0.16667e^{-1.5t}\,u(t)$$

4-7 SOLVING THÉVENIN AND NORTON CIRCUITS

Thévenin and Norton equivalent circuits are frequently used to simplify analysis of networks. As with all Laplace circuits, Thévenin and Norton equivalent circuits are worked exactly as if they were dc circuits. The following procedure works in both the time domain and the Laplace domain, so it may be used to calculate values on paper as well as physically applied to a circuit in the lab.

Thévenin or Norton Equivalent Impedance

1. Remove the component or network for which you are finding the Thévenin or Norton equivalent circuit. The point of removal must have two and only two terminals.
2. Make all independent sources zero. A zero-volt voltage source is a short circuit, and a zero-ampere current source is an open circuit.
3. Drive the terminal where the component or network was removed with a voltage source V_z, and determine the impedance that V_z sees. For a circuit in the lab, we measure I_z and divide it into V_z to determine the impedance.

4. (a) To find the Thévenin voltage, find the open-circuit voltage with the component or network and V_z removed.
   (b) To find the Norton current, short the terminals and find the current in the short with the component or network and V_z removed.

The impedance we will find will be a function of s. The function may not be an easily recognizable combination of resistance, inductance, and/or capacitance. We therefore put the function next to a box and treat the box as a component. Do not get this confused with block diagrams, which are entirely different.

EXAMPLE 4-19

Using a Thévenin equivalent circuit, find $I_1(s)$ in Fig. 4-50 when $R_1 = 1 \, \Omega$ and $4 \, \Omega$.

Solution: Figure 4-51a shows the circuit with the current source at zero, the component removed, and a voltage source, V_z, driving the two terminals where the component was removed. The source sees a series network of a capacitor and resistor, which is the Thévenin equivalent impedance.

$$Z_{\text{TH}}(s) = \frac{1}{5s} + 4 = \frac{4(s + 0.05)}{s}$$

Using Fig. 4-51b, we can find the open-circuit voltage at the terminals a–b, which is the Thévenin equivalent voltage. The terminals are marked a–b so that we know which terminal is which when we draw the Thévenin equivalent circuit. We usually measure the voltage at a with respect to b. When we measure this way, the plus side of the source in the equivalent circuit will be closest to the a terminal.

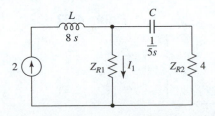

Figure 4-50

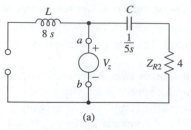

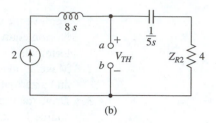

(a)   (b)

Figure 4-51

Using Kirchhoff's voltage law, the voltage across the open terminals is

$$V_{TH} = V_C + V_{R2}$$

and using Ohm's law, the voltage across the capacitor and resistor is

$$V_C = 2\left(\frac{1}{5s}\right) = \frac{2}{5s} \quad \text{and} \quad V_{R2} = 2(4) = 8$$

Substituting into the Kirchhoff's voltage law equation gives

$$V_{TH} = \frac{2}{5s} + 8 = \frac{8(s + 0.05)}{s}$$

The equivalent circuit, and the component for which the Thévenin equivalent circuit was found, is shown in Fig. 4-52. Since it is not a pure resistance, capacitance, or inductance, a block is used to indicate a component, but it is still treated as any other Laplace impedance.

We can now easily find the current through the resistor for any value it may become. We could put in a value and solve, or we could use the variable Z_{R1} and solve for a general equation. Using the second choice and Ohm's law, we obtain

$$I(s) = \frac{\left[\dfrac{8(s + 0.05)}{s}\right]}{\left[\dfrac{4(s + 0.05)}{s} + Z_{R1}\right]}$$

Figure 4-52

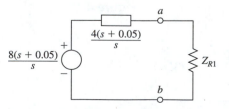

Because Z_{R1} is a constant, we can put this in a form so that when Z_{R1} is substituted in the function, it will be in a standard form.

$$I(s) = \frac{\left(\dfrac{8}{4 + Z_{R1}}\right)(s + 0.05)}{s + \left(\dfrac{0.2}{4 + Z_{R1}}\right)}$$

Substituting in $Z_{R1} = 1$ yields

$$I(s) = \frac{1.6(s + 0.05)}{s + 0.04}$$

and $i(t)$ is

$$i(t) = 1.6\,\delta(t) + 0.016e^{-0.04t}\,u(t)$$

Substituting in $Z_{R1} = 4$, we have

$$I(s) = \frac{s + 0.05}{s + 0.025}$$

and $i(t)$ is

$$i(t) = \delta(t) + 0.025e^{-0.025t}\,u(t)$$

When we have dependent sources in the circuit, we cannot remove them. With dependent sources involved, we must use one of two methods to solve the Thévenin or Norton equivalent impedance.

For both methods, we still remove all independent sources, remove the network for which we want to find the Thévenin or Norton equivalent circuit, and drive that point with a source, V_z. The first method is to find an equation having only the variable V_z and I_z, and solve for V_z/I_z, which is the equivalent impedance. The second method is to find an equation for V_z and I_z in terms of the same variable, divide V_z by I_z, and cancel out the variable that V_z and I_z are in terms of.

EXAMPLE 4-20

Find the Norton equivalent circuit for the network C_2 and L in Fig. 4-53.

Solution: Figure 4-54 shows the independent source removed, the network removed, and the V_z source inserted. Normally, one method is used, but we will use both methods and solve for the equivalent impedance twice.

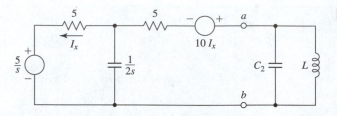

Figure 4-53

Method 1. There are a number of good approaches to this, but let's use Kirchhoff's voltage law around the outside loop.

$$V_z - 10I_x - 5I_z - 5I_x = 0$$

$$V_z = 5I_z + 15I_x$$

This equation has three variables in it, but we want only V_z and I_z. We can use current divider rule to find I_x in terms of I_z.

$$I_x = \frac{I_z\left(\dfrac{1}{2s}\right)}{5 + \dfrac{1}{2s}} = \frac{0.1I_z}{s + 0.1}$$

When this is substituted into the Kirchhoff's voltage law equation, we have

$$V_z = 5I_z + 15\left(\frac{0.1\,I_z}{s + 0.1}\right)$$

$$= \left[5 + \left(\frac{1.5}{s + 0.1}\right)\right]I_z$$

$$= \frac{5(s + 0.4)I_z}{s + 0.1}$$

Solving for V_z/I_z gives

$$\frac{V_z}{I_z} = \frac{5(s + 0.4)}{s + 0.1}$$

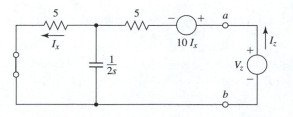

Figure 4-54

Method 2.　　In this method we need to find I_z in terms of a variable, and find V_z in terms of the same variable. This variable can be any variable. Let's use I_x so that we can see the similarities between this method and the previous method.

First we will use the current divider to find I_x in Fig. 4-54.

$$I_x = \frac{I_z\left(\dfrac{1}{2s}\right)}{5 + \dfrac{1}{2s}} = \frac{0.1\,I_z}{s + 0.1}$$

Next we use Kirchhoff's voltage law around the outside loop.

$$V_z - 10I_x - 5I_z - 5I_x = 0$$

This time we need to find I_z in terms of I_x. This can be done by solving the current divider rule for I_z:

$$I_z = 10I_x(s + 0.1)$$

and substituting into the Kirchhoff's voltage law equation.

$$V_z - 10I_x - 5[10I_x(s + 0.1)] - 5I_x = 0$$

$$V_z = I_x(50s + 20)$$

Finally, we divide V_z by I_z and cancel the I_x.

$$\frac{V_z}{I_z} = \frac{I_x(50s + 20)}{10I_x(s + 0.1)} = \frac{5(s + 0.4)}{s + 0.1}$$

which gives us the same result. One or the other method may be easier depending on the particular circuit, but when done properly, the same result can be found with either method.

The next step is to find the Norton short-circuit current. We remove the source we added and short the terminals, as shown in Fig. 4-55. This is a multiple-source circuit, so we will use mesh.

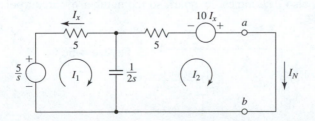

Figure 4-55

Using Fig. 4-55, we can find the short-circuit current as indicated at the terminals *a*–*b*, which is the Norton equivalent current. The terminals are marked *a*–*b* so that we know which terminal is which in the equivalent circuit. We usually measure the current from *a* to *b*. When we measure this way, the current source in the equivalent circuit points toward the *a* terminal. In this case, I_N will be equal to I_2.

The mesh equations from the circuit are

$$\left(5 + \frac{1}{2s}\right)I_1 - \frac{1}{2s} \qquad I_2 = \frac{5}{s}$$

$$-\frac{1}{2s}I_1 + \left(5 + \frac{1}{2s}\right)I_2 = 10I_x \quad \text{or} \quad = 10(-I_1)$$

Notice that $-I_1$ is substituted for I_x.

Moving the I_1 term to the left and removing fractions, we get

$$(10s + 1)I_1 - \qquad\qquad I_2 = 10$$

$$(20s - 1)I_1 + (10s + 1)I_2 = 0$$

We only need to solve for I_2, which is

$$I_2 = I_N = \frac{-2(s - 0.05)}{s(s + 0.4)}$$

Figure 4-56 shows the Norton equivalent circuit.

4-8 CIRCUIT ORDER

The circuit order is the order of the time-domain differential equation required to solve the circuit. Therefore, a first-order circuit generates a first-order differential equation, a second-order circuit generates a second-order differential equation, and so on. When using Laplace transforms, it is unnecessary to write the differential equation because different orders of circuit are easily identified in the Laplace domain or from the circuit.

The order of the circuit is determined by the number of nonredundant reactive components (inductors and capacitors) in the circuit. The number of nonredundant reactive components also determines the poles, so the number of circuit poles is equal to the circuit

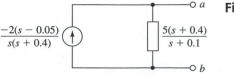

Figure 4-56

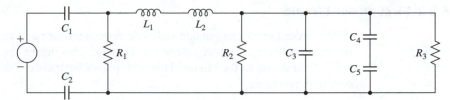

Figure 4-57

order. The order of the differential equation, the circuit order, and the poles are, therefore, due to the nature of the circuit, which is how and where the R, L, and Cs are used.

To help distinguish the circuit poles from the source poles, either an impulse source or step source with a magnitude of 1 is used. When an impulse source is used, there will be no poles due to the source in the Laplace equation because the Laplace transform of the impulse is a constant. When the impulse is used, the response is usually called the *impulse response*.

In the lab it is easier to generate a step function. The response is therefore called the *step response*. The step function will put a pole value of zero in the Laplace function, making it easy to identify.

4-8-1 Redundancy

Redundant components are two or more of the same type of component (R, L, or C) that can be combined into a single component of equivalent value. An example of this is two 1 μF capacitors connected in parallel to create 2 μF of capacitance. There are two physical capacitors, but in the equations they appear as one 2 μF capacitor. One of the capacitors was a redundant component. Therefore, to determine the circuit's order, first replace all redundant components with a single equivalent component. This is only necessary for the reactive components. The circuit order can then be determined by counting the number of nonredundant reactive components.

EXAMPLE 4-21

Find the order of the circuit in Fig. 4-57.

Solution: C_1 and C_2 are in series because the same current flows through them, and they can therefore be combined into a single value. L_1 and L_2 are in series and can be combined. C_4 and C_5 are in series and when combined the resultant capacitor can be combined with C_3, which is in parallel. Figure 4-58 shows the circuit with the redundancy removed. From this circuit we count the number of reactive components, which is three. This is, therefore, a third-order circuit.

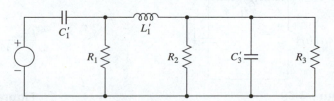

Figure 4-58

4-8-2 First-Order Circuits

A first-order circuit can contain only one nonredundant reactive component. The circuit can contribute only one real pole to the response. This means that the inverse will contain a $Ke^{-\alpha t}$ term due to the circuit. This will always be transient unless the circuit contains only pure reactance.

EXAMPLE 4-22

For the first-order circuit shown in Fig. 4-59, find the Laplace voltage across the capacitor and the current through it. Identify the term that is due to the nature of the circuit.

Solution: The Laplace circuit is shown in Fig. 4-60. Using the voltage divider rule, we obtain

$$V_C(s) = \frac{\left(\dfrac{4}{s^2 + 16}\right)\left(\dfrac{3}{s}\right)}{2 + \dfrac{3}{s}}$$

$$= \frac{6}{(s^2 + 16)(s + 1.5)}$$

The pole due to the factor $(s^2 + 16)$ is from the source, and the pole due to the factor $(s + 1.5)$ is from the nature of the circuit.

4-8-3 Second-Order Circuits

A second-order circuit will have two nonredundant reactive components. Consequently, the circuit response will have two poles due to the circuit. With two poles there are three possible cases for the inverse Laplace transform. Each of the three cases may have different responses when combined with the effects of the other poles due to the sources. Figure 4-61a shows the impulse response where the poles are due only to the circuit. Figure 4-61b shows the step response where the circuit poles are combined with a pole of $s = 0$ (a step function). Each of the three cases has a name associated with it.

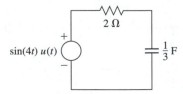

Figure 4-59　　　　　　　　　　　　　　　**Figure 4-60**

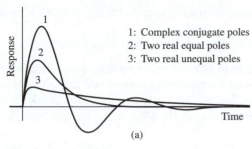

Figure 4-61

1. *Underdamped.* The two poles are complex conjugates, and the inverse of the complex conjugates has the form

$$Ke^{-\alpha t} \sin(\omega t + \theta)\, u(t)$$

2. *Critically damped.* The two poles are real values and equal. The inverse has the form

$$(K_1 t + K_2)e^{-\alpha t}\, u(t)$$

3. *Overdamped.* The two poles are real values and unequal. The inverse has the form

$$K_1 e^{-\alpha_1 t} + K_2 e^{-\alpha_2 t}$$

EXAMPLE 4-23

For the equation shown, is the quadratic in the denominator underdamped, critically damped, or overdamped?

$$V(s) = \frac{s + 4}{s(s^2 + 6s + 9)}$$

Solution: The factors of the quadratic are $s = -3, -3$. Therefore, we may write the equation

$$V(s) = \frac{s + 4}{s(s + 3)^2}$$

Because the quadratic has two equal real roots, the system is critically damped.

When designing second-order circuits we may only need to determine if the circuit is underdamped, critically damped, or overdamped without factoring. To do this, we use another form of the quadratic:

$$s^2 + 2\zeta\omega_n s + \omega_n^2 \qquad\qquad (4\text{-}5)$$

where ω_n = the natural resonant frequency, in rad/sec (ω is the Greek lowercase letter omega)

ζ = the damping ration (ζ is the Greek lowercase letter zeta)

This form of the quadratic is also used in control systems design. It shows different aspects of the second-order function to give us a better perspective of how components in a system or circuit affect the overall response. We will take only a limited look at this form.

If we solve for the roots of the function in Eq. (4-5) using the quadratic formula, we have

$$s = -\frac{b}{2} \pm \sqrt{\left(\frac{b}{2}\right)^2 - c} \tag{4-6}$$

$$= -\frac{2\zeta\omega_n}{2} \pm \sqrt{\left(\frac{2\zeta\omega_n}{2}\right)^2 - \omega_n^2}$$

$$= -\zeta\omega_n \pm \omega_n \sqrt{\zeta^2 - 1}$$

Notice that ζ determines which of the three possible cases we have.

1. $\zeta < 1$: *underdamped.* $\zeta^2 - 1$ will be negative, so we may rewrite the roots as

$$s = -\zeta\omega_n \pm \omega_n \sqrt{(-1)(1 - \zeta^2)} \tag{4-7}$$

$$s = -\zeta\omega_n \pm j\omega_n \sqrt{1 - \zeta^2}$$

2. $\zeta = 1$: *critically damped.* $\zeta^2 - 1 = 0$, so we may rewrite the roots as

$$s = -\omega_n, - \omega_n \tag{4-8}$$

3. $\zeta > 1$: *overdamped.* $\zeta^2 - 1$ will be positive, so we will have two unequal real roots.

$$s = -\zeta\omega_n + \omega_n \sqrt{\zeta^2 - 1}, \qquad -\zeta_n - \omega_n \sqrt{\zeta^2 - 1} \tag{4-9}$$

We notice that ω_n appears in Eqs. (4-8) and (4-9) where the inverse does not have a sinusoidal term. This frequency does not apply directly to this inverse, but in Eq. (4-8) it applies to the gain and phase response curves in Chapter 5. In the overdamped case, ω_n does not have any significance here or in Chapter 5, but is used simply to avoid introduction of another variable.

Previously when we factored a complex quadratic in the denominator, the poles were called $s = -\alpha \pm j\omega$. The ω became the frequency of the sinusoidal function in the inverse Laplace transform. This is actually called the *damped resonant frequency,* and we should add a d subscript to distinguish it from the natural resonant frequency ω_n.

Only in the underdamped case can we show a relationship of ω_d, the damped resonant frequency, and α, the damping constant, to Eq. (4-7). Because the poles are at

$$s = -\alpha \pm j\omega_d$$

and because Eq. (4-7) is

$$s = -\zeta\omega_n \pm j\omega_n \sqrt{1 - \zeta^2}$$

we can find the relation by setting the real parts of each equation equal to each other and setting the imaginary parts equal to each other.

$$\alpha = \zeta\omega_n \tag{4-10a}$$

$$\omega_d = \omega_n \sqrt{1 - \zeta^2} \tag{4-10b}$$

EXAMPLE 4-24

For the second-order pole, find ζ, ω_n, α, and ω_d.

$$\frac{N(s)}{s^2 + 4s + 13}$$

Solution: We set the third term, 13, equal to the third term in Eq. (4-5) to find ω_n:

$$\omega_n^2 = 13 \Rightarrow \omega_n = \sqrt{13} = 3.6056$$

and then set the second terms equal to solve for ζ.

$$2\zeta\omega_n s = 4s$$

$$\zeta = \frac{4}{2\omega_n} = \frac{4}{2(3.6056)} = 0.55470$$

We can find α and ω_d by factoring the quadratic or by using Eq. (4-10).

$$\alpha = \zeta\omega_n = (0.55470)(3.6056) = 2$$

$$\omega_d = \omega_n \sqrt{1 - \zeta^2} = (3.6056)\sqrt{1 - 0.55470^2} = 3$$

This means we could write the quadratic in the form

$$(s + 2)^2 + 9$$

EXAMPLE 4-25

For Fig. 4-62, write the equation for the Laplace voltage across the capacitor and find ζ, ω_n, α, and ω_d.

Figure 4-62

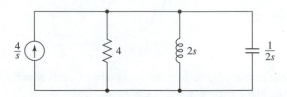

Solution: This is a current source driving a parallel circuit, so we can find the Laplace impedance of R, L, and C and then use Ohm's law to find the voltage across the network, which is also the voltage across the capacitor.

$$Z(s) = \frac{1}{\dfrac{1}{4} + \dfrac{1}{2s} + 2s} = \frac{0.5s}{s^2 + 0.125s + 0.25}$$

$$V(s) = Z(s)I(s) = \frac{0.5s}{s^2 + 0.125s + 0.25}\frac{4}{s}$$

$$= \frac{2}{s^2 + 0.125s + 0.25}$$

Using Eq. (4-5), we find that

$$\omega_n = \sqrt{0.25} = 0.5$$

and

$$2\zeta\omega_n = 0.125 \Rightarrow \zeta = \frac{0.125}{(2)(0.5)} = 0.125$$

To find α and ω_d, we can use Eq. (4-10) or we can just factor. This time, let's factor.

$$s = -0.0625 \pm j0.49608$$

Therefore,

$$\alpha = 0.0625$$

$$\omega_d = 0.49608$$

EXAMPLE 4-26

For the circuit shown in Fig. 4-63, find the value of R in order for the circuits current to be over-, under-, and critically damped.

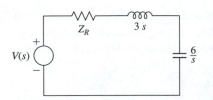

Figure 4-63

Solution: This can be solved by finding how ζ varies with changes in R. Let's first find the equation for the current.

$$I(s) = \frac{V(s)}{Z(s)} = \frac{V(s)}{Z_R + sL + \dfrac{1}{sC}}$$

$$= \frac{\left(\dfrac{s}{L}\right) V(s)}{s^2 + \dfrac{Z_R}{L} s + \dfrac{1}{LC}}$$

Using Eq. (4-5), we obtain

$$\omega_n^2 = \frac{1}{LC} \Rightarrow \omega_n = \frac{1}{\sqrt{LC}} = \frac{1}{\sqrt{3\left(\dfrac{1}{6}\right)}} = 1.4142$$

$$2\zeta\omega_n = \frac{Z_R}{L} \Rightarrow \zeta = \frac{Z_R}{2\omega_n L} = \frac{Z_R}{2(1.4142)(3)}$$

Therefore,

$$Z_R = \zeta(8.4853)$$

From this result we can determine R by substituting different values of ζ. Therefore,

$$\zeta = 1 \Rightarrow R = 8.4853 \Rightarrow \text{critically damped}$$

$$\zeta > 1 \Rightarrow R > 8.4853 \Rightarrow \text{overdamped}$$

$$\zeta < 1 \Rightarrow R < 8.4853 \Rightarrow \text{underdamped}$$

4-9 USING MATLAB

MATLAB has several operations in the Symbolic toolbox that allow us to manipulate symbolic polynomials so that we can mimic the steps that would be taken to work a problem with pencil and paper. MATLAB itself has many operations that perform polynomial manipulation on matrices that contain only the polynomial's coefficient. However, the real power of using MATLAB for solving the circuits in this chapter is its ability to solve the simultaneous equations having coefficients that are functions of *s*.

EXAMPLE 4-27

The following set of equations was generated from an electronic circuit using the mesh current method. Find the Laplace current for I_1.

$$(2s + 4)I_1 - (2s)\ I_2 - (4)\ I_3 = \frac{3}{s}$$

$$-(2s)\ I_1 + (3s + 5)I_2 - (5)\ I_3 = 4$$

$$-(4)\ I_1 - (5)\ I_2 + \left(\frac{2}{s} + 9\right)I_3 = 0$$

Solution: From the Symbolic toolbox we will use the solve operation. This operation allows us to enter the simultaneous equations and explicitly list what variables to solve. Listing what variables to solve for is important because we will be using an *s* as part of the coefficients. If we did not indicate which variables to solve for, then MATLAB could easily assume that we need to solve for *s* as well.

Besides allowing the entry of a very long command line, MATLAB allows the line to be broken up into several lines. To continue typing a command on the next line down, three periods in a row at the end of the current line are added. This tells MATLAB that we are going to continue the command on the next line. This is a very convenient feature that we will use to make the entry of the three equations easier. Enter the following in the Command Window.

```
x = solve (...
'(2*s+4)*I1  - 2*s *I2      - 4*I3           = 3/s', ...
'-2*s* I1    + (3*s+5)*I2   - 5*13           = 4', ...
'-4*I1       - 5*I2         + (2/s +9)*I3 = 0', ...
'I1', 'I2',  'I3')
```

Notice that the equations are entered first and the variables to solve for are listed last.

Instead of an answer, MATLAB prints a list of the variables for which it has solved a value. In this case, the results will be stored in the locations x.I1, x.I2, and x.I3. To display the result for I1, enter the following command in the Command Window.

```
pretty(x.I1)
```

Figure 4-64 shows the MATLAB Command Window with the result of this calculation. Notice that the function is not in a very good form because the powers of *s* are out of order and the function is not completely factored. Using MATLAB, we can factor the numerator and denominator to rewrite this answer in a much better form.

```
        x = solve (. . .
        '(2*s+4) *I1 - 2*s *I2    - 4*I3          = 3/s', . . .
        '-2*s* I1    +(3*s+5) *I2 - 5*I3          = 4', . . .
        '-4*I1         - 5* I2     + (2/s +9)*I3 = 0', . . .
        'I1', 'I2',  'I3')
x =

        I1: [1x1 sym]
        I2: [1x1 sym]
        I3: [1x1 sym]
>> pretty (x.I1)
                            3        2
                  24 s + 59 s + 26 s + 10
            3/2 ─────────────────────────────
                        2           3
                  s (12 s + 22 s + 9 s + 20)
```

Figure 4-64   MATLAB Command Window for Example 4-27

First we determine the multiplying constant for our final form by multiplying the beginning multiplier, 3/2, by the coefficient of the highest power of *s* in the numerator, 24, and then dividing this result by the coefficient of the highest power of *s* in the denominator, 9. Therefore, in this case, the final multiplying factor will be 4.

To find the roots of the numerator and denominator, we must first separate the numerator polynomial and denominator polynomial by entering the following command.

$$[n,d] = numden (x.I1);$$

This stores the numerator polynomial into *n* and the denominator polynomial into *d*. We then set the printout for 5 significant digits by entering the following command in the Command Window.

$$format \ short \ e$$

Next, the numerator roots are found by entering the following command.

$$roots \ (sym2poly(n))$$

and the denominator roots are found by entering

$$roots \ (sym2poly(d))$$

Figure 4-65 shows the Command Window after entering the previous four command lines. If these last few commands are going to be used often, then it would be advisable to combine them in an M-file program. From the values calculated in Fig. 4-65, we can construct the answer.

```
>> [n,d] = numden (x.I1);
>> format short e
>>   roots (sym2poly(n))

ans =

-2.0250e+000
-2.1669e-001 +3.9851e-001i
-2.1669e-001 -3.9851e-001i

>> roots (sym2poly (d))

ans =
                                      0
-1.4714e-001 +1.4550e+000i
-1.4714e-001 -1.4550e+000i
-1.0391e+000
```

Figure 4-65   MATLAB Command Window for Example 4-27

$$I_1(s) = \frac{4(s + 2.0250)[(s + 0.21669)^2 + (0.39851)^2]}{s(s + 1.0391)[(s + 0.14714)^2 + (1.4550)^2]}$$

To return MATLAB to its normal printout mode, enter `Format` on the Command line.

The last few steps in the previous example could be written into an M-file to automate the process of restructuring the output format. This problem could be continued by finding the inverse and graphing the result, as shown in previous MATLAB sections.

4-10 USING THE TI-89

The TI-89 is very powerful in its ability to manipulate symbolic functions. It can duplicate almost any calculus or algebraic step required to solve a problem by hand. There are many built-in operations in the TI-89 that manipulate not only polynomials but also polynomial fractions. However, one of the best uses of the TI-89 is in solving simultaneous equations having coefficients that are functions of s.

EXAMPLE 4-28

The following set of equations was generated from an electronic circuit using the mesh current method. Find the Laplace current for I_1.

$$(2s + 4) I_1 \quad - (2s) \ I_2 \quad - (4) \quad I_3 = \frac{3}{s}$$

$$-(2s) \quad I_1 + (3s + 5) I_2 \quad - (5) \quad I_3 = 4$$

$$-(4) \quad I_1 \quad - (5) \ I_2 + \left(\frac{2}{s} + 9\right) I_3 = 0$$

Solution: We will use the simult() operation to solve the simultaneous equations. This operation works for coefficients containing symbolic expressions, such as s, as well as for numeric coefficients. To access this operation, press the following keystrokes.

2nd [MATH] (scroll to 4:Matrix) ENTER (scroll to 5:simult( ) ENTER

Figure 4-66a shows the screen just before the last Enter key is pressed.

Next, a left square bracket is entered followed by each coefficient of the first equation separated by commas. Then a semicolon is entered, and each coefficient of the second equation is entered. The third equation's coefficients are entered in a similar manner, but a right square bracket is entered after the last coefficient instead of a semicolon. Next,

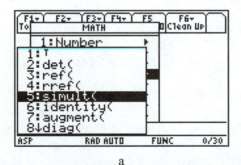

a

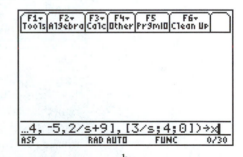

b

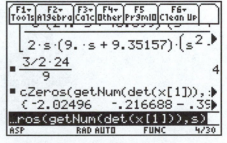

c

d

Figure 4-66

a comma and a left square bracket are entered. Then the equation's constants are entered, but the constants are separated by a semicolon. To end the simult( ) operation, a right square bracket and right parenthesis are entered. At the end of this operation, we may store the result in a variable. For our particular case, the Entry line should be as follows.

$$simult([2s+4,-2s,-4;-2s,3s+5,-5;-4,-5,2/s+9],[3/s;4;0]) \rightarrow x$$

Figure 4-66b shows the screen after the Enter key is pressed.

When the Enter key is pressed, the answer to all three variables will be displayed and stored in x as an array, but we are interested only in first variable I_1. The answer for I_1 can be accessed by specifying the first element in the x array. This is done by using x[1]. The second and third element are accessed by using x[2] and x[3] respectively. At the same time that we access x[1], we can factor it as well by using the factor( ) operation. To do this enter the following on the Entry line.

$$factor(x[1],s)$$

The s at the end of the command specifies that x[1] is factored using s as the variable in the expression. Figure 4-66c shows the screen after the Enter key is pressed. Notice that the answer is not in a very good form. However, we can use some of the calculator's built in functions to come up with a better form.

First we can determine an overall multiplying constant for this fraction. We begin with the current numerator constant, 3, that is shown on calculator's display divided by the denominator constant, 2, also shown on the display. Then we multiply this by the coefficients of the highest power of s in each factor in the numerator and divide this by the coefficients of the highest power of s in each denominator factor. This results in

$$(3/2) * (24)/(9) = 4$$

Next we find the roots of the numerator. To find the roots of I_1, we must remove I_1 from the x array and convert it to an expression. This can be accomplished by using the det( ) operation, which is normally used to find the determinant of an array. However, when a particular element of an array is specified when using this operation, it removes the element and converts the element to an expression. With this expression we then strip out the numerator using the getNum( ) operation, and determine the real and complex roots using the cZeros( ) operation. These can be combined and entered on the Entry line as follows.

$$cZeros(getNum(det(x[1]),s)$$

This returns a list of the roots for the numerator. Figure 4-66d shows the screen after this command is executed. To find the roots of the denominator, we only need to change the command slightly to

$$cZeros(getDenom(det(x[1]),s)$$

With the values from these calculations we can construct the final answer in a much better form.

$$I_1 = \frac{4(s + 2.0250)[(s + 0.21669)^2 + (0.39851)^2)]}{s(s + 1.0391)[(s + 0.14714)^2 + (1.4550)^2]}$$

The last few steps could be written as a function and called from the Entry line to easily put the answer into our desired form. This problem can be continued to find the inverse Laplace transform and to plot a graph, as shown in previous TI-89 sections.

PROBLEMS

Section 4-1

1. Redraw the circuit shown in Fig. 4-67 without switches and show the initial conditions for the reactive component.

2. Redraw the circuit shown in Fig. 4-68 without switches and show the initial conditions for the reactive components.

3. Redraw the circuit shown in Fig. 4-69 without switches and show the initial conditions for the reactive components.

4. Redraw the circuit shown in Fig. 4-70 without switches and show the initial conditions for the reactive components.

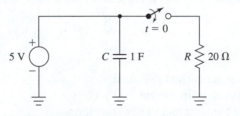

Figure 4-67

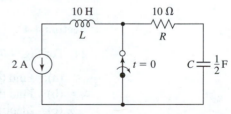

Figure 4-68

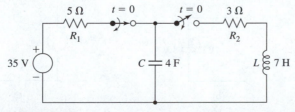

Figure 4-69

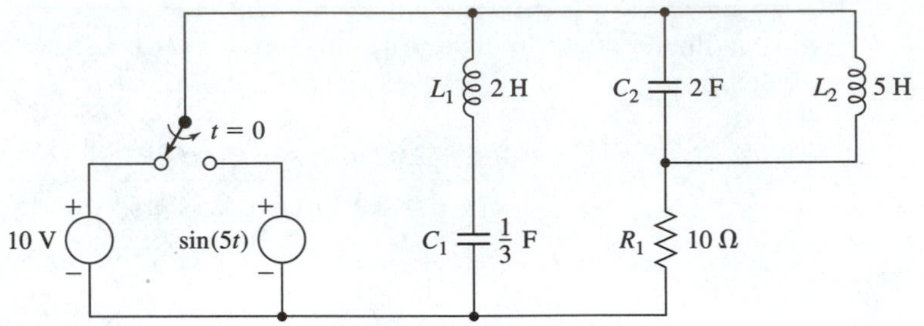

L1 2 H
C2 2 F
L2 5 H
t = 0
10 V
sin(5t)
C1 1/3 F
R1 10 Ω

Figure 4-70

Section 4-2

5. Draw the Laplace Norton equivalent circuit (natural equivalent circuit) for Fig. 4-71a, and then using source conversion, show the Thévenin equivalent form.

6. Draw the Laplace Thévenin equivalent circuit (natural equivalent circuit) for Fig. 4-71b, and then using source conversion, show the Norton equivalent form.

Section 4-3

7. Convert Fig. 4-67 to the Laplace equivalent circuit with the reactive component in its natural equivalent form.

8. Convert Fig. 4-69 to the Laplace equivalent circuit with the reactive components in their natural equivalent form.

9. Convert Fig. 4-70 to the Laplace equivalent circuit with the reactive components in their natural equivalent form.

10. Draw the Laplace equivalent circuit for Fig. 4-72 using:

    (a) All voltage sources.
    (b) All current sources. (Don't forget the applied voltage source.)

Section 4-4

11. For Fig. 4-73:

    (a) Find the Laplace current indicated.
    (b) Find the time-domain current.
    (c) Identify the transient and steady-state terms.

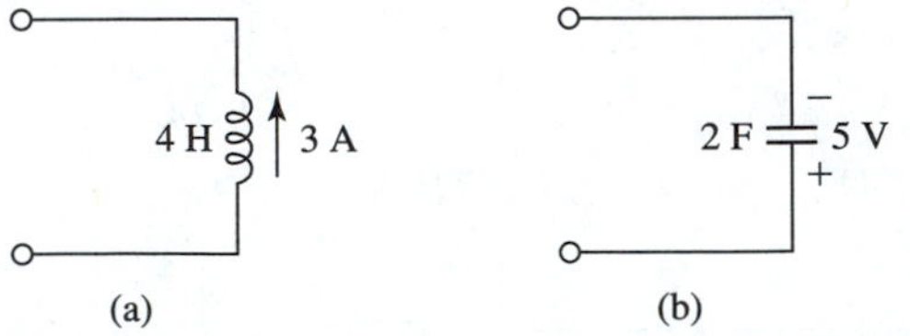

4 H
3 A
(a)
2 F
5 V
(b)

Figure 4-71

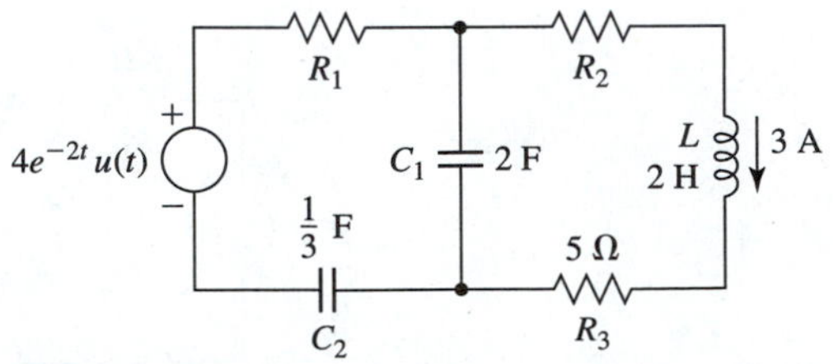

5 Ω
R1
10 Ω
R2
4e^{-2t} u(t)
C1 2 F
L 2 H
3 A
1/3 F
C2
5 Ω
R3

Figure 4-72

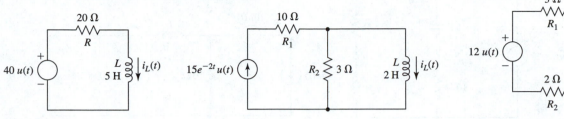

Figure 4-73   **Figure 4-74**   **Figure 4-75**

12. For Fig. 4-74:

    (a) Find the Laplace current indicated.
    (b) Find the inverse Laplace transform of the current.
    (c) Find the steady-state current.

13. For Fig. 4-75:

    (a) Find the Laplace voltage indicated.
    (b) Find the inverse Laplace transform of the voltage.
    (c) Find the steady-state response for this output.

14. For Fig. 4-76:

    (a) Find the Laplace voltage indicated.
    (b) Find the time-domain voltage.
    (c) Find the steady-state voltage.

15. For Fig. 4-77:

    (a) Find the Laplace current indicated.
    (b) Find the time-domain current.

16. For Fig. 4-78:

    (a) Find the Laplace voltage V_{a-b}.
    (b) Find the time-domain voltage.

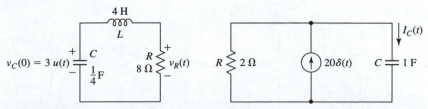

Figure 4-76   **Figure 4-77**

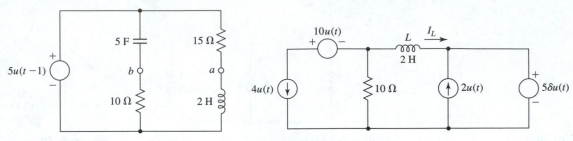

Figure 4-78   **Figure 4-79**

Section 4-5

17. For Fig. 4-79, using superposition, find the time-domain current through the inductor in the direction indicated.

18. Using superposition, find the Laplace voltage as indicated in Fig. 4-80.

19. For Fig. 4-81:

    (a) Write the mesh current equations.
    (b) Remove the fractions from the equations found in part (a) so that the equations are in a form suitable for solving by determinants.

20. Solve the simultaneous equations for $I_1(s)$ and $I_2(s)$

$$(2s + 2)I_1 \qquad - (2s) \quad I_2 = \frac{10}{s + 5}$$

$$-(2s) \quad I_1 + \left(2s + 3 + \frac{1}{s}\right)I_2 = 0$$

21. For Fig. 4-82:

    (a) Use the mesh current method to find the Laplace current $I_R(s)$ and $I_C(s)$.
    (b) Using the results in part (a), find the Laplace current $I_S(s)$.

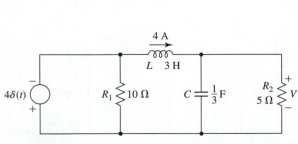

Figure 4-80

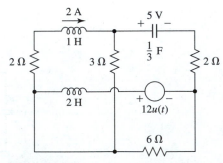

Figure 4-81

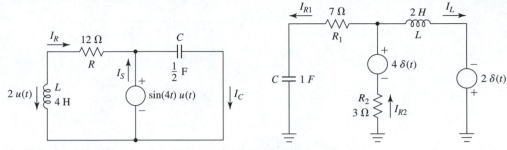

Figure 4-82   **Figure 4-83**

22. For Fig. 4-83:

    (a)   Use the mesh current method to find the Laplace current $I_{R1}(s)$ and $I_L(s)$.

    (b)   Using the results in part (a), find the Laplace current $I_{R2}(s)$.

23. For Fig. 4-84:

    (a)   Write the node voltage equations.

    (b)   Remove the fractions from the equations found in part (a) so that the equations are in a form suitable for solving by determinants.

    (c)   Solve for the voltage V_a

24. For Fig. 4-85:

    (a)   Write the node voltage equations for the circuit.

    (b)   Remove the fractions from the equations found in part (a) so that the equations are in a form suitable for solving by determinants.

    (c)   Solve for the voltage across R_3.

25. For Fig. 4-84:

    (a)   Write the simultaneous equations using the modified node voltage technique described in Section 4-5-3.

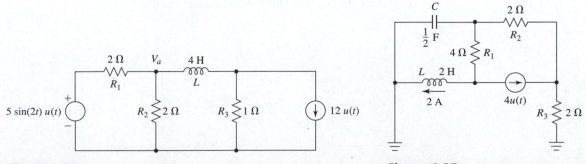

Figure 4-84   **Figure 4-85**

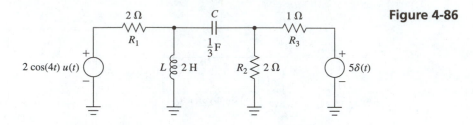

Figure 4-86

(b) Remove the fractions from the equations found in part (a) so that the equations are in a form suitable for solving by determinants.

(c) Solve for the voltage V_a.

26. For Fig. 4-86:

(a) Write the simultaneous equations using the modified node voltage technique described in Section 4-5-3.

(b) Remove the fractions from the equations found in part (a) so that the equations are in a form suitable for solving by determinants.

Section 4-6

27. For Fig. 4-87, convert the VCVS to a

(a) CCVS

(b) VCCS

(c) CCCS

28. For Fig. 4-87, find the current (measure in the downward direction through the capacitor in terms of the input voltage).

29. Use the mesh current method to find the Laplace current through R_2 in Fig. 4-88.

30. For Fig. 4-88, use the node voltage method to find the Laplace voltage across R_2.

Section 4-7

31. For Fig. 4-89:

(a) Find the Laplace Thévenin equivalent impedance for the inductor L_2.

(b) Find the Laplace Thévenin equivalent voltage for the inductor L_2.

(c) Draw the Laplace Thévenin equivalent circuit.

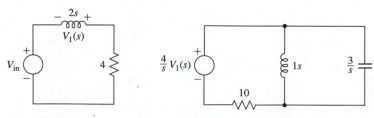

Figure 4-87

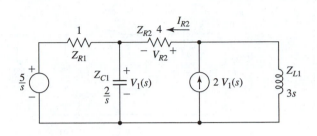

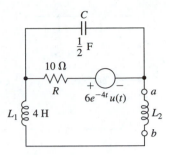

Figure 4-88   **Figure 4-89**

32. For Fig. 4-90:

   (a)   Find the Laplace Thévenin equivalent impedance for the network L_2 and C_2.
   (b)   Find the Laplace Thévenin equivalent voltage for the network L_2 and C_2.
   (c)   Draw the Laplace Thévenin equivalent circuit.

33. For Fig. 4-89:

   (a)   Find the Laplace Norton equivalent impedance for the inductor L_2.
   (b)   Find the Laplace Norton equivalent current for the inductor L_2.
   (c)   Draw the Laplace Norton equivalent circuit.
   (d)   Use source conversion on the Norton equivalent circuit and verify that it is the same as the Thévenin equivalent circuit found in Problem 31.

34. For Fig. 4-90:

   (a)   Find the Laplace Norton equivalent impedance for the network L_2 and C_2.
   (b)   Find the Laplace Norton equivalent current for the network L_2 and C_2.
   (c)   Draw the Laplace Norton equivalent circuit.

Section 4-8

35. Find the order of the circuits in Fig. 4-91.

36. Find the step response for the voltage across R_2 in Fig. 4–92, and find the value of the pole due to the circuit.

37. Find the impulse response for the current through C_1 in Fig. 4-92.

Figure 4-90

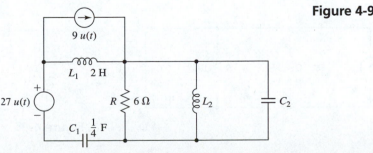

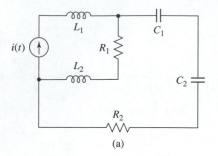

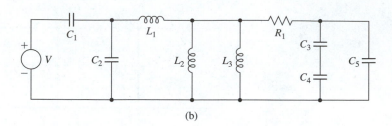

(a)   (b)

Figure 4-91

38. For the following second-order equations, find ζ, ω_n, α, and ω_d.

    (a) $s^2 + 10s + 25$
    (b) $4s^2 + 6s + 36$

39. For the following second-order equations, find ζ, ω_n, α, and ω_d.

    (a) $s^2 + 29s + 100$
    (b) $s^2 + 9$

40. Find the Laplace current in Fig. 4-93. Is the circuit under- , over- , or critically damped?

41. Find the values of C in Fig. 4-94 to cause the voltage, $v(t)$, to be under- , over-, and critically damped.

42. Is the current through R in Fig. 4-95 under- , over- , or critically damped?

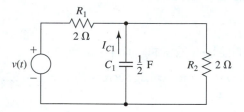

Figure 4-92

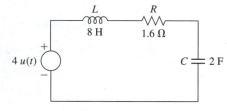

Figure 4-93

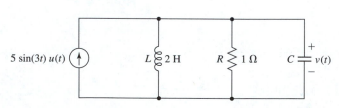

Figure 4-94

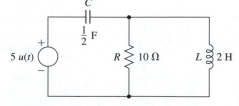

Figure 4-95

Sinusoidal Steady State

OBJECTIVES

Upon successful completion of this chapter, you should be able to:

- Define the concept of a transfer function and identify the general variables $G(s)$, $C(s)$, and $R(s)$ that are used in it.
- Determine the transfer function of an electronic circuit.
- Calculate the step response and the impulse response from a transfer function.
- Interpret stability of a transfer function from its pole–zero plot on an s-plane and the speed of decay for each pole.
- Discuss how a plot of magnitude and phase is used to predict the output response of a complex input signal.
- Calculate the exact magnitude and exact phase shift of an input signal to a circuit at a particular frequency from the transfer function of the circuit.
- Describe the difference between the Bode plot of magnitude and phase and the exact plot of magnitude and phase.
- Draw the Bode plot of frequency and phase given the transfer function.
- Predict the frequency location and the magnitude value of any peaks caused by second-order poles and second-order zeros in a transfer function.
- Calculate the transfer function from a given Bode plot of magnitude and express it in standard form.

5-0 INTRODUCTION

Sinusoidal steady-state analysis starts with the transfer function. For an electronic circuit, the transfer function is the ratio of the output voltage or current to the input voltage or current. There are a number of ways to use the transfer function in analyzing a circuit. We can use it for stability analysis or network analysis, but the most obvious way to use the transfer function is to find the output for a given input. If we find the ratio of the output magnitude and phase to the input magnitude and phase, we can graph the gain (magnitude function) and the phase (phase function) for a range of frequencies. These graphs can be used to determine which frequencies the circuit amplifies and which frequencies it attenuates.

In this chapter, we concentrate on the magnitude function and phase function of the sinusoidal steady-state response. To do this analysis, the system must be linear.

A circuit is linear if it exhibits amplitude linearity and the superposition principle. Amplitude linearity means that if the input signal's amplitude is changed by a multiplying factor, the output signal's amplitude will change by the same multiplying factor. Therefore, if an input voltage $v_{in}(t)$ produces an output voltage of $v_{out}(t)$, then an input of $Kv_{in}(t)$ will produce an output of $Kv_{out}(t)$. The superposition principle means that if an input signal of $v_{1\,in}(t)$ produces an output signal of $v_{1\,out}(t)$ and an input signal of $v_{2\,in}(t)$ produces an output signal of $v_{2\,out}(t)$, then when the input signal is $v_{1\,in}(t) + v_{2\,in}(t)$ the output will be equal to $v_{1\,out}(t) + v_{2\,out}(t)$. For our circuits to exhibit amplitude linearity and the superposition principle, all the components—R, L, and C—must have either fixed values or values that vary in time only.

5-1 TRANSFER FUNCTION

In general, the transfer function is the ratio of the output to the input. The transfer function is not restricted to voltage and current. In control systems it may represent the ratio of the output rpm of a motor to the input voltage, but the transfer function is always the ratio of the output function and the input function.

The ratio of output voltage to input voltage is the voltage gain, but this is a special case of the transfer function and has no units associated with it. In electronics, the input might be current and the output a voltage that causes the transfer function to have units. In this case, it is not a gain. So we should avoid calling the ratio a *gain,* and instead refer to it as a *transfer function.*

In Laplace transform notation, the transfer function is usually denoted by $G(s)$, the output is denoted by $C(s)$, and the input is denoted by $R(s)$. Stated in an equation,

$$G(s) = \frac{C(s)}{R(s)} \tag{5-1}$$

These letter names are not chosen arbitrarily. $G(s)$ stands for gain, $C(s)$ stands for controlled variable or controlled output, and $R(s)$ stands for the reference input.

Once we have the transfer function we can calculate the output for any given input by multiplying the transfer function by the input.

$$C(s) = R(s)\, G(s) \tag{5-2}$$

We must remember that multiplication in the Laplace domain does not correspond to multiplication in the time domain. In the time domain this calculus procedure is called *convolution.*

In a case where we have an unknown circuit, we can apply an input and measure the output. From the output, we can determine the transfer function. Two types of inputs are typically used to do this: the step input and the impulse input. The output is called the *step response* and the *impulse response,* respectively.

Both the step and impulse responses have advantages and disadvantages. The step response is easiest to generate in the lab; however, we must ignore the zero-value pole generated by the input to determine the transfer function. Therefore, the output from the step is the transfer function with an added pole. For the impulse response, the output is the complete transfer function; however, we can only approximate the impulse function in the lab.

An impulse does not add any poles or zeros, so the impulse response is the inverse of the transfer function. To show this, we start with Eq. (5-2):

$$C(s) = R(s)\, G(s)$$

and substitute the impulse input

$$R(s) = \mathcal{L}[\delta(t)] = 1$$

into it:

$$C(s) = 1 \, G(s)$$

and find the inverse

$$c(t) = g(t) \qquad\qquad \textbf{(5-3)}$$

The impulse response is therefore the inverse Laplace of the transfer function.

There are two mathematical methods used to find the transfer function. These two methods are similar to the methods used in Section 4-7 when we were finding the Thévenin impedance for circuits with dependent sources. The only difference is that instead of using V_z/I_z, we will use "output/input," where the "output" or "input" may be voltage or current.

The first method is to find an equation having only the output and input variable and solve for output/input, which is the transfer function. The second method is to find an equation for the output and input in terms of the same variable, divide the output by the input, and cancel out the variable that the output and input are in terms of.

EXAMPLE 5-1

For the circuit in Fig. 5-1, find the transfer function where $V_{in}(s)$ is the input and $V(s)$ is the output. Find the unit step response.

Solution: Let's use the first method. In this circuit we can use the voltage divider rule, which will give us one equation involving only the two variables.

$$V(s) = \frac{V_{in}(s)\dfrac{1}{s}}{1 + \dfrac{1}{s}}$$

Simplifying and solving for output over input gives

$$G(s) = \frac{V(s)}{V_{in}(s)} = \frac{1}{s + 1}$$

To find the step response, we must first find the Laplace transform of a step input, which is

$$R(s) = \mathcal{L}[1 \, u(t)] = \frac{1}{s}$$

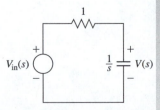

Figure 5-1

Then using Eq. (5-2), we have

$$C(s) = R(s)\, G(s)$$

$$V(s) = \left(\frac{1}{s}\right)\left(\frac{1}{(s+1)}\right)$$

$$= \frac{1}{s(s+1)}$$

Last, we find the inverse Laplace.

$$c(t) = v(t) = 1\, u(t) - e^{-t}\, u(t)$$

EXAMPLE 5-2

For Fig. 5-2, find the transfer function using I as the output and V_{in} as the input. Find the impulse response.

Solution: Let's use the second method to solve this circuit. We can find the input and output in terms of any variable. Let's use I_{in}. First we will find V_{in} in terms of I_{in} using Ohm's law.

$$V_{in}(s) = I_{in}(s)\, Z_T = I_{in}(s)\left(1 + \frac{(s)\left(\dfrac{2}{s}\right)}{\left(s + \dfrac{2}{s}\right)}\right)$$

simplifying this yields

$$V_{in} = \frac{I_{in}\left[(s+1)^2 + 1\right]}{s^2 + 2}$$

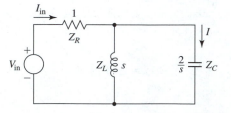

Figure 5-2

Next we need to find I in terms of I_{in}. Using the current divider rule, we find

$$I = \frac{I_{in}\, s}{\left(s + \dfrac{2}{s}\right)}$$

Simplifying this, we have

$$I = \frac{I_{in}\, s^2}{s^2 + 2}$$

Now we divide the output by the input to find the transfer function.

$$\frac{I}{V_{in}} = \frac{\left[\dfrac{I_{in}\, s^2}{s^2 + 2}\right]}{\left[\dfrac{I_{in}\,[(s + 1)^2 + 1]}{s^2 + 2}\right]}$$

Canceling terms, we have

$$G(s) = \frac{I(s)}{V_{in}(s)} = \frac{s^2}{[(s + 1)^2 + 1]}$$

Now that we have the transfer function, we can find the inverse of it to get the impulse response.

$$g(t) = \mathcal{L}^{-1}[G(s)] = \mathcal{L}^{-1}\left[\frac{s^2}{[(s + 1)^2 + 1]}\right]$$

The inverse Laplace transform is therefore

$$g(t) = \delta(t) + 2e^{-t}\sin(t - 90°)$$

5-2 POLE–ZERO PLOT AND STABILITY

Once the transfer function is found, we can plot the poles and zeros on a Cartesian plane. A lot of material has been written for analyzing systems using pole–zero plots. The pole–zero plot is also used extensively in active-filter texts. We will look at an introduction to this approach.

5-2-1 Pole–Zero Plot

The poles and zeros of the transfer function are plotted on the s-plane. The s-plane is a Cartesian plane with the x-axis representing real values (σ-axis) and the y-axis representing imaginary values ($j\omega$-axis). We plot the values of s that make the transfer function zero (zeros) or infinity (poles). Poles are indicated by an $\times$ and zeros are indicated with a $\circ$.

EXAMPLE 5-3

Draw the pole–zero plot for the transfer function shown.

$$G(s) = \frac{15\,(s+2)[(s+1)^2 + 9]}{s^2[(s+4)^2 + 4]}$$

Solution: First we find the roots of the numerator, which are the zeros.

$$s = -2, -1 + j3, -1 - j3$$

These are the ○'s plotted on the graph in Fig. 5-3.

Next we find the roots of the denominator, which are the poles.

$$s = 0, 0, -4 + j2, -4 - j2$$

These are the ✕'s plotted on the graph.

Notice that there is a second-order pole at the origin and that its order is noted by using a superscript of 2. Also notice that the constant multiplier in the numerator of the transfer function does not show up anywhere in the pole–zero plot.

5-2-2 Stability Using Pole–Zero Plot

For an amplifier to work properly, the output must respond to the input. When the output does not, the system is unstable. We can determine the stability from the pole–zero plot. There are three possible cases for stability: stable, unstable, or marginally stable. These can be determined by the output when the system's input is an impulse function.

1. The system is stable if the impulse response goes to zero in a finite amount of time.
2. The system is unstable if the impulse response increases without bound.
3. The system is marginally stable if the impulse response goes to a finite value or moves between finite limits.

Because the poles of a Laplace function determine the form of the time response, the stability depends on the poles and not the zeros. Figure 5-4 shows the three regions: to the left of the $j\omega$-axis, on the $j\omega$-axis, and to the right of the $j\omega$-axis.

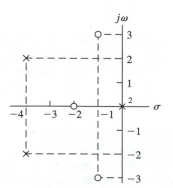

Figure 5-3

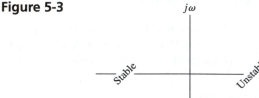

Figure 5-4

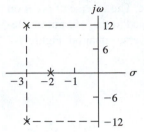

Figure 5-5

The rightmost pole of the transfer function determines the stability of the entire circuit or system. This is similar to a chain being as strong as its weakest link. When all poles of the transfer function are to the left of the $j\omega$-axis, the circuit is stable. When the rightmost pole is single order and on the $j\omega$-axis, the circuit is marginally stable. A multiple-order pole on the $j\omega$-axis makes the circuit unstable. When there are any poles to the right of the $j\omega$-axis, the circuit is unstable.

When we observe the form of the inverse Laplace transform for each type of pole, it becomes obvious why the s-plane is divided into these three regions. The key to understanding why these regions exist is in the exponential multiplier's power. The exponential multiplier part of an inverse Laplace transform controls whether an inverse term will decay or not, and as we move from left to right along the x-axis, we will see three different forms of the exponential multiplier—$e^{-\alpha t}$, e^{0t}, $e^{+\alpha t}$. These three forms of the exponential control the stability of each inverse Laplace transform term.

When real or complex poles are on the left of the $j\omega$-axis, the time response always contains an $e^{-\alpha t}$ multiplier. For example, if the transfer function is

$$G(s) = \frac{N(s)}{(s + 2)[(s + 3)^2 + 144]}$$

then it has all poles to the left of the $j\omega$-axis, as shown in Fig. 5-5. The inverse due to the $(s + 2)$ factor is

$$f_1(t) = Ke^{-2t}u(t)$$

This part of the inverse is transient and will go away in steady state, as shown in Fig. 5-6a. The inverse due to the factor $[(s + 3)^2 + 144]$ is

$$f_2(t) = Ke^{-3t}\sin(12t + \theta)u(t)$$

This is also transient and will go to zero in steady state, as shown in Fig. 5-6b.

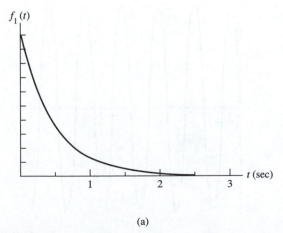

(a)

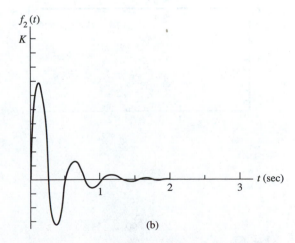

(b)

Figure 5-6

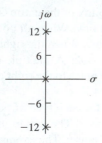

Figure 5-7

Single-order poles on the $j\omega$-axis are called marginally stable. The inverse of poles on the origin may be considered to contain an exponential factor raised to the $-0t$ power. Because this term is equal to 1, we usually do not write e^{-0t} in the inverse equation. Figure 5-7 shows the poles for the transfer function

$$G(s) = \frac{N(s)}{s(s^2 + 144)}$$

The pole due to s gives us a response of

$$f_1(t) = K\,u(t)$$

which is shown in Fig. 5-8a, and the pole due to $(s^2 + 144)$ gives us

$$f_2(t) = K\sin(12t + \theta)\,u(t)$$

which is shown in Fig. 5-8b. This circuit may or may not be stable. If the circuit is supposed to be an oscillator and we have only the complex poles, it will be stable, but it will not be useful as an amplifier. When there is a pole at the origin and if the input is also a step function, the inverse will be

$$f_1(t) = (K_1 t + K_2)\,u(t)$$

which is unstable.

This instability will also occur for multiple-order complex poles on the $j\omega$-axis. Therefore, real or complex poles will be unstable if they appear as a multiple-order pole on the $j\omega$-axis.

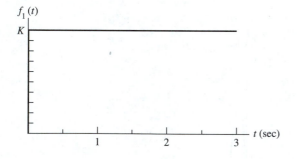

(a)

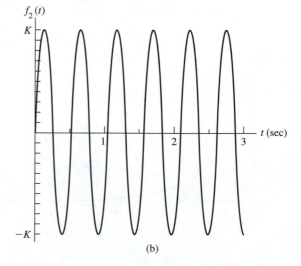

(b)

Figure 5-8

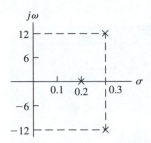

Figure 5-9

Poles on the right of the $j\omega$-axis will make the system unstable. The inverse of these poles will contain the factor $e^{+\alpha t}$, which will cause the output to increase without bound. The transfer function

$$G(s) = \frac{N(s)}{(s - 0.2)[(s - 0.3)^2 + 144]}$$

has all poles on the right of the $j\omega$-axis, as shown in Fig. 5-9. The inverses of the poles due to $(s - 0.2)$ and $[(s - 0.3)^2 + 144]$ are

$$f_1(t) = Ke^{+0.2t} u(t)$$

$$f_2(t) = Ke^{+0.3t} \sin(12t + \theta) u(t)$$

respectively, and are shown in Fig. 5-10. These functions increase without bound and are therefore unstable.

Another point of interest is the speed of decay for the inverse function and its relationship to the $j\omega$-axis. If we start to the far left of the $j\omega$-axis, the function's exponential multiplier will have a very large negative power and will decay very rapidly. The closer we move to the $j\omega$-axis, the slower the function will decay because the exponential multiplier, a, becomes smaller. When we reach the $j\omega$-axis, the function will not decay at all and will become a steady-state value. When we move to the right, the function will increase without bound. The farther to the right of the $j\omega$-axis we go, the faster this increase will become.

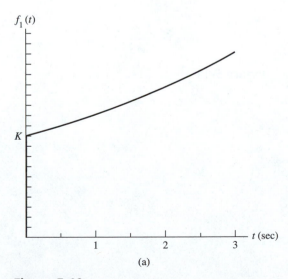

(a)

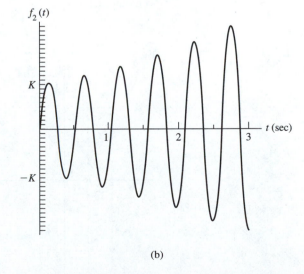

(b)

Figure 5-10

EXAMPLE 5-4

Draw the pole–zero plot and determine the stability of the circuit given the following transfer function.

$$G(s) = \frac{(s - 3)[(s + 1)^2 + 16]}{s(s + 1)^2 [(s + 2)^2 + 9]}$$

Solution: The pole–zero plot is shown in Fig. 5-11. This system is marginally stable because the rightmost pole is on the $j\omega$-axis and is single order. There is a zero in the right half of the s-plane, but because the zeros do not determine the form of the inverse, they do not affect the stability of the system.

5-3 STEADY-STATE FREQUENCY RESPONSE AND THE BODE PLOT

An amplifier or filter will have many different types of input signals. These inputs may be complex and constantly changing. It would be impossible to analyze the output response for each different signal. Fortunately, most signals can be approximated by a combination of sinusoidal components. If we have a graph of how our circuit responds to each sinusoidal frequency and we can write the input signal as a summation of sinusoidal functions, then we can determine the output response of our circuit.

For example, assume that we have the graph of how a circuit's output responds to an input signal. Furthermore, assume that at 100 Hz the graph shows the circuit will produce an output that is 20 times the input signal shifted by 30°, and at 400 Hz the output will be 5 times the input signal shifted by 60°. From this information we can conclude that if we apply a two-tone signal of $v_{in}(t) = 12 \sin(2\pi(100) + 15°) + 12 \sin(2\pi(400) + 15°)$ to the input, we will measure $v_{out}(t) = 240 \sin(2\pi(100) + 45°) + 60 \sin(2\pi(400) + 75°)$ on the output.

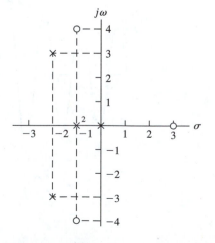

Figure 5-11

Normally, we draw two graphs—the magnitude function and the phase function—on semilog graph paper. (A review of semilog graph paper is given in Appendix D.) There are two popular methods used to draw the graphs. One is the exact plot, which usually requires a computer. The other way is the Bode plot, which is an approximation. We will look at both of these methods.

5-3-1 Converting the Transfer Function to the Time Domain

In the time domain, the magnitude function and phase function are found by the ratio of the ac steady-state output to input. The transfer function is also the ratio of the output to the input as stated in Eq. (5-1), but this is in Laplace and not in the time domain. To convert this Laplace ratio to the time-domain ac steady-state ratio, we simply replace s with $j\omega$.

To understand why this works, we need to look at what we did to the circuit component values when going to Laplace. For the inductor we had

$$L \Rightarrow sL$$

If we replace s by $j\omega$, we have

$$sL \Rightarrow j\omega L$$

The j is a phase shift of 90° and ω is equal to $2\pi f$. We also know that the inductive reactance is $X_L = 2\pi f L$. Therefore, we have

$$L \Rightarrow sL \Rightarrow j\omega L = X_L \underline{/90°} \tag{5-4}$$

With a similar procedure, the capacitor is

$$C \Rightarrow \frac{1}{sC} \Rightarrow \frac{1}{j\omega C} = X_C \underline{/-90°} \tag{5-5}$$

The resistance is the same in each step.

$$R \Rightarrow R \Rightarrow R = R \tag{5-6}$$

When we let s become $j\omega$, we have, in a sense, been doing steady-state ac circuit analysis, but this is true if, and only if, we let $s = j\omega$ in the final step and we assume that the input is a pure sinusoidal function. When we use a Laplace transfer function in this way, we have a function in terms of frequency. By substituting in different frequencies, we can graph the magnitude and phase outputs as frequency changes without doing the steady-state ac circuit analysis at each individual frequency.

The procedure to find the time-domain magnitude function and phase function is as follows:

1. Find the Laplace transfer function.
2. Replace s by $j\omega$ to find the magnitude function, $M(j\omega)$, and the phase function, $P(j\omega)$.

$$M(j\omega) \underline{/P(j\omega)} = G(s)\,|_{s\,=\,j\omega} \tag{5-7}$$

3. Plug in different values of ω. This will result in a numerical magnitude and phase value.
4. On semilog graph paper, plot the dB value of the magnitude, $M_{dB}(j\omega) = 20 \log[M(j\omega)]$, and the phase $P(j\omega)$.

From our experience with the transfer function, we know that we will have either real or complex conjugate roots in the denominator and numerator along with a constant multiplier. Because this is true, we can break up the frequency analysis of the transfer function into

1. A gain constant
2. Real poles
3. Real zeros
4. Complex poles
5. Complex zeros

We will look at these five components separately to understand how they affect the magnitude and the phase angle of a signal. Then we will see how they combine in a transfer function so that we will be able to predict the output response.

To make the analysis easier, we need to rewrite the transfer function into a different form. We want a single numerator constant, K, and all factors having the constant part equal to 1. Don't be disturbed about not understanding why this is easier because this point will not become apparent until we go through Section 5-3-7.

Therefore, real poles and zeros of the form

$$(s + \omega_c)^r \tag{5-8a}$$

become

$$\omega_c^r\left(\frac{s}{\omega_c} + 1\right)^r \tag{5-8b}$$

We use ω_c because this value tells us the critical frequency (also called the *corner* or *break frequency*) of this factor. The ω^r multiplier will then become part of the gain constant K.

The complex conjugate poles and zeros in the form of

$$(s^2 + 2\zeta\omega_c s + \omega_c^2)^r \tag{5-9a}$$

become

$$\omega_c^{2r}\left(\frac{s^2}{\omega_c^2} + \frac{2\zeta}{\omega_c}s + 1\right)^r \tag{5-9b}$$

Here again, ω_c tells us the critical frequency. This value is still ω_n, but we will use ω_c because in this context we are looking at magnitude functions and phase functions. The ω_c^{2r} multiplier will become part of the gain constant K.

EXAMPLE 5-5

Write the following transfer function in a form suitable for frequency analysis.

$$G(s) = \frac{87(s + 2)[(s + 1)^2 + 4]}{(s^2 + 6s + 9)(s^2 + 4s + 29)}$$

Solution: We must first factor everything into the standard form as described in Section 3-1-2 so that we can determine which roots are real, which are complex, and if any factors will cancel out.

$$G(s) = \frac{87(s + 2)[(s + 1)^2 + 4]}{(s + 3)^2[(s + 2)^2 + 25]}$$

Once all the root types are identified and any factors canceled, we put the factors into the forms shown in Eq. (5-8a) and (5-9a).

$$G(s) = \frac{87(s + 2)(s^2 + 2s + 5)}{(s + 3)^2(s^2 + 4s + 29)}$$

The next step is to make each factor's constant part equal to 1 by putting each factor in the form of Eq. (5-8b) or (5-9b).

$$G(s) = \frac{87(2)\left(\dfrac{s}{2} + 1\right)(5)\left(\dfrac{s^2}{5} + \dfrac{2}{5}s + 1\right)}{(3)^2\left(\dfrac{s}{3} + 1\right)^2(29)\left(\dfrac{s^2}{29} + \dfrac{4}{29}s + 1\right)}$$

And finally, we combine all the constant multipliers in the numerator to form the gain constant K.

$$G(s) = \frac{\dfrac{10}{3}\left(\dfrac{s}{2} + 1\right)\left(\dfrac{s^2}{5} + \dfrac{2}{5}s + 1\right)}{\left(\dfrac{s}{3} + 1\right)^2\left(\dfrac{s^2}{29} + \dfrac{4}{29}s + 1\right)}$$

5-3-2 Gain Constant

When we put the transfer function in a form suitable for drawing the gain function and phase function, there will be a constant in the numerator, called the *gain constant,* which will be denoted by the variable K.

$$M_{dB}(j\omega)\ \underline{/P(j\omega)} = G(s)\bigg|_{s=j\omega} = \frac{KN(s)}{D(s)}\bigg|_{s=j\omega}$$

The gain constant can be expressed as a magnitude and phase.

$$M(j\omega) \underline{/P(j\omega)} = |K| \underline{/0°} \quad \text{or} \quad |K| \underline{/-180°} \qquad \textbf{(5-10)}$$

The phase angle will be 0° when K is positive, and the phase angle will be $-180°$ when K is negative. We typically choose $-180°$ instead of using $+180°$.

The magnitude function is almost exclusively a plot on semilog graph paper with the vertical axis in dB. (Appendix D reviews semilog graph paper.) The gain constant, unaffected by changes in frequency, will have a constant dB value given by

$$M_{dB}(j\omega) = 20 \log|K| \qquad \textbf{(5-11)}$$

The magnitude in dB caused by the gain constant will be positive for $|K| > 1$, and the magnitude in dB will be negative for $|K| < 1$. Figure 5-12a shows both cases for $|K|$.

For the phase function, we look at the sign of K instead of the magnitude of K. If K is positive, the phase angle is 0°, and if K is negative, the phase angle is $-180°$. Figure 5-12b shows the two possible cases for the phase angle.

5-3-3 Real Poles

We must separate real poles into nonzero and zero poles. The analysis of these two value types is different. First we will look at nonzero poles.

The nonzero pole in the correct form for frequency analysis using Eq. (5-8b) with $r = 1$ is

$$\frac{1}{\omega_c\left(\dfrac{s}{\omega_c} + 1\right)}$$

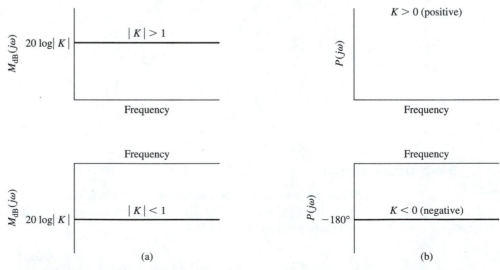

(a) **(b)**

Figure 5-12

The multiplier ω_c is part of the constant, K. Therefore, we will work with

$$\frac{1}{\dfrac{s}{\omega_c} + 1}$$

We can determine the magnitude and phase by letting $s = j\omega$:

$$M(j\omega)\ \underline{/P(j\omega)} = \left(j\frac{\omega}{\omega_c} + 1 \right)^{-1} \tag{5-12}$$

Keep in mind that ω_c is a fixed constant value and ω is the variable we will substitute into with different frequency values. Using rectangular-to-polar conversion, we obtain

$$M(j\omega) = \left[\sqrt{\left(\frac{\omega}{\omega_c}\right)^2 + 1^2} \right]^{-1} \tag{5-13a}$$

$$M_{dB}(j\omega) = -20 \log \sqrt{\left(\frac{\omega}{\omega_c}\right)^2 + 1^2} \tag{5-13b}$$

$$P(j\omega) = \frac{1}{\tan^{-1}\left(\dfrac{\omega/\omega_c}{1}\right)}$$

$$P(j\omega) = -\tan^{-1}\left(\frac{\omega}{\omega_c}\right) \tag{5-14}$$

Substituting values of ω into Eqs. (5-13b) and (5-14) that range from below ω_c to above ω_c, we can plot the graphs shown in Fig. 5-13. In Fig. 5-13a we get approximately 0 dB below ω_c, because in Eq. (5-13b) ω/ω_c becomes much less than 1 and we are essentially finding the logarithm of 1. When ω is above ω_c, in Eq. (5-13b), then ω/ω_c becomes much greater than 1 and we essentially have $-20 \log(\omega/\omega_c)$. When $\omega = \omega_c$, we calculate $-20 \log(2^{1/2})$, which is equal to -3.0103 dB.

Roll-off is the rate or slope of the magnitude as frequency changes. For the magnitude function, when we are above ω_c, we find that the magnitude approaches $-20 \log(\omega/\omega_c)$. The more we increase the frequency, the closer we get to this asymptote. Using this asymptote, the magnitude changes by 20 dB for each change in the frequency by a factor of 10 (which is called a *decade*). Therefore, we have a roll-off rate of -20 dB/decade.

Similarly, we can find significant points on the phase function. In Fig. 5-13b, using Eq. (5-14), we see that when $\omega = \omega_c$, we calculate $-\tan^{-1}(1)$. This value is exactly $-45°$. Once ω is less than $0.1\omega_c$, the phase shift is approximately $0°$ because we are calculating the arctangent of approximately 0. When ω is greater than $10\omega_c$, we are finding the arctangent of a large number, which is approximately $-90°$.

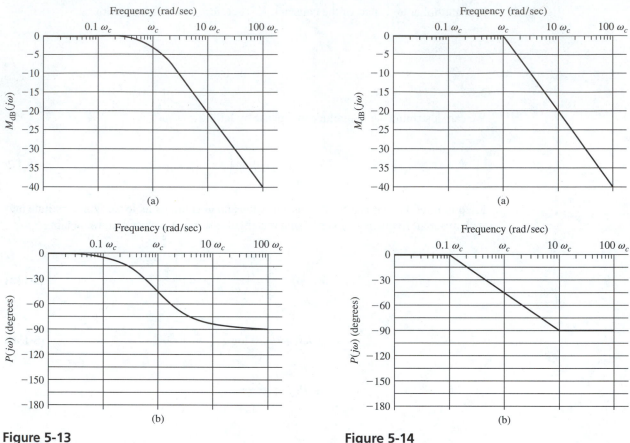

Figure 5-13

Figure 5-14

These observations lead us to the Bode plot, which is an approximation using straight lines. Figure 5-14 shows the Bode plot for the single pole:

$$\frac{1}{\dfrac{s}{\omega_c} + 1}$$

For the Bode magnitude function in Fig. 5-14a, the gain is 0 dB up to and including the break frequency ω_c. At ω_c we must remember that the exact value is -3.0103 dB, even though we draw it at 0 dB. Above ω_c we draw a line with a slope of -20 dB/decade. So we draw the asymptote instead of the exact values.

For the Bode phase function, we assume that there is no phase shift for frequencies less than a decade below the break frequency, and we assume $-90°$ for frequencies greater than a decade above the break frequency. Between a decade below and a decade above the break frequency, we draw a line at a slope of $-45°$ per decade. The only place where this approximation is exactly equal to the exact plot is at the break frequency, where both types of plots are $-45°$.

Figure 5-15 shows both the Bode approximation and the exact plot of the single pole. The frequency plot in Fig. 5-15a shows that the maximum error occurs at ω_c, which is only 3.0103 dB.

Frequency (rad/sec)

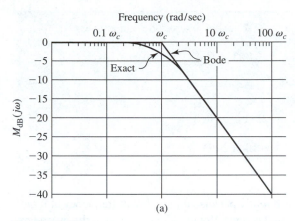

(a)

Frequency (rad/sec)

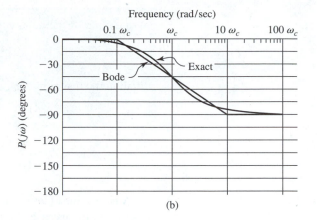

(b)

Figure 5-15

EXAMPLE 5-6

Draw the Bode plot with the frequency scale in hertz. Find the exact value of magnitude and phase at 10 Hz.

$$\frac{1}{\dfrac{s}{20} + 1}$$

Solution: The break frequency is 20 rad/sec. Using

$$\omega = 2\pi f$$

we can solve for the break frequency in hertz.

$$f = \frac{\omega}{2\pi} = \frac{20}{2\pi} = 3.1831 \text{ Hz}$$

We must include at least one decade below 3.1831 Hz (which is 0.31831 Hz) and one decade above (which is 31.831 Hz) in order to show all the phase shift. Figure 5-16 shows the magnitude function and phase function.

To find the exact values at a frequency we substitute $s = j\omega$ into the transfer function. We were given a frequency in hertz, so we must first convert it to rad/sec.

$$\omega = 2\pi f = 2\pi(10) = 62.832 \text{ rad/sec}$$

Substituting into the function, we have

$$\frac{1}{j\dfrac{62.832}{20} + 1}$$

Using polar–rectangular conversions, we solve for the magnitude and phase.

$$M(j\omega) \; \underline{/P(j\omega)} = 0.30331 \; \underline{/-72.343°}$$

Finding the magnitude in dB, we get

$$M_{\text{dB}}(j\omega) = 20 \log(0.30331)$$

$$= -10.362 \text{ dB}$$

A pole value of zero is dealt with differently because it has a break frequency at 0 rad/sec or at dc. In this case, the exact plot and the Bode plot are identical. Using the procedure for nonzero poles, we have

$$M(j\omega) \; \underline{/P(j\omega)} = \left. \frac{1}{s} \right|_{s=j\omega}$$

$$M(j\omega) \; \underline{/P(j\omega)} = (j\omega)^{-1}$$

Splitting this into a gain function and a phase function, we have

$$M(j\omega) = \omega^{-1} \tag{5-15a}$$

$$M_{\text{dB}}(j\omega) = -20 \log \omega \tag{5-15b}$$

$$P(j\omega) = \frac{1}{\tan^{-1}\left(\dfrac{\omega}{0}\right)} = -\tan^{-1} \infty = -90° \tag{5-16}$$

The magnitude function is shown in Fig. 5-17a. Notice that it has a value of 0 dB at 1 rad/sec and the slope is -20 dB/decade.

Figure 5-17b shows the phase function. Because for any value of ω we are finding the arctangent of ∞, we will always calculate $-90°$.

5-3-4 Real Zeros

Real zeros are very similar to real poles so we will not spend as much time on them. We will look at nonzero real zeros first, and then look at zero-value zeros.

The nonzero zero in proper form is

$$\frac{s}{\omega_c} + 1 \tag{5-17}$$

Substituting in $j\omega$ for s gives

$$M(j\omega) \; \underline{/P(j\omega)} = \frac{j\omega}{\omega_c} + 1$$

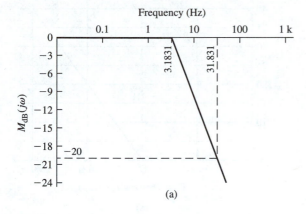

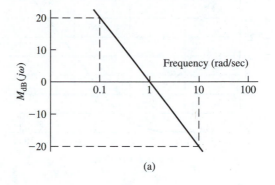

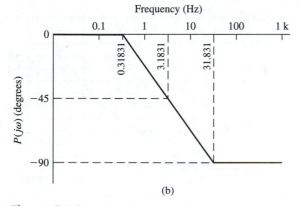

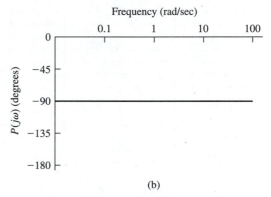

Figure 5-16 **Figure 5-17**

and separating the magnitude and phase using rectangular-to-polar conversion yields

$$M(j\omega) = \sqrt{\left(\frac{\omega}{\omega_c}\right)^2 + 1^2} \tag{5-18a}$$

$$M_{\text{dB}(j\omega)} = 20 \log \sqrt{\left(\frac{\omega}{\omega_c}\right)^2 + 1^2} \tag{5-18b}$$

$$P(j\omega) = \tan^{-1}\left(\frac{\omega}{\omega_c}\right) \tag{5-19}$$

Notice that these are the same equations as real nonzero poles except that the minus sign is missing. The procedure for plotting the exact plot and the Bode plot is the same as for the real nonzero poles. Figure 5-18 shows the exact plot, and Fig. 5-19 shows the Bode plot. Everything is exactly the same except that the slopes for the magnitude function and phase function are positive.

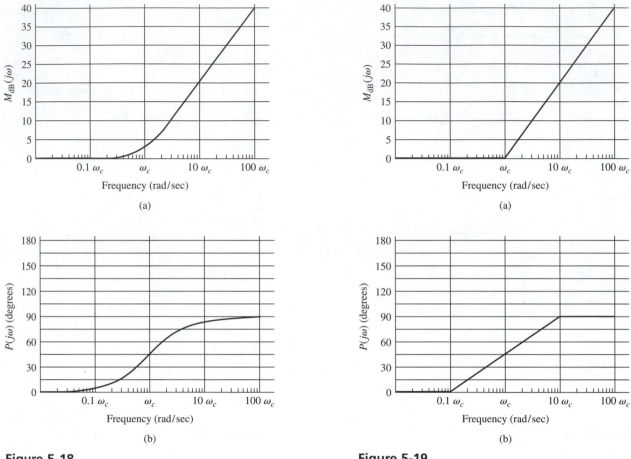

Figure 5-18

Figure 5-19

A real zero-value zero will be exactly like a zero-value pole (break frequency at 0 rad/sec) except that the slope of the magnitude will be positive and the phase shift will be positive. The exact plot and Bode plot for a zero-value zero is the same graph. Following the same approach as for the zero-value pole, the equations are

$$M(j\omega) \underline{/P(j\omega)} = s \mid_{s = j\omega}$$

$$M(j\omega) \underline{/P(j\omega)} = j\omega$$

$$M(j\omega) = \omega \qquad \qquad \text{(5-20a)}$$

$$M_{dB}(j\omega) = 20 \log \omega \qquad \qquad \text{(5-20b)}$$

$$P(j\omega) = \tan^{-1}\left(\frac{\omega}{0}\right) = \tan^{-1}(\infty) = 90° \qquad \qquad \text{(5-21)}$$

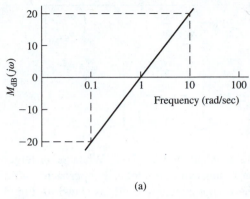

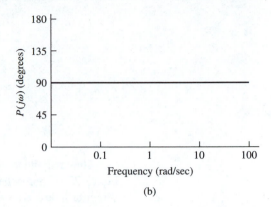

(a)

(b)

Figure 5-20

Notice that these are the same equations as the real zero-value pole except that these equations are positive. The exact and Bode plot are the same graphs and are shown in Fig. 5-20.

5-3-5 Complex Poles

A complex pole, with $r = 1$, in the transfer function will appear as

$$\frac{1}{s^2 + 2\zeta\omega_c s + \omega_c^2}$$

This function must have $\zeta < 1$ since $\zeta \geq 1$ indicates that the quadratic equation can be factored into two real values. From here on in this section we will assume that ζ is less than 1. Arranging the function in the correct form for frequency analysis gives

$$\frac{1}{\omega_c^2\left(\dfrac{s^2}{\omega_c^2} + \dfrac{2\zeta}{\omega_c}s + 1\right)}$$

The ω_c^2 factor will become part of the gain constant. Therefore, we will work with

$$\frac{1}{\dfrac{s^2}{\omega_c^2} + \dfrac{2\zeta}{\omega_c}s + 1}$$

Substituting in $j\omega$ for s, we find the magnitude function and the phase function.

$$M(j\omega) \; \underline{/P(j\omega)} = \frac{1}{\dfrac{(j\omega)^2}{\omega_c^2} + \dfrac{2\zeta}{\omega_c}j\omega + 1} \tag{5-22}$$

$$M(j\omega) = \left\{\sqrt{\left(\frac{2\zeta\omega}{\omega_c}\right)^2 + \left[1 - \left(\frac{\omega}{\omega_c}\right)^2\right]^2}\right\}^{-1} \tag{5-23a}$$

$$M_{dB}(j\omega) = -20 \log \sqrt{\left(\frac{2\zeta\omega}{\omega_c}\right)^2 + \left[1 - \left(\frac{\omega}{\omega_c}\right)^2\right]^2} \tag{5-23b}$$

$$P(j\omega) = -\tan^{-1}\left[\frac{\dfrac{2\zeta\omega}{\omega_c}}{1 - \left(\dfrac{\omega}{\omega_c}\right)^2}\right] \tag{5-24}$$

The magnitude and phase are graphed in Fig. 5-21. When ω is larger than ω_c, Eq. (5-23b) approaches $20 \log(\omega/\omega_c)^2$ and will asymptotically approach -40 dB/decade. When ω is less than ω_c, we will have approximately 0 dB, as shown in Fig. 5-21a. The phase function shown in Fig. 5-21b is similar to real poles except that it goes through 180° phase change.

Figure 5-21 shows a graph where $\zeta = 0.70711$ (exact value is $\sqrt{2}/2$). We can see from Eqs. (5-23) and (5-24) that the function is different for different values of ζ. Figure 5-22 shows the magnitude function and phase function for different values of ζ.

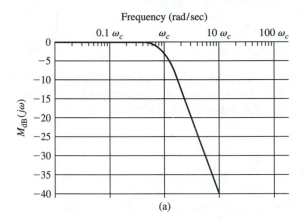

(a)

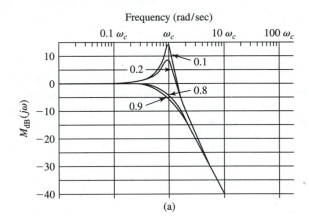

(a)

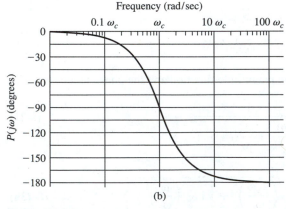

(b)

Figure 5-21

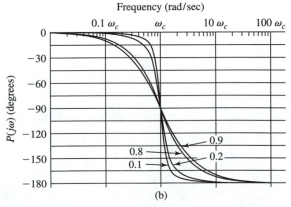

(b)

Figure 5-22

By observation, the magnitude function has a peak for $\zeta < 0.70711$, and the frequency where this peak occurs is slightly less than ω_c. Even though the peak is lower than ω_c, the asymptote will start at ω_c.

We can find at what frequency, ω_{pk}, this peak occurs by setting the derivative of Eq. (5-23a) equal to zero. (Actually, we take the derivative of the function under the radical because when this is at maximum, the square root will also be at maximum.) The result is

$$\omega_{pk} = \omega_c \sqrt{1 - 2\zeta^2} \tag{5-25}$$

This will have real values only for ζ less than 0.70711; for values greater than or equal to 0.70711 there is no peak.

By substituting the value of ω_{pk} from Eq. (5-25) into Eq. (5-23b), the peak value of the magnitude function at this frequency is

$$\Delta M_{dB} = -20 \log \left(2\zeta \sqrt{1 - \zeta^2}\right) \tag{5-26}$$

This equation does not really show the peak value, but it shows the difference between the magnitude value below ω_c and the peak.

Even though the complex poles can give an infinite peak, the Bode plot is based on $\zeta = 0.70711$, as shown in Fig. 5-21. We straighten the curve so that it is 0 dB up to and including ω_c and then follows the asymptote. Figure 5-23 shows the Bode plot. If we need to know what this peak value is, we use Eq. (5-26). If we need to know at what frequency the peak occurs, we use Eq. (5-25). However, when this accuracy is required, we need to use a computer and plot the exact functions.

5-3-6 Complex Zeros

We will find the same parallels between complex poles and complex zeros that we did between real poles and real zeros. A complex zero in the proper form is

$$\frac{s^2}{\omega_c^2} + \frac{2\zeta}{\omega_c} s + 1$$

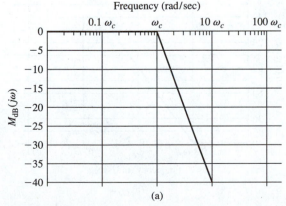
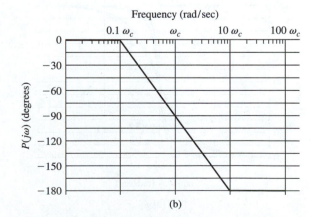

Figure 5-23

This is the reciprocal of a complex pole, and the procedure to find the magnitude function and phase function is exactly the same.

$$M(j\omega) \ \underline{/P(j\omega)} = \frac{(j\omega)^2}{\omega_c^2} + \frac{2\zeta}{\omega_c}j\omega + 1$$

$$M(j\omega) = \sqrt{\left(\frac{2\zeta\omega}{\omega_c}\right)^2 + \left[1 - \left(\frac{\omega}{\omega_c}\right)^2\right]^2} \qquad \textbf{(5-27a)}$$

$$M_{dB}(j\omega) = 20 \log \sqrt{\left(\frac{2\zeta\omega}{\omega_c}\right)^2 + \left[1 - \left(\frac{\omega}{\omega_c}\right)^2\right]^2} \qquad \textbf{(5-27b)}$$

$$P(j\omega) = \tan^{-1}\left[\frac{\dfrac{2\zeta\omega}{\omega_c}}{1 - \left(\dfrac{\omega}{\omega_c}\right)^2}\right] \qquad \textbf{(5-28)}$$

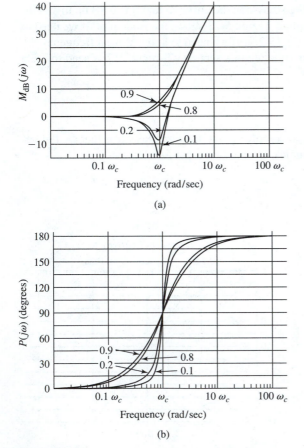

Figure 5-24

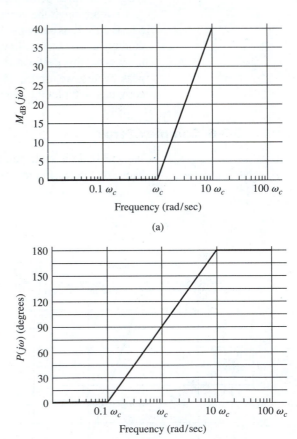

Figure 5-25

Notice that Eqs. (5-27b) and (5-28) are the negative of Eqs. (5-23b) and (5-24).

Figure 5-24 shows the exact plot using different values of ζ. Notice that this is upside down to Fig. 5-22. These are so similar it stands to reason that the peak occurs at the same frequency as the poles.

$$\omega_{\text{pk}} = \omega_c \sqrt{1 - 2\zeta^2} \tag{5-29}$$

As with the complex poles, this has real values only for $\zeta < 0.70711$.

The difference between the function's value at a frequency much less than ω_c and the peak value is given by the negative of Eq. (5-26).

$$\Delta M_{\text{dB}} = 20 \log \left(2\zeta \sqrt{1 - \zeta^2} \right) \tag{5-30}$$

The Bode plot for complex zeros is also based on $\zeta = 0.70711$, as shown in Fig. 5-25.

EXAMPLE 5-7

Draw the Bode plot for the transfer function. What is the peak value, and at what frequency does it occur?

$$\frac{25}{s^2 + 4s + 25}$$

Solution: First we put the function in the proper form for frequency analysis

$$\frac{1}{\dfrac{s^2}{25} + \dfrac{4}{25}s + 1}$$

From this we see that ω_c^2 is 25 and ω_c is 5. The Bode plot is shown in Fig. 5-26.

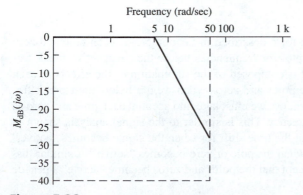
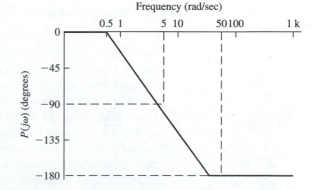

Figure 5-26

Using the center coefficient, we see that

$$\frac{2\zeta}{\omega_c} = \frac{4}{25}$$

Therefore,

$$\zeta = \frac{4\omega_c}{2(25)} = \frac{4(5)}{2(25)} = 0.4$$

Using Eq. (5-26), we obtain

$$\Delta M_{dB} = -20 \log \left(2\zeta \sqrt{1 - \zeta^2}\right)$$

$$= -20 \log \left(2(0.4)\sqrt{1 - (0.4)^2}\right)$$

$$= 2.6954 \text{ dB}$$

In this case, ΔM_{dB} is also equal to the value of the peak magnitude because the function is at 0 dB below the break frequency.

Using Eq. (5-25), the peak occurs at

$$\omega_{pk} = \omega_c \sqrt{1 - 2\zeta^2}$$

$$= 5 \sqrt{1 - 2(0.4)^2}$$

$$= 4.1231 \text{ rad/sec}$$

5-3-7 Combining Poles and Zeros

When the transfer function is in the form described in Section 5-3-1, each pole and zero contributes 0 dB to the magnitude plot for frequencies below their respective break frequency. Because the magnitude plot is composed of the summation of the effects of each pole and zero, we may ignore these poles and zeros when we are below their break frequency. As we move to higher frequencies, we must take into account each pole and zero as it becomes "active" at its break frequency. This is similar to the signal analysis shown in Section 2-8, but instead of looking at the time shift for when the signal becomes "active," we look at the break frequency for when the pole or zero becomes "active." A similar discussion applies to the phase plot, except that the poles and zeros become "active" a decade below the break frequency.

We can summarize how the poles and zeros affect the magnitude and phase graphs as follows.

1. Zeros contribute positive slopes, and poles contribute negative slopes.
2. For the magnitude plot,

   a. Real poles and zeros of the form $\left(\dfrac{s}{\omega_c} + 1\right)$ begin their roll-off at ω_c and have a roll-off rate of ± 20 dB/decade (use $+$ for zeros and $-$ for poles).

   b. Real poles and zeros of the form s are always rolling off at a rate of ± 20 dB/decade (use $+$ for zeros and $-$ for poles) and have a value of 0 dB at 1 rad/sec, or 0.15915 Hz.

   c. Complex poles and zeros of the form $\left(\dfrac{s^2}{\omega_c^2} + \dfrac{2\zeta}{\omega_c}s + 1\right)$ begin their roll-off at ω_c and have a roll-off rate of ± 40 dB/decade (use $+$ for zeros and $-$ for poles). If ζ is less than $\sqrt{2}/2$, then the pole or zero will cause a peak at $\omega_{pk} = \omega_c \sqrt{1 - 2\zeta^2}$ with a magnitude of $\Delta M_{dB} = \pm 20 \log\left(2\zeta\sqrt{1 - \zeta^2}\right)$ (use $+$ for zeros and $-$ for poles).

   d. A gain constant, K, in the numerator will shift the entire magnitude graph by $K_{dB} = 20 \log K$.

3. For the phase angle plot,

   a. Real poles and zeros of the form $\left(\dfrac{s}{\omega_c} + 1\right)$ begin their roll-off at a decade below ω_c and end at a decade above ω_c. Their roll-off rate is $\pm 45°$/decade (use $+$ for zeros and $-$ for poles).

   b. Real poles and zeros of the form s contribute a fixed phase shift of $\pm 90°$ (use $+$ for zeros and $-$ for poles) at all frequencies.

   c. Complex poles and zeros of the form $\left(\dfrac{s^2}{\omega_c^2} + \dfrac{2\zeta}{\omega_c}s + 1\right)$ begin their roll-off at a decade below ω_c and end at a decade above ω_c. Their roll-off rate is $\pm 90°$/decade (use $+$ for zeros and $-$ for poles). When ζ becomes less than $\sqrt{2}/2$ and approaches zero, the points where the roll-off begins and ends move toward the break frequency, causing the roll-off rate to increase.

   d. A $-$ sign in the numerator will shift the entire phase graph by 180°.

The general steps for graphing the magnitude function and phase function are

1. Put the transfer function in the proper form, as described in Section 5-3-1.
2. Find the break frequencies of the poles and zeros.
3. Start at least one decade below the lowest break frequency.
4. Move up in frequency, adding the effects of each break frequency until you are at least one decade above the highest break frequency. For multiple breaks occurring at the same frequency, the magnitude will actually be off by more than 3 dB, but we will ignore this because the Bode plot is an approximation. This type of error will also occur with the phase angles, and we will ignore it as well.
5. For the magnitude plot, find the dB value of the gain constant, K, and shift the graph by that amount.

EXAMPLE 5-8

Draw the magnitude function and the phase function for the transfer function shown.

$$\frac{170,000(s + 10)}{(s + 100)(s + 10000)}$$

Solution: First we put the transfer function in the proper form.

$$\frac{1.7\left(\dfrac{s}{10} + 1\right)}{\left(\dfrac{s}{100} + 1\right)\left(\dfrac{s}{10000} + 1\right)}$$

Next we list the break frequencies (the lowest break frequency first) and the gain constant in dB.

$$\omega_{C1Z} = 10 \text{ rad/sec}$$

$$\omega_{C2P} = 100 \text{ rad/sec}$$

$$\omega_{C3P} = 10000 \text{ rad/sec}$$

$$K_{dB} = 20 \log K$$

$$= 20 \log(1.7) = 4.6090 \text{ dB}$$

Notice the subscripting of the break frequencies (the critical frequencies). The subscript *C1Z* means that the first critical frequency is a zero, and the subscript *C2P* means that the second critical frequency is a pole.

Before drawing the graph, we must determine the appropriate range of frequencies for our graphs. In order to include all the changes that occur, we must start at least one decade below the lowest nonzero break frequency and go to at least one decade above the highest break frequency. In this case, the lowest break frequency is 10 rad/sec and the highest break frequency is 10000 rad/sec. Therefore, we will graph from 1 rad/sec to 100 k rad/sec.

The magnitude plot is now drawn. The left column of Fig. 5-27 shows the step-by-step procedure for drawing the magnitude. Figure 5-27a shows ω_{C1Z} and Fig. 5-27b shows ω_{C2P} drawn separately. These are then added together to form Fig. 5-27c. We then draw the next break frequency, ω_{C3P}, as shown in Fig. 5-27d. This is then added to Fig. 5-27c to form the result of all the break frequencies added together in Fig. 5-27e. Finally, the gain constant is taken into account by adding its value of $k_{dB} = 4.6090$ dB to Fig. 5-27e to give the final magnitude plot shown in Fig. 5-27f.

A similar procedure is followed to construct the phase angle plot shown in the right column of Fig. 5-27. We have to be a little more careful because the phase angle slopes

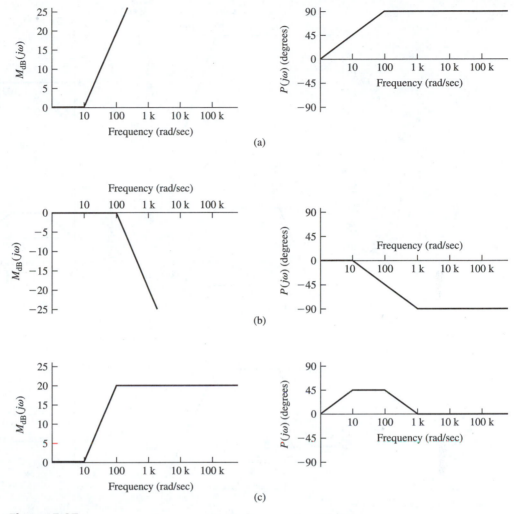

Figure 5-27

start *and* stop. Figure 5-27a shows the phase shift due to ω_{C1Z}, and Fig. 5-27b shows the phase shift due to ω_{C2P}. Figure 5-27c shows these two added together. Notice that when these two roll-offs overlap between 10 rad/sec and 100 rad/sec, they cancel each other effectively, making the slope 0. Then between 100 rad/sec and 1 k rad/sec the graph is pulled down because the phase shift due to ω_{C1Z} has leveled off while ω_{C2P} continues to pull the graph down. After 1 k rad/sec both phase shift slopes level off and therefore so does the composite in Fig. 5-27c. The next step is to add the effect of ω_{C3P}. The phase shift due to ω_{C3P} is shown in Fig. 5-27d, and it is added to the graph in Fig. 5-27c to construct the final graph in Fig. 5-27e. This graph is then copied without modification for Fig. 5-27f so that the final drawings of both the magnitude and phase angle can be seen side-by-side.

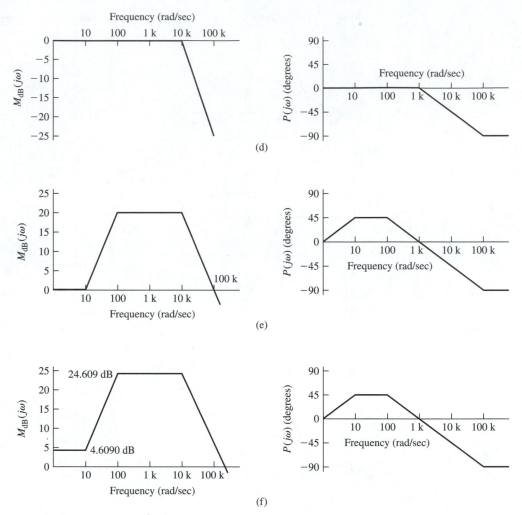

Figure 5-27 *(continued)*

In the last example we see that the slopes add and subtract just like they did in Section 2-8 using ramps. With this in mind, we could have started with Fig. 5-27e and drawn from left to right, adding the effect of each break frequency's slope as it starts and/or stops. Then we would have added in K_{dB} to construct the final drawing in Fig. 5-27f. We really only need to draw individual break frequencies when we are unfamiliar with how they affect the overall graph.

Example 5-8 was relatively simple because the break frequencies occurred only at major divisions on the log-scale. For example, when we went from 10 to 100 rad/sec, as seen in the left column of Fig. 5-27c, it was easy to determine that the magnitude was 20 dB because we were one decade above 10 rad/sec with a roll-off rate of 20 dB/decade. The same was true for the phase angle graph—everything occurred on even decade increments. This is most often not the case.

When the break frequencies are uneven decade increments, we must calculate the magnitude or phase using the equations from Appendix D. In this type of problem, the overall procedure stays the same as shown in Example 5-8, but we must calculate the significant points where new slopes start and stop. Let's look at an example where the break frequencies are not at even decades.

EXAMPLE 5-9

Draw the Bode plot for the transfer function shown. Calculate the crossing of the 0-dB line.

$$\frac{-7.5 \times 10^6 s}{(s + 15)(s^2 + 400s + 1 \times 10^6)}$$

Solution: In the correct form for frequency analysis,

$$\frac{-0.5s}{\left(\dfrac{s}{15} + 1\right)\left[\dfrac{s^2}{(1 \times 10^3)^2} + \dfrac{0.4}{1 \times 10^3}s + 1\right]}$$

Listing the break frequencies and the gain constant in dB, we have

$$\omega_{C1Z} = 0 \text{ rad/sec}$$

$$\omega_{C2P} = 15 \text{ rad/sec}$$

$$\omega_{C3P} = 1 \times 10^3 \text{ rad/sec}$$

$$K_{dB} = 20 \log K$$

$$= 20 \log 0.5 = -6.0206 \text{ dB}$$

Figure 5-28a shows the magnitude function without the gain constant, K_{dB}, taken into account. Construction of this graph begins by determining the range of frequencies. The lowest nonzero break frequency is 15 rad/sec, and a decade below this is 1.5 rad/sec. Therefore, we will begin at the next major division below this, which is 1 rad/sec. The highest break frequency is 1 k rad/sec and, therefore, we will end at 10 k rad/sec.

To draw the magnitude function, we begin with the lowest break frequency; in this case, it will be due to ω_{C1Z}. This will cause a constant roll-off of 20 dB/dec that will pass through 1 rad/sec. The graph will continue this until it reaches 15 rad/sec, where the next break frequency affects the graph. The magnitude at this point is calculated using Eq. (D-2), Eq. (D-4), and Eq. (D-5a). Using Eq. (D-2), we calculate the distance between 1 rad/sec and 15 rad/sec.

$$ND = \log\left(\frac{f_2}{f_1}\right)$$

$$= \log\left(\frac{15}{1}\right) = 1.1761 \text{ dec}$$

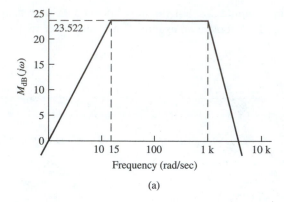

(a)

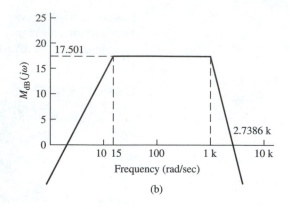

(b)

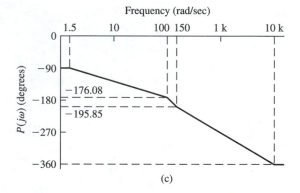

(c)

Figure 5-28

The change in magnitude between these two points using Eq. (D-5a) is then

$$RR = \frac{\Delta A}{ND}$$

$$\Delta A = (ND)(RR)$$

$$= (1.1761 \text{ decades})(20 \text{ dB/dec}) = 23.522 \text{ dB}$$

We can calculate the value this obtains at the point where ω_{C2P} starts by using Eq. (D-4).

$$\Delta A_{dB} = A_{2\,dB} - A_{1\,dB}$$

$$A_{2\,dB} = A_{1\,dB} + \Delta A_{dB}$$

$$= 0 \text{ dB} + 23.522 \text{ dB}$$

$$= 23.522 \text{ dB}$$

The break frequency, ω_{C2P}, that occurs at 15 rad/sec causes a slope of -20 dB/dec, which is equal and opposite to the slope caused by ω_{C1Z}. Therefore, the graph levels off with a slope of 0 dB until the next break frequency, ω_{C3P}, at 1 k rad/sec. Because ω_{C3P} is a complex–conjugate pole, it will contribute a roll-off rate of -40 dB/dec to the current slope of 0 dB/dec. Therefore, the graph will roll off at a rate of -40 dB/dec for frequencies above 1 k rad/sec.

Next the gain constant, $K_{dB} = -6.0206$ dB, is taken into account. We simply add this value to significant points on our graph in Fig. 5-28a. Our only significant point is at 23.522 dB. This point will become 17.501 dB—23.522 dB + $(-6.0206$ dB) = 17.501 dB. The completed graph is shown in Fig. 5-28b.

With the graph shifted to its final location, we can determine the two x-axis crossings. This is just a matter of applying the three equations from Appendix D. We will first find the lower-frequency x-axis crossing. We begin with Eq. (D-4).

$$\Delta A_{dB} = A_{2\,dB} - A_{1\,dB}$$

$$= 17.501 - 0$$

$$= 17.501 \text{ dB}$$

The slope to the left is 20 dB/dec, and the number of decades between the two frequencies can be found using Eq. (D-5a).

$$RR = \frac{\Delta A}{ND}$$

Solving for the number of decades gives

$$ND = \frac{\Delta A}{RR}$$

$$= \frac{17.501 \text{ dB}}{20 \text{ dB/dec}}$$

$$= 0.87506 \text{ decade}$$

Next we find this lower frequency by solving Eq. (D-2) for f_1.

$$ND = \log\frac{f_2}{f_1} \Rightarrow f_1 = \frac{f_2}{10^{ND}}$$

$$f_1 = \frac{15}{10^{0.87506}} = 2 \text{ rad/sec}$$

Keeping in mind that the roll-off rate on the upper frequency side of the graph is -40 db/dec, we follow the same procedure. The change in the dB value is calculated using Eq. (D-4)

$$\Delta A_{dB} = A_{2\,dB} - A_{1\,dB}$$

$$= 0 - 17.501$$

$$= -17.501 \text{ dB}$$

Next we find the number of decades using Eq. (D-5a).

$$\text{RR} = \frac{\Delta A}{\text{ND}}$$

$$\text{ND} = \frac{\Delta A}{\text{RR}}$$

$$= \frac{-17.501 \text{ dB}}{-40 \text{ dB/dec}}$$

$$= 0.43753 \text{ decade}$$

Solving for the high-frequency f_2 in Eq. (D-2) yields

$$\text{ND} = \log\left(\frac{f_2}{f_1}\right) \Rightarrow f_2 = 10^{\text{ND}} f_1$$

$$f_2 = 10^{0.43753} (1 \times 10^3) = 2.7386 \text{ k rad/sec}$$

With the magnitude graph drawn we turn our attention to drawing the graph of the phase shift. The phase function shown in Fig. 5-28b starts at $-90°$ due to the combined effect of the ω_{C1Z} contributing $90°$ and the $-$ sign in the transfer function contributing $\pm 180°$. We will choose to use $-180°$ because this will yield a value closer to $0°$ to begin the graph with.

The break frequency at $\omega_{C2P} = 15$ rad/sec will start affecting the phase graph a decade below, at 1.5 rad/sec. It will cause the graph to roll off at a rate of $-45°$/dec. If there were no other break frequencies, the graph would have leveled off to a value of $-180°$ at 150 rad/sec, but ω_{C3P} affects the graph before this happens.

The break frequency at $\omega_{C3P} = 1$ k rad/sec starts rolling off at 100 rad/sec, and the break frequency at $\omega_{C2P} = 15$ rad/sec does not stop until 150 rad/sec. Therefore, the two overlap between 100 rad/sec and 150 rad/sec. We need to find the value of the graph at 100 rad/sec using Eq. (D-2), Eq. (D-6), and Eq. (D-7a) in Appendix D. Using Eq. (D-2), we calculate the distance between the 1.5 rad/sec and 100 rad/sec.

$$ND = \log\left(\frac{f_2}{f_1}\right)$$

$$= \log\left(\frac{100}{1.5}\right) = 1.8240 \text{ dec}$$

From Eq. (D-7a), we find the change in degrees between the two points.

$$RR = \frac{\Delta D}{ND} \Rightarrow \Delta D = (RR)(ND)$$

$$= (-45°/\text{dec})(1.8240 \text{ dec})$$

$$= -82.076°$$

The value of the phase shift at this point in the graph is then calculated by using Eq. (D-6) and solving for D_2.

$$\Delta D = D_2 - D_1 \Rightarrow D_2 = D_1 + \Delta D$$

$$= -90° + (-82.076°)$$

$$= -172.08°$$

This value is then indicated on the graph at 100 rad/sec.

From 100 rad/sec to 150 rad/sec both slopes combine. The roll-off rate due to ω_{C2P} is $-45°/\text{dec}$ because it is a real pole, and the roll-off rate due to ω_{C3P} is $-90°/\text{dec}$ because it is complex. This gives a total roll-off rate of $-135°/\text{dec}$ in this frequency interval. The procedure for finding the phase angle at 150 rad/sec is the same as was used to find the phase shift at 100 rad/sec except that we use the $-135°/\text{dec}$ roll-off rate. First the number of decades is found between 100 rad/sec and 150 rad/sec using Eq. (D-2).

$$ND = \log\left(\frac{f_2}{f_1}\right)$$

$$= \log\left(\frac{150}{100}\right) = 0.17609 \text{ dec}$$

From Eq. (D-7a) we find the change in degrees between the two points.

$$RR = \frac{\Delta D}{ND} \Rightarrow \Delta D = (RR)(ND)$$

$$= (-135°/\text{dec})(0.17609 \text{ dec})$$

$$= -23.772°$$

The value of the phase shift at this point in the graph is then calculated by using Eq. (D-6) and solving for D_2.

$$\Delta D = D_2 - D_1 \Rightarrow D_2 = D_1 + \Delta D$$

$$= -172.08° + (-23.772°)$$

$$= -195.85°$$

This value is then indicated on the graph at 150 rad/sec.

The phase shift roll-off rate then goes to $-90°$/dec after 150 rad/sec because ω_{C2P} levels off. We can then use the same procedure to find the final value when ω_{C3P} levels off. However, at this point we know that all of the break frequencies have stopped rolling off and that the final value will be equal to the sum of the total phase shifts due to each break frequency. Therefore, we add the total phase shift due to the minus sign in the transfer function, due to ω_{C1Z}, due to ω_{C2P}, and due to ω_{C3P}.

$$\text{Final value} = -180° + 90° - 90° - 180° = -360°$$

EXAMPLE 5-10

For the transfer function shown, draw the Bode plot.

$$\frac{2 \times 10^5(s + 1)}{s(s + 100)(s + 200)}$$

Solution: The correct form is

$$\frac{10\left(\dfrac{s}{1} + 1\right)}{s\left(\dfrac{s}{100} + 1\right)\left(\dfrac{s}{200} + 1\right)}$$

Listing the break frequencies and the dB gain, we have

$$\omega_{C1P} = 0 \text{ rad/sec}$$

$$\omega_{C2Z} = 1 \text{ rad/sec}$$

$$\omega_{C3P} = 100 \text{ rad/sec}$$

$$\omega_{C4P} = 200 \text{ rad/sec}$$

$$K_{dB} = 20 \log K$$

$$= 20 \log 10 = 20 \text{ dB}$$

Because our lowest nonzero break frequency is 1 rad/sec, we will start the graph at 0.1 rad/sec, and because a decade above our highest break frequency is 2 k rad/sec, we will end at 10 k. rad/sec.

Figure 5-29a shows the magnitude without K_{dB} being taken into account. Our lowest break frequency, ω_{C1P}, is always rolling off and passes through 1 rad/sec, but at 1 rad/sec ω_{C2Z} starts, and because it is a numerator factor, a zero, it causes the roll-off rate to go to 0 dB/dec. Then at 100 rad/sec ω_{C3P} adds a roll-off rate of -20 dB/dec. This rate continues for only a short distance because ω_{C4P} becomes active at 200 rad/sec. Here we need to determine the value of the graph using the equations from Appendix D. Starting with Eq. (D-2),

$$ND = \log\left(\frac{f_2}{f_1}\right)$$

$$= \log\left(\frac{200}{100}\right) = 0.30103 \text{ dec}$$

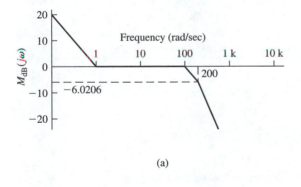

(a)

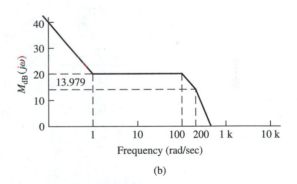

(b)

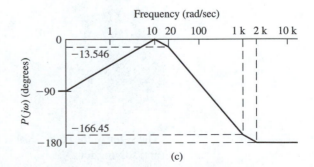

(c)

Figure 5-29

From Eq. (D-5a), we find the change in dB between the two points.

$$RR = \frac{\Delta A_{dB}}{ND} \Rightarrow \Delta A_{dB} = (RR)(ND)$$

$$= (-20 \text{ dB/dec})(0.30103 \text{ dec})$$

$$= -6.0206 \text{ dB}$$

The dB value of this point is then calculated by using Eq. (D-4) and solving for $A_{2\,dB}$.

$$\Delta A_{dB} = A_{2\,dB} - A_{1\,dB} \Rightarrow A_{2\,dB} = A_{1\,dB} + \Delta A_{dB}$$

$$= 0 \text{ dB} + (-6.0206 \text{ dB})$$

$$= -6.0206 \text{ dB}$$

After 200 rad/sec, the roll-off rate becomes -40 dB/dec.

The graph is then shifted up by the 20 dB due to K_{dB}. The final graph is shown in Fig. 5-29b.

The graph of the phase shift is shown in Fig. 5-29c. It starts at $-90°$ because of the break frequency $\omega_{C1P} = 0$ rad/sec. The break frequency of $\omega_{C2Z} = 1$ rad/sec starts affecting the graph at 0.1 rad/sec. Because it is a factor in the numerator, it will add a phase shift rate of $45°$/dec. Just as it completes at 10 rad/sec at a value of $0°$, ω_{C3P} starts affecting the graph by adding a roll-off rate of $-45°$/dec. This roll-off rate continues only until 20 rad/sec, where ω_{C4P} starts affecting the graph. At 20 rad/sec we must use the equation from Appendix D to determine the value. From Eq. (D-2),

$$ND = \log\left(\frac{f_2}{f_1}\right)$$

$$= \log\left(\frac{20}{10}\right) = 0.30103 \text{ dec}$$

From Eq. (D-7a) we find the change in degrees between the two points.

$$RR = \frac{\Delta D}{ND} \Rightarrow \Delta D = (RR)(ND)$$

$$= (-45°/\text{dec})(0.30103 \text{ dec})$$

$$= -13.546°$$

The value of the phase shift at this point in the graph is then calculated by using Eq. (D-6) and solving for D_2.

$$\Delta D = D_2 - D_1 \Rightarrow D_2 = D_1 + \Delta D$$

$$= 0° + (-13.546°)$$

$$= -13.546°$$

At this point, the roll-off rate is $-90°/\text{dec}$ until the break frequency ω_{C3P} ends at 1 k rad/sec. We now need to calculate the value of the graph at 1 k rad/sec. From Eq. (D-2),

$$\text{ND} = \log\left(\frac{f_2}{f_1}\right)$$

$$= \log\left(\frac{1000}{20}\right) = 1.6990 \text{ dec}$$

From Eq. (D-7a) we find the change in degrees between the two points.

$$\text{RR} = \frac{\Delta D}{\text{ND}} \Rightarrow \Delta D = (\text{RR})(\text{ND})$$

$$= (-90°/\text{dec})(1.6990 \text{ dec})$$

$$= -152.91°$$

The value of the phase shift at this point in the graph is then calculated by using Eq. (D-6) and solving for D_2.

$$\Delta D = D_2 - D_1 \Rightarrow D_2 = D_1 + \Delta D$$

$$= -13.546° + (-152.91°)$$

$$= -166.45°$$

After 1 k rad/sec, the slope is back to $-45°/\text{dec}$ until 2 k rad/sec. Here we know that all of the break frequencies have stopped rolling off and that the final value will be equal to the sum of the total phase shifts due to each break frequency. Therefore, we add the total phase shift due to ω_{C1P}, due to ω_{C2Z}, due to ω_{C3P}, and due to ω_{C4P}.

$$\text{Final value} = -90° + 90° - 90° - 90° = -180°$$

When break frequencies are very close to each other, as in the last example, the accuracy of the Bode plot may increase or decrease. This is because a Bode plot is an approximation of the exact plot, with its largest inaccuracy at the break frequencies. If two break frequencies are close to each other and both are due to poles or both are due to zeros, then the Bode plot will be further from the exact plot. However, if two break frequencies that are close to each other are due to a pole and a zero, then the Bode plot will be closer to the exact plot. Therefore, when we observe a Bode plot with two sudden slope changes, if the second slope change increases the magnitude of the slope rate, then this location will be further from the exact plot, but if the second slope change decreases the magnitude of the slope rate, then this location will be closer to the exact plot.

After we have graphed a few functions, it becomes relatively easy to write the transfer function from a given magnitude function. The process can be briefly described as follows:

1. List the break frequencies. Indicate whether it is a pole or zero and the number of roots required.
2. Draw the graph from this list to make sure that it has the correct shape.
3. Determine the amount of shift in dB's between the graph drawn in step 2 and the given graph. This value will be K_{dB}. Then from K_{dB}, solve for K.
4. From this information, write the function in the Bode plot form. Then rearrange the function into the standard form, as described in Sec. 3-1-2.

One problem is where we have a roll-off rate of 40 dB/dec or more. From the exact plot we could measure the peak to determine ζ, but from a Bode plot this is not possible. We will make the following assumption.

RULE: When a roll-off rate exceeds 40 dB/sec, we will use quadratics with $\zeta = \sqrt{2}/2$ and, if required, one real pole or zero to meet the required roll-off rate.

EXAMPLE 5-11

Write the transfer function from the magnitude function shown in Fig. 5-30.

Solution: First we must find all the slope rates on the given graph. This is done by using the equations form Appendix D. For the slope at the beginning we find

$$\Delta A_{dB} = A_{2\,dB} - A_{1\,dB}$$

$$= 10\ dB - 0\ dB = 10\ dB$$

$$ND = \log\left(\frac{f_2}{f_1}\right)$$

$$= \log\left(\frac{0.1}{0.0562}\right) = 0.25026\ dec$$

$$RR = \frac{\Delta A_{dB}}{ND}$$

$$= \frac{(10\ dB)}{(0.25026\ dec)} = 39.958\ dB/dec$$

We can only have roll-off rates in 20 dB/dec increments, so we always round to the nearest 20 dB/dec value. Therefore, in this case we will use 40 dB/dec.

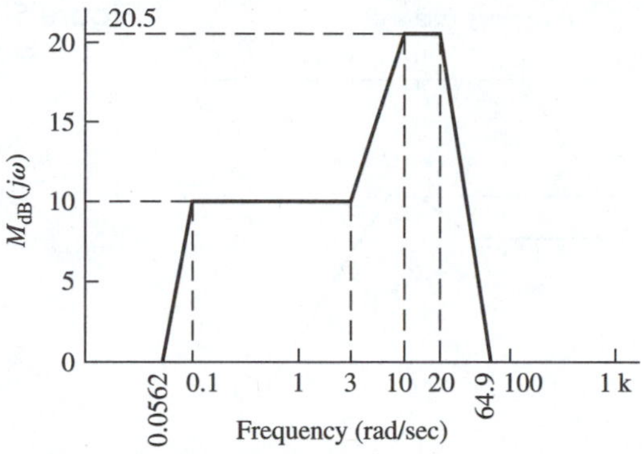

Figure 5-30

Using a similar process, we find that the slope between 3 rad/sec and 10 rad/sec is 20 dB/dec, and the slope between 20 rad/sec and 64.9 rad/sec is -40 dB/dec. These calculations are left to the reader to verify.

From knowing the slopes and the break frequencies, we can determine the poles and zeros that will be needed for this transfer function. The beginning has a slope of 40 dB/dec, so we will need two s factors in the numerator. Then at 0.1 rad/sec we see that the slope goes to 0 dB/dec. Therefore, we will require a quadratic in the denominator with a break frequency at 0.1 rad/sec in order to counteract the 40 dB/dec from the s^2 factor in the numerator. Next, at 3 rad/sec a slope of 20 dB/dec starts. This will require a first-order zero. Then in order to stop the positive slope at 10 rad/sec, a first-order pole will be required. Finally, at 20 rad/sec a quadratic pole will be needed to cause the slope to become -40 dB/dec. This can be summarized in the following list.

$$2 \text{ at } \omega_{C1Z} = 0 \text{ rad/sec}$$

$$2 \text{ at } \omega_{C2P} = 0.1 \text{ rad/sec}$$

$$\omega_{C3Z} = 3$$

$$\omega_{C4P} = 10 \text{ rad/sec}$$

$$2 \text{ at } \omega_{C5P} = 20 \text{ rad/sec}$$

Using this list, we can draw a graph that does not take into account K_{dB}. This graph is shown in Fig. 5-31. Using any significant point in Fig. 5-31 and comparing it to its location in Fig. 5-30, we can determine the amount of shift that is required by K_{dB}. We see that in Fig. 5-31 the area between 0.1 rad/sec and 3 rad/sec is at -40 dB, and in Fig. 5-30 this area is at 10 dB. Therefore, the shift between these points is 50 dB. From this we can solve for the K value by

$$K_{dB} = 20 \log K$$

$$50 = 20 \log K$$

$$K = 10^{(50/20)} = 316.23$$

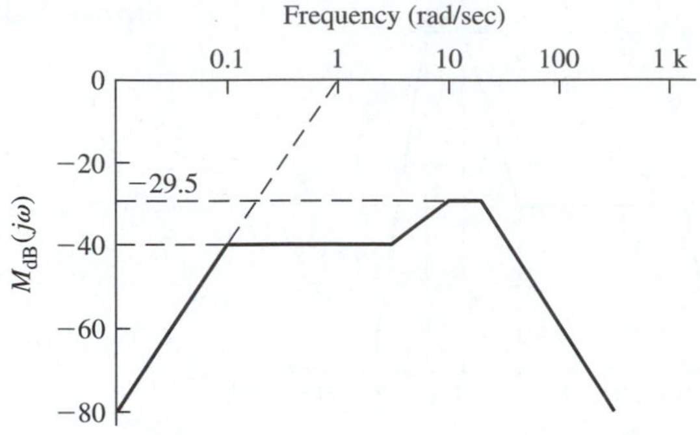

Figure 5-31

From this information, and remembering to use $\zeta = \sqrt{2}/2$ for the quadratics, we can write the transfer function in the form suitable for drawing the Bode plot.

$$\frac{316.23\, s^2 \left(\dfrac{s}{3} + 1\right)}{\left[\dfrac{s^2}{0.1^2} + \dfrac{2(0.70711)}{0.1}s + 1\right]\left(\dfrac{s}{10} + 1\right)\left[\dfrac{s^2}{20^2} + \dfrac{2(0.70711)}{20}s + 1\right]}$$

Next we put this in a proper transfer function form by making the coefficients of the highest power of s in each factor equal to 1.

$$\frac{(0.1^2)(10)(20^2)\left(\dfrac{1}{3}\right)(316.23)s^2(3)\left(\dfrac{s}{3} + 1\right)}{(0.1)^2\left[\dfrac{s^2}{0.1^2} + \dfrac{2(0.70711)}{0.1}s + 1\right](10)\left(\dfrac{s}{10} + 1\right)(20)^2\left[\dfrac{s^2}{20^2} + \dfrac{2(0.70711)}{20}s + 1\right]}$$

Evaluating the constants, the final result is

$$\frac{4216.4s^2(s + 3)}{(s^2 + 0.14142s + 0.01)(s + 10)(s^2 + 28.284s + 400)}$$

5-4 USING MATLAB

MATLAB can be used to calculate the exact magnitude in dB and phase angle at a particular frequency, but more important it can graph the exact magnitude and phase of a function. Example 5-12 in this section demonstrates how to do this using an M-file.

EXAMPLE 5-12

Write an M-file that will draw the exact plot of magnitude and phase for a given transfer function. The magnitude and phase should be drawn in separate Figure Windows and then together in the same Figure Window. Use the following transfer function to test the M-file. Plot the graphs from 0.1 rad/sec to 100 rad/sec.

$$G(s) = \frac{10s}{[(s + 1)^2 + 25]}$$

Solution: Figure 5-32 shows the completed M-file. This file can be entered by using the built-in text editor. To open the editor, select **File -> New -> M-File** from the MAT-LAB desktop. Once the text is entered, it is saved with the filename Frq.M. Let's look at the lines of code.

The first line defines that this is a function and that it is called by using Frq with three elements in its argument—f, Bdec, and Ndec. The next few lines that begin with a % sign are comments that will be printed when the command "help Frq" is entered. This help section defines the arguments and gives an example of how to call the function.

The first command is the echo off command. This makes it so that the commands after this one are not echoed onto the MATLAB Command Window as they are executed.

The next line defines s as a symbol so that the rest of the commands in the M-file know that s is being used as a symbol. Even though we define s as a symbol in the MATLAB Command Window before using this M-file, we must also define it as a symbol in the M-file. This is because when an M-file is executed, it creates its own set of variables that are separate from the variables entered in the Command Window. When the M-file has finished executing all the commands, these variables that were created by the M-file are deleted.

The next command sets up a variable for the number of points to use across a graph. This M-file can be edited to increase this number of points for better resolution on the graph, but this will also increase the computer memory required and slow the execution.

The following group of three command lines calculates the magnitude and phase for the graphs. The first line in this group uses the logspace command to create a vector of frequencies that are equally spaced across a log axis. In the argument of this command the logarithm of Bdec is taken because this command requires the value to be an exponent of 10 rather than the actual value of the frequency. The second command line in this group calculates the magnitude by taking 20 times the log base 10 of the absolute value of the function, f, with j.*w substituted in for s. The last command line of the group calculates the phase by taking the angle of the function, f, with j.*w substituted in for s.

The next group of four command lines causes the magnitude graph to be plotted in figure 1. The commands here are very similar to the commands shown in Section 2-11 for plotting time-domain functions. The big difference is that the `semilogx` command is used instead of the `plot` command. These two commands work exactly alike except that the `semilogx` command makes the *x*-axis logarithmic instead of linear. The `pause` command at the end of this group stops the execution of the M-file and waits for any key on the keyboard to be pressed to continue the execution. This allows users to look at the graph as long as they want before the next Figure Window is shown.

The next group of four command lines plots the phase angle. The `figure(2)` command causes a second Figure Window to open instead of clearing and redrawing over

```
function Frq(f,Bdec,Ndec)
%Frequency Response plot of Magnitude in dB and Phase in degrees
%Frq(f,Bdec,Ndec)
%       where f is a function of the variable s
%              Bdec is the beginning decade
%              Ndec is the number of decades above Bdec to show
%Example:
%      s = sym ('s');
%      f = 10/(s + 10);
%      Frq(f, 0.1, 4)
%

echo off
s = sym('s'); %let the M-file know that s is the variable
NumPoints = 200; %set the number of points to plot across the graph

%Calculate the values of the magnitude in dB and the phase in degrees
w = logspace (log10(Bdec),log10(Bdec)+Ndec,NumPoints); %get the frequencies
m = 20 .* log10 (abs(subs(f,s,j .* w))); %calculate the magnitude in dB
p = angle(subs(f,s,j.*w)).*(180 ./ pi); %calculate the phase angle in degrees

% Draw the Magnitude of the function on a full Figure window
figure (1) %start at figure 1
semilogx(w,m); grid on;
xlabel ('Frequency in rad/sec'); ylabel ('Magnitude in dB'); figure(gcf)
pause%wait until the they are done looking at the graph

% Draw the Phase of the function on a full Figure window
figure(2) %open another figure window
semilogx(w,p); grid on;
xlabel ('Frequency in rad/sec'); ylabel ('Phase in Degrees'); figure(gcf)
pause %wait until the they are done looking at the graph

% Draw the Magnitude and Phase on a split Figure window
figure(3) %open another figure window
subplot(2,1,1) %split figure for two plots
semilogx(w,m); grid on;
xlabel ('Frequency in rad/sec'); ylabel ('Magnitude in dB'); figure(gcf)
subplot(2,1,2) %point to lower location to plot on
semilogx(w,p); grid on;
xlabel ('Frequency in rad/sec'); ylabel ('Phase in Degrees'); figure(gcf)
```

Figure 5-32

MATLAB M-file for Example 5-12

the first Figure Window. In this way, when the M-file has finished, we can examine either the figure with the magnitude or the figure with the phase.

The last group of seven command lines causes the magnitude to be plotted in the top half of a Figure Window and the phase to be plotted in the bottom half of the same Figure Window. The command that makes this possible is the `subplot` command. The first two numbers in the argument of this command set the Figure Window for 2 rows of 1 graph per row, and the last number indicates whether to plot in the upper half, 1, or the lower half, 2, of the Figure Window.

With the M-file entered and saved, we can use it to plot the exact graph of magnitude and phase for any transfer function. To run this M-file for the function given in this example, we enter the following in the Command Window.

```
s = sym('s');
f = 10 * s / ((s + 1)^2 + 25);
Frq (f, 0.1, 3)
```

Figure 5-33 shows the plot of the last Figure Window that has both magnitude and phase together.

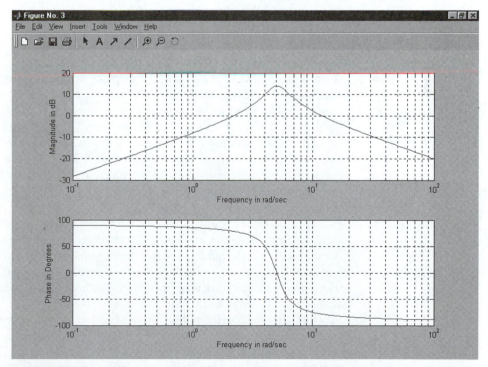

Figure 5-33

The M-file in the previous example can be varied in many ways. It could be broken into two M-files—one just for magnitude and one just for phase. Some of the command lines could be removed so that it shows only the split screen. The number of points to calculate the graph could be incorporated into the argument Frq. The M-file could be written to draw the Bode plot version along with the exact plot.

The use of MATLAB in this chapter can be extended beyond drawing the magnitude and phase graphs. For example, we can use MATLAB to multiply the transfer function by any input signal, find the inverse, and then display the output time-domain signal.

5-5 USING THE TI-89

The TI-89 can easily calculate the exact magnitude in dB and the exact phase angle at a particular frequency, and by programming the calculator, it can graph the exact magnitude and phase of a function. The example in this section demonstrates how to program the TI-89 to draw the exact plot on its screen.

EXAMPLE 5-13

Write a program that will draw the exact plot of magnitude and phase for a given transfer function. Draw the magnitude on the full screen, then draw the phase on the full screen, and then draw both graphs on a split screen. Use the following transfer function to test the program. Plot the graphs from 0.1 rad/sec to 100 rad/sec.

$$G(s) = \frac{10s}{[(s + 1)^2 + 25]}$$

Solution: Figure 5-34 shows the completed program. Although this program can be entered directly into the calculator, it is more practical to use the TI-GRAPH-LINK cable and software. In this way, the program can be typed on a computer keyboard and downloaded into the the calculator. For more information about the TI_GRAPH_LINK, see your guidebook or visit the TI Web site at http://www.ti.com/calc.

The very first line of the program states the elements of the argument that are passed to this program and the names that the elements will be called. The second line indicates that this is a program and not a function. The next few lines are comments that describe what the elements of the argument will be used for. Notice that s must be undefined before this program is called.

The next three lines of code set up the local variables. An advantage of defining the variables in the program as local is that they will be automatically removed from the calculator's memory after the program has finished running. Another advantage of using local variables is that if the variable name is already being used elsewhere, the variable being used outside of the program will not be changed by the program because local variables are stored in a different section in the memory.

Just after the local variables are defined, two constants are defined. The first one is the number of pixels in the x-direction that are available across the calculator's screen, and the second variable is a speed factor. The speed factor is the number of pixels to move across the x-axis screen before calculating a new y-axis value. If this is

```
(Fs,fb,nd)
Prgm
©Frq(Fs,fb,nd)
©Be sure the s does not exist before calling this program.
©Fs must be a function of s
©fb is the beginning frequency in rad/sec
©nd is the number of decades to show

Local graphit, ym, yp, md
Local temp, scnPix, speed, i
Local ymax1, ymax2, ymin1, ymin2
154→scnPix© screen size in pixels
2→speed© speed factor 1=best resolution but uses the most memory and
runs the slowest

©save the current mode settings in md
  getMode("All")→md

©+++++++++++++++++++++++++
©SUBROUTINE - Graph the function y
©+++++++++++++++++++++++++
Define Graphit(y,fb,nd,scnPix,speed,ymx,ymm)=Prgm
 Local xnew,xold,ynew,yold,y,i,i2,offset

©init the graph's range for x and y
 If ymx > 0 then
    1.1*ymx→ymax
 else
    0.9*ymx→ymax
EndIf
If ymn > 0 then
    0.9*ymn→ymin
else
    1.1*ymn→ymin
EndIf

©beginning frequency location on the graph
 log(fb)→xmin

©ending frequency location on the graph
 log(fb) + nd →xmax
 0→xscl ©no tick marks on x-axis
 0→yscl ©no tick marks on y-axis
 ClrDraw ©clear last graph

©init beginning point to draw
 log(fb)→xold
©Get the first element of y. det is used to convert it to an expression
 det(y[2])→yold
```

Figure 5-34

TI-89 program for Example 5-13

```
©draw the line from point to point
 For i,speed,scnPix+speed,speed
  i/speed + 2→offset ©point to next entry
  i→xnew
  log(fb) + (i* nd/scnPix)→xnew
  det(y[offset])→ynew
  Line xold,yold,xnew,ynew ©draw the line
  xnew→xold
  ynew→yold
EndFor

©Mark the frequency axis
.For i, log(fb), log(fb)+nd
  For i2, 1, 10
    lineVert i + log(i2) ©draw a line
  EndFor
EndFor

EndPrgm
©+++++++++++++++++++++++


©+++++++++++++++++++++++
©   Main
©+++++++++++++++++++++++

©init y as a matrix.
©Since this is not a valid entry we will
© not use it when drawing the graph
 [0]→ym
 [0]→yp

©Init top and bottom graph
©Value to display
 0→ymin1
 0→ymax1
 0→ymin2
 0→ymax2

©set the modes for degrees
 setMode ({"Angle","Degree"})

©Calculate Magnitued and Phase
 For i,0,scnPix+speed,speed
 ©sub in s = jw
  fs|s= i *fb * 10^(i* nd/scnPix)→temp
```

Figure 5-34 *(Continued)*

```
©store magnitued in ym and phase in yp
augment(ym;[20*log(abs(temp))])→ym
augment(yp;[angle(temp)])→yp

©locate max and min
©magnitude values in y direction
 If 20*log(abs(temp))>ymax1
  20*log(abs(temp))→ymax1
 If 20*log(abs(temp))<ymin1
  20*log(abs(temp))→ymin1
©Phase angle values in y direction
 If angle(temp) > ymax2
  angle(temp)→ymax2
 If angle(temp) < ymin2
  angle(temp)→ymin2
EndFor

©graph the magnitude on full screen
 Graphit(ym,fb,nd,scnPix,speed,ymax1,ymin1)
©showdisplay until the Enter is pressed
 Pause

©graph the phase angle on full screen
 Graphit(yp,fb,nd,scnPix,speed,ymax2,ymin2)
©show where 180 degrees is.
 LineHorz-180: LineHorz 180
©showdisplay until the Enter is pressed
 Pause

©Show both on a split screen
©set the Mode for a split screen
 setMode({"Graph","Sequence","Split Screen",
 "Top-Bottom","number of graphs","2","Graph2","Sequence",
 "split 1 App","home","split 2 App","Graph"})
©Graph the Magnitude
 Graphit(ym,fb,nd,scnPix,speed,ymax1,ymin1)
 Switch(2)©go to other graph
©Graph the Phase graph
 Graphit(yp,fb,nd,scnPix,speed,ymax2,ymin2)
 LineHorz-180: LineHorz 180
©showdisplay until the Enter is pressed
 Pause

©go back to original mode with Full screen
 setMode(md)
EndPrgm
```

Figure 5-34 *(Continued)*

set to 1, then a new calculation is performed for each pixel we move across the *x*-axis. If this is set to 10, then a new value is calculated for every 10 pixels that we move across the screen. In this way, when we want a general shape quickly, we can set this value to 10 or 20, and then when we want a more accurate shape, we can set this value to 1 or 2.

The next function is the getMode( ) function. This function returns a list of the current mode setting. This list is stored in the variable md. Because several of the mode quantities will be changed in this program, we will want to return the calculator to the mode that it was in before this program was called. The last instruction that is executed in this program restores the mode setting from the list saved in the md variable.

The next several lines of code are a subroutine that does the actual printing on the calculator's screen. This subroutine is called from the main program, which appears later in the code. It will be easier to understand this subroutine if we first examine the main routine, and then discuss this subroutine later.

Moving past the subroutine, we find the beginning of the main routine. The first two instructions in the main routine initialize two matrices—one for storing the magnitude values calculated, ym, and one for storing the phase values, yp. The zero that is stored in them will not be used. The only reason for storing a value in the variables is to create the two matrices. This is because the augment( ) function that must be used later requires that these matrices already exist.

The next four instructions initialize locations for storing the values of the maximum value and minimum value that magnitude and phase will have in the specified frequency range. The actual values used will be found in the following For loop using If instructions.

The setMode( ) function sets the calculator in the degree mode. If the calculator is already in this mode, then this instruction will have no effect. This is done because it is simpler to always set the calculator to the degree mode in the program than it is to remember to put the calculator in this mode before using this program.

The next group of instructions inside the For loop not only calculate the maximum and minimum *y*-values, they also create a matrix of values for the magnitude and phase of the function. The loop counter i increments by pixels across the *x*-axis of the screen. The speed variable determines how many pixels are skipped between calculations. If the speed variable is set to 1, then a value is calculated at each *x*-axis pixel. Just under the For instruction is where the values for both the magnitude and phase are calculated. The function fs is evaluated with the s variable set to $i*fb * 10^\wedge(i* nd/scnPix)$. The first *i* in this expression is the $\sqrt{-1}$, and the second i is the counter from the For instruction. The value calculated is put in a temporary variable, temp. The next two functions pull out the magnitude and the phase angle separately and add them to the matrix ym and yp, respectively, using the augment ( ) function. As discussed previously, the group of If instructions at the end of the For loop determines the maximum and minimum *y*-values in the given frequency range.

The next function executed is the Graphit( ) function. This is a call to the subroutine at the top of this program. Notice that the magnitude matrix is being passed as the first argument. The beginning frequency, number of decades, the speed variable, the maximum value in the *y*-direction, and the minimum value in the *y*-direction are passed in this call. This subroutine plots the calculated values on the calculator's screen.

After Graphit( ) is executed, there is a Pause instruction. This will cause the execution of this program to be suspended until the user presses the Enter key. In this way the user can observe the graph as long as desired.

The phase matrix is then passed to the Graphit( ) function. This will cause the magnitude graph to be erased and the phase graph to be drawn. Just after this function call, there are two LineHorz instructions. These will draw a horizontal line at -180 and $+180$ degrees on the calculator's screen.

The next group of instructions splits the screen and displays the magnitude in the upper portion of the screen and the phase in the lower portion of the screen. To split the screen, the setMode function, having several mode settings specified, is used. Notice how the Graphit( ) function is exactly the same as when we called it for full screen. However, the Switch(2) function needs to be used to direct the output to the bottom screen. If we had needed to go back to the top screen, we would have used the Switch(1) function.

Finally, at the end of the program is the SetMode(md) function. This restores the calculator's mode back to what it was just before this program was executed. The list of parameters for the original mode was stored in the variable md at the beginning of this program.

At this point we can go back to the top of the program and discuss the Graphit( ) subroutine. The first group of If instructions increases the maximum y-direction value and the minimum y-direction value passed to the subroutine by 10%. The values are stored in the variables reserved for the screen parameters of the display. These reserved variables are ymax and ymin for the y-direction and xmax and xmin for the x-direction.

The Line instruction will be used to draw the passed values on the calculator's screen. This instruction plots only linearly, so we must convert our x-values to logarithmic values before plotting, as discussed in Appendix D. Therefore, we must take the log of every x-axis value to determine where it goes on the screen. With this in mind, the beginning x-axis value and the ending x-axis value are stored into xmin and xmax, as shown in the program.

The next three instructions complete the setting of the graph screen parameters. First the tick marks for the x-axis and y-axis are turned off by storing zero in xscl and yscl. Then the graph is cleared using the ClrDraw instruction.

Because the Line instruction draws from one set of x-y points to another set of x-y points, we need to initialize a set of points to start from. This is done in the next two instructions. The first x-value will be at the beginning decade, and the first y-value will be in the y matrix at the second location. We access the element in the second row of the y matrix by using $y[2]$. The second value in the y matrix is used because the first value was put into the matrix to initialize the variable so that the augment( ) function could be used. Therefore, we skip this first value. Notice that we are taking the determinant of $y[2]$. The reason for this is that the Line instruction must have an expression, and when we access one element by using $y[2]$, it returns a 1×1 matrix. To convert this single-element matrix to an expression, we calculate the determinant, and the determinant of a single-element matrix is the value of that one element.

After setting the initial xold and yold values, a For loop is set up to step through the rest of the values. In this loop, new values for x and y are determined and stored into

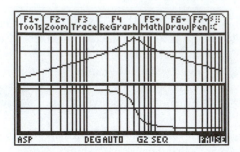

Figure 5-35

xnew and ynew. The line is then drawn between the xold and yold to the xnew and ynew location. Then the current xnew and ynew values are stored in the xold and yold. The program loops back and gets the next values for xnew and ynew. This process continues until the graph is plotted across the calculator's screen.

The final group of instructions in this subroutine prints vertical lines using the LineVert instruction. The loop is set up so that it will draw vertical lines at the locations of 1, 2, 3, 4, 5, 6, 7, 8 and 9 that are between the decades, as seen on regular semilog graph paper.

Even though the program is a little complicated, it is quite simple to use. To plot the given function, we begin by entering the transfer function on the Entry line

$$f = 10*s/((s+1)^\wedge 2+25)$$

and then we enter

$$Frq(f,0.1,3)$$

Figure 5-35 shows the split screen image of both the magnitude and phase.

This program can be varied in several ways. Tick marks along the axis for the dB values and degree values could be added. Elements could be added to the argument of this program allowing the caller to specify the x and y range so that closeup views of the graph could be observed.

There is much more that the TI-89 could be used for in this chapter. For example, a transfer function could be multiplied by an input signal, the inverse found, and this resulting time-domain function graphed.

PROBLEMS

Section 5-1

1. For the circuit in Fig. 5-36:

   (a) Find the transfer function V_L/V_{in}.
   (b) Find the step and impulse response.

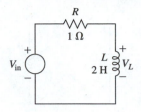

Figure 5-36

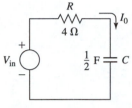

Figure 5-37

2. For the circuit in Fig. 5-37:

   (a) Find the transfer function I_O/V_{in}.
   (b) Find the step and impulse response.

3. For the circuit in Fig. 5-38:

   (a) Find the transfer function V_O/I_{in}.
   (b) Find the step response.

4. For the circuit in Fig. 5-39:

   (a) Find the transfer function V_O/V_{in}.
   (b) Find the impulse response.

5. For the circuit in Fig. 5-40:

   (a) Find the transfer function I_L/V_{in}.
   (b) Using the transfer function, find $I_L(s)$ when $v_{in}(t) = 5e^{-2(t-1)} u(t-1)$.

6. For the circuit in Fig. 5-41:

   (a) Find the transfer function V_L/V_{in}. (*Hint:* Find V_{in} in terms of I_1, and V_L in terms of I_1.)
   (b) Find the impulse response.

Section 5-2

7. For the transfer function shown:

   (a) Draw the pole–zero plot.
   (b) For the poles that are stable, find the transient and steady-state poles.
   (c) Is the system stable, unstable, or marginally stable?
   (d) For the poles that are in the stable region, write them in order of their speed of decay, the fastest-decaying pole first.

$$G(s) = \frac{s + 2}{s(s + 3)[(s + 1)^2 + 36]}$$

8. Use the directions in Problem 7.

$$G(s) = \frac{25 s^2(s - 3)}{(s + 2)(s + 4)^2}$$

Figure 5-38

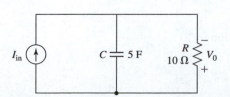

Figure 5-39

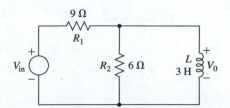

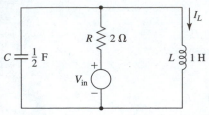

Figure 5-40

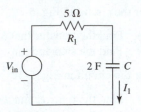

Figure 5-41

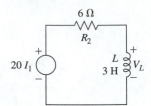

9. Use the directions in Problem 7.

$$G(s) = \frac{8[(s - 2)^2 + 25]}{(s + 1)(s - 3)(s + 4)^2}$$

10. Use the directions in Problem 7.

$$G(s) = \frac{(s - 3)(s + 4)}{(s^2 + 4)(s + 5)}$$

Section 5-3

Section 5-3-1

11. Write the transfer function in a form suitable for frequency analysis.

$$G(s) = \frac{5s^2(s + 10)}{(s + 2)[(s + 3)^2 + 9]}$$

12. For the transfer function shown:

(a) Write it in a form suitable for frequency analysis.
(b) Find the ac steady-state magnitude in dB and phase in degrees at a frequency of 10 rad/sec.

$$G(s) = \frac{15(s + 5)^2}{s^2(s^2 + 5s + 6)}$$

Section 5-3-2

13. Draw the magnitude function and phase function for:

(a) $K = 10$
(b) $K = -10$

14. Draw the magnitude function and phase function for:

    (a) $K = 0.1$
    (b) $K = -0.1$

Section 5-3-3

15. For the transfer function shown:

    (a) Draw the Bode plot.
    (b) Find the exact value of the magnitude in dB and phase in degrees at a frequency of 10 Hz.

$$G(s) = \frac{100}{s + 100}$$

16. Draw the Bode plot with the frequency scale in hertz.

$$G(s) = \frac{1}{s}$$

Section 5-3-4

17. Draw the Bode plot with the frequency scale in hertz.

$$G(s) = 0.1(s + 10)$$

18. For the transfer function shown:

    (a) Draw the Bode plot.
    (b) Calculate the exact magnitude in dB and phase in degrees at 12 rad/sec.

$$G(s) = s$$

Section 5-3-5

19. For the transfer function shown:

    (a) Draw the Bode plot.
    (b) Calculate ΔM_{dB} and the frequency of the peak.

$$G(s) = \frac{400}{s^2 + 16s + 400}$$

20. For the transfer function shown:

    (a) Draw the Bode plot.
    (b) Calculate ΔM_{dB} and the frequency of the peak.

$$G(s) = \frac{225}{s^2 + 225}$$

Section 5-3-6

21. For the transfer function shown:

    (a) Draw the Bode plot.
    (b) Calculate ΔM_{dB} and the frequency of the peak.

    $$G(s) = 0.04(s^2 + 8s + 25)$$

22. For the transfer function shown:

    (a) Draw the Bode plot.
    (b) Calculate ΔM_{dB} and the frequency of the peak.

    $$G(s) = 0.25(s^2 + 4)$$

Section 5-3-7

23. For the transfer function shown, draw the Bode plot when:

    (a) $K = -10$
    (b) $K = 0.1$

    $$G(s) = Ks$$

24. Draw the Bode plot for the transfer functions and calculate the exact value of the magnitude in dB and phase in degrees at a frequency of 20 rad/sec.

    (a) $\dfrac{120}{s + 150}$
    (b) $12.5(s + 0.2)$

25. Draw the Bode plot for the transfer functions and calculate the magnitude at the peaking frequency. (This will be the maximum magnitude in one case and the minimum magnitude in the other case.)

    (a) $\dfrac{5000}{s^2 + 4s + 1000}$
    (b) $7.5 \times 10^{-8}(s^2 + 400s + 4 \times 10^6)$

26. Draw the Bode plot for the transfer function.

    $$G(s) = \frac{-7.168 \times 10^7(s + 1)}{s(s + 800)^2}$$

27. With the frequency scale in hertz:

    (a) Draw the Bode plot.
    (b) Calculate the frequency of any x-axis crossing on the magnitude graph.
    (c) Calculate the exact magnitude in dB and phase in degrees at 500 Hz.

    $$G(s) = \frac{2 \times 10^4 s}{(s + 5)(s + 2000)}$$

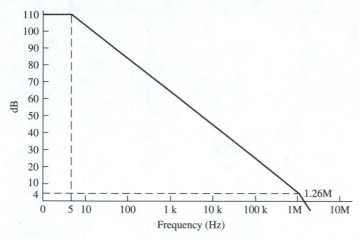

Figure 5-42

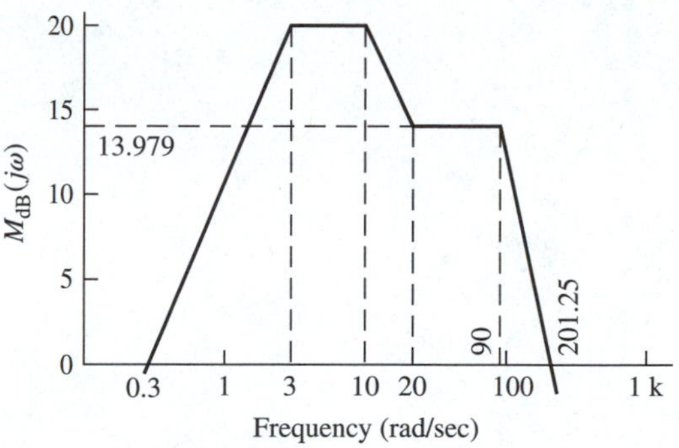

Figure 5-43

28. For the transfer function:

    (a)   Draw the Bode plot.
    (b)   Find the exact value of magnitude in dB and phase in degrees at 0.3 rad/sec.

$$G(s) = \frac{(s + 0.3)^2}{s^2}$$

29. Draw the Bode plot for the transfer function.

$$G(s) = \frac{6.4 \times 10^7 s}{(s + 100)(s^2 + 800s + 6.4 \times 10^5)}$$

30. For the transfer function:

    (a)   Draw the Bode plot.
    (b)   Calculate any x-axis crossing on the magnitude graph.

$$\frac{4 \times 10^{12} s^2}{(s + 1 \times 10^3)(s + 4 \times 10^3)(s + 5 \times 10^4)(s + 2 \times 10^5)}$$

31. Figure 5-42 shows the magnitude plot of an op-amp. Write the transfer function for this op-amp.

32. From the magnitude plot shown in Fig. 5-43, write the transfer function.

6

Introduction to Filters

OBJECTIVES

Upon successful completion of this chapter, you should be able to:

- Define the relationship between a gain function and a loss function.
- Identify whether a signal is amplified or attenuated given the magnitude in either a dB loss value or in a dB gain value.
- Normalize and denormalize a function in both frequency and magnitude.
- Identify and discuss the significance of a filter specification's minimum loss, maximum loss, pass band, transition band, and stop band.
- Draw and define the low-pass, high-pass, band-pass, and Notch filter specifications.
- Define Q and discuss how it affects filter complexity.
- Redefine an unsymmetrical band-pass or notch filter specification into a symmetrical filter specification by either increasing the Q of the filter or modifying the center frequency.
- Define the difference between passive and active filters.
- Write the transfer function for an op-amp given the frequency response characteristics of the op-amp.
- Calculate an op-amp's required open-loop gain, slew rate, and power supply magnitude required to use a simplified model for the op-amp in filter analysis.
- Discuss the types of components that should be used in a filter design for best results.

6-0 INTRODUCTION

In this chapter we define the terminology used when working with filters and also typical assumptions used for the op-amp. This chapter is a transition from using Laplace transforms to applying Laplace transforms. The real power of Laplace can be demonstrated by its application in design work, and in this book we will use active filters.

6-1 FILTER GRAPHS

Filter graphs are graphs of transfer functions. We can plot the response of a filter or we can define what type of filter response that we want. For the type of filters used in this book, the phase is usually ignored. In advanced texts on filters, when phase is of concern, other types of filters are used.

6-1-1 Loss Function

When the transfer function is the ratio of the output voltage to the input voltage, it is called a *gain function,* and when the ratio is the input voltage to the output voltage, it is called a *loss function.* When working with a filter, we typically use the loss function instead of the gain function.

There are two reasons for using the loss function instead of the gain function. First we are concerned with how much a signal is attenuated. Second, because most of the function's response is due to poles, it is easier to work with the poles when they are in the numerator.

From this point on we must be careful about our use of the words *gain* and *loss,* since they do not refer to the signal level. The gain can be either an increase (amplification) or a decrease (attenuation) in signal level. Similarly, the loss function does not mean a decrease in the signal. It means that if the loss function magnitude, in dB, is positive, there is a decrease (attenuation) in signal level, and if the loss function magnitude, in dB, is negative, there is an increase (amplification) in signal level. We must use *attenuation* and *amplification* to describe a decrease or increase in the signal level, and use *gain* and *loss* to indicate which type of function we are using.

As we saw in the last chapter, when we let $s = j\omega$ in the gain function, $G(s)$, we can split the result into a magnitude function, $M(j\omega)$, and a phase function, $P(j\omega)$.

$$M(j\omega) \; \underline{/P(j\omega)} = G(s) \Big|_{s = j\omega} \tag{6-1}$$

In a similar manner, we can define the loss function as $L(s)$, and since the loss function is the reciprocal of the gain function we will also be able to split it into a magnitude function, $ML(j\omega)$, and a phase function, $PL(j\omega)$.

$$ML(j\omega) \; \underline{/PL(j\omega)} = L(s) \Big|_{s = j\omega} = \frac{1}{G(s)} \Big|_{s = j\omega} \tag{6-2}$$

In this book we will be concentrating on the magnitude function and ignore the phase shift when working with filters. The magnitude functions, in dB, for the gain function and loss function are

$$M_{\text{dB}}(j\omega) = 20 \log |G(j\omega)| \tag{6-3}$$

$$ML_{\text{dB}}(j\omega) = 20 \log |L(j\omega)| \tag{6-4}$$

Because the gain function, $G(j\omega)$, and loss function, $L(j\omega)$, are reciprocals,

$$ML_{\text{dB}}(j\omega) = -M_{\text{dB}}(j\omega) \tag{6-5}$$

When looking at a loss function graph we need to redesign our thinking. The larger the positive dB value, the more the signal is attenuated; and the smaller the dB value, the less the signal is attenuated. If the loss is a negative dB value, the signal is amplified. We could think in terms of stepping over the function. The higher the graphs go, the more the effort required to step over, or the more difficult it is for the signal to get through.

6-1-2 Normalized Graphs

The magnitude of the loss function typically has the frequency and the magnitude normalized. A gain function could also be normalized, but because we are concentrating on filters, we will only examine normalizing loss functions. We will start with frequency normalization.

To normalize the frequency,

1. Choose a frequency to normalize to. This is some significant frequency such as a break frequency or a center frequency.
2. Divide all frequencies by this chosen frequency.

When we normalize, the chosen frequency becomes the value 1. Because we are dividing a frequency by a frequency, the frequency axis will be unitless and will usually be indicated by Ω. Frequency normalization expressed in an equation form is

$$\Omega = \frac{\omega}{\omega_x} \tag{6-6a}$$

or expressed with the frequency in hertz:

$$\Omega = \frac{f}{f_x} \tag{6-6b}$$

where Ω = the normalized frequency (unitless)
ω or f = the frequency to be normalized
ω_x or f_x = the frequency we normalize to

EXAMPLE 6-1

Normalize the loss function shown in Fig. 6-1 to 3 krad/sec. What is the frequency of 25 krad/sec on the normalized graph?

Solution: Each frequency is divided by 3 krad/sec using Eq. (6-6) and plotted as shown in Fig. 6-2. Notice that this graph has the same shape as Fig. 6-1 but that the break occurs at 1.

Using Eq. (6-6), we can calculate the normalized value for 25 krad/sec.

$$\Omega = \frac{\omega}{\omega_x}$$

$$= \frac{25 \text{ k}}{3 \text{ k}}$$

$$= 8.333$$

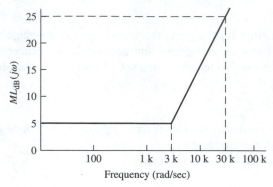

Figure 6-1

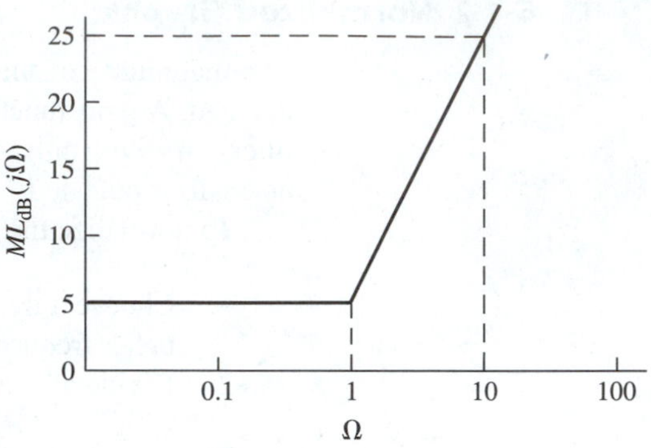

Figure 6-2

From the last example we see that the shape of the graphs is exactly the same. A big advantage of normalized graphs is that they will always have the same shape when denormalized to any frequency. To denormalize a normalized graph,

1.   Choose a frequency to denormalize the normalized frequency to.
2.   Multiply each normalized frequency by this value.

This will change the unitless axis back to an axis with units. We can put this in equation form by solving Eq. (6-6) for ω or f.

$$\omega = \Omega\omega_x \qquad \text{(6-7a)}$$

$$f = \Omega f_x \qquad \text{(6-7b)}$$

where Ω =   the normalized frequency (unitless)
    ω or f =   the denormalized frequency
    ω_x or f_x =   the frequency we are denormalizing to

The frequency ω_x does not have to be the same value as that used in Eq. (6-6). This allows us to shift the function to any frequency we want. Because the new graph must still have major divisions marked with 1×10 to some integer power, the normalized locations on major divisions may or may not be on the major divisions of the denormalized graph.

EXAMPLE 6-2

Denormalize Fig. 6-2 to a frequency of 250 rad/sec. Find the denormalized frequency of 0.13.

Solution: We multiply each frequency by 250 rad/sec using Eq. (6-7). The denormalized graph is shown in Fig. 6-3.
    Using Eq. (6-7) gives

$$\omega = \Omega\omega_x$$

$$= (0.13)(250)$$

$$= 32.5 \text{ rad/sec}$$

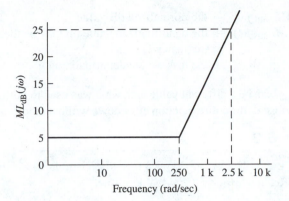

Figure 6-3

A loss function can also have its magnitude normalized. To normalize a loss function's magnitude,

1. Choose a dB value to normalize to. On a loss function this is typically the lowest dB value of the graph.
2. Subtract it from each magnitude value.

We will notice that when this is done, the shape of the graph will remain the same, and the bottom of the graph will be 0 dB. In equation form,

$$ML_{NdB}(j\omega) = ML_{dB}(j\omega) - M_{xdB} \tag{6-8}$$

where $ML_{NdB}(j\omega)$ = the normalized dB value
$ML_{dB}(j\omega)$ = the value to be normalized (ω may be the normalized frequency Ω)
M_{xdB} = the loss we are normalizing to

Substraction of logarithms corresponds to a division of the values, so we could express Eq. (6-8) as

$$ML_N(j\omega) = \frac{ML(j\omega)}{M_x(j\omega)}$$

This equation is the same as Eq. (6-6), thus we see that magnitude normalization is the same as frequency normalization. Therefore, magnitude denormalization is also the same as frequency denormalization.

To denormalize a loss function,

1. Choose a gain to denormalize the normalized gain to.
2. Add this value to each normalized value.

We can state this in equation form by solving Eq. (6-8) for $ML_{dB}(j\omega)$.

$$ML_{dB}(j\omega) = ML_{NdB}(j\omega) + M_{xdB} \tag{6-9}$$

where $ML_{NdB}(j\omega)$ = the normalized dB value

$ML_{dB}(j\omega)$ = the denormalized value (ω may be the normalized frequency Ω)

M_{xdB} = the loss we are denormalizing to

This M_{xdB} can be a different value than what was used to normalize the function in Eq. (6-8). This will shift the entire function up or down without modifying the shape of the curve.

EXAMPLE 6-3

Normalize the 5-dB level in Fig. 6-4 to 0 dB. Then denormalize the function to 3 dB with the frequency denormalized to 4 krad/sec.

Solution: Using Eq. (6-8) at significant points on the graph,

$$ML_{NdB}(j\omega) = ML_{dB}(j\omega) - M_{xdB}$$

$$= 5 - 5 = 0 \text{ dB}$$

$$ML_{NdB}(j\omega) = ML_{dB}(j\omega) - M_{xdB}$$

$$= 25 - 5 = 20 \text{ dB}$$

The function normalized in loss and frequency is shown in Fig. 6-5a.

When denormalizing, we may denormalize the loss and the frequency in any order. We will denormalize the loss first.

To denormalize the loss to 3 dB, we use Eq. (6-9) on the significant points of the normalized graph

$$ML_{dB}(j\omega) = ML_{NdB}(j\omega) + M_{xdB}$$

$$= 0 + 3 = 3 \text{ dB}$$

$$ML_{dB}(j\omega) = ML_{NdB}(j\omega) + M_{xdB}$$

$$= 20 + 3 = 23 \text{ dB}$$

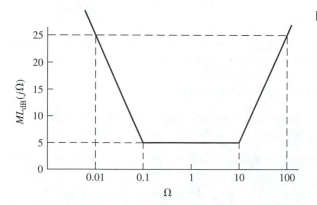

Figure 6-4

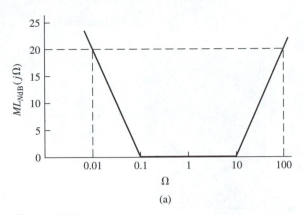

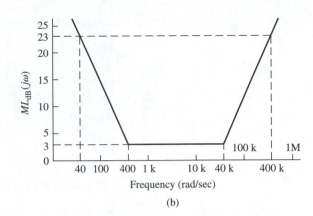

Figure 6-5

> This will shift Fig. 6-5a up by 3 dB.
> We denormalize the frequency using Eq. (6-7) at significant points on the graph.
>
> $$\omega = \Omega\omega_x$$
>
> $$= (10)(4 \text{ krad/sec}) = 40 \text{ krad/sec}$$
>
> $$\omega = \Omega\omega_x$$
>
> $$= (0.1)(4 \text{ krad/sec}) = 400 \text{ rad/sec}$$
>
> This will shift Fig. 6-5a to the right by 4 krad/sec. Figure 6-5b shows the denormalized function.

6-1-3 Graph Specifications for Filters

To design a filter, it must first be specified, which is typically done on a graph. The graph is made up of bands of frequencies with loss areas blocked out. We want certain bands of frequencies passed with a limit on the maximum amount they can be attenuated and a band of frequencies stopped by a minimum amount of attenuation.

An ideal filter would be one where, at one frequency, the loss would go from no attenuation to complete attenuation. Currently, this is impossible. We must therefore have a band of frequencies to give the filter a transition from the pass band to the stop band.

Let's define these terms and associate a variable name to them.

ω_s: the frequency of the edge of a stop band
ω_p: the frequency of the edge of a pass band
Transition band: the band of frequencies between ω_s and ω_p
L_{max}: the normalized maximum dB loss in the pass band
L_{min}: the normalized minimum dB loss in the stop band

Figure 6-6 shows a typical filter specification. A transition band will always exist between a pass-band edge frequency, ω_p, and a stop-band edge, ω_s. The filter to meet

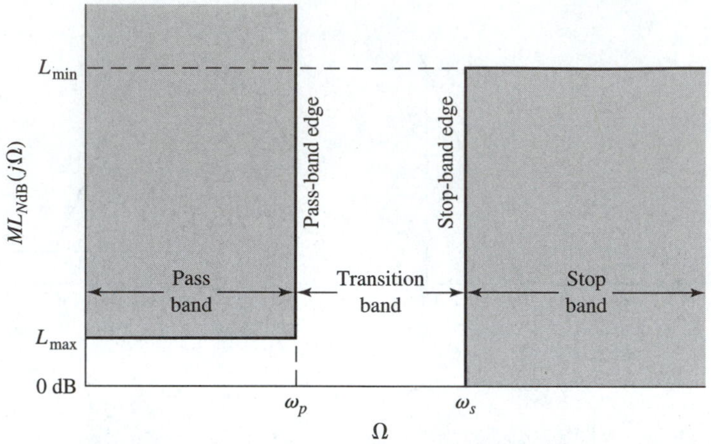

Figure 6-6

this specification must remain under L_{max} up to ω_p, and the filter must be above L_{min} in the stop band.

As ω_p and ω_s get closer, and as the difference between L_{min} and L_{max} gets larger, we approach the ideal filter. The closer we get to the ideal filter, the more complex and expensive the filter becomes.

6-2 FILTER DEFINITIONS

There are many types of filters. We will concentrate on a few. These are the basic building blocks by which more complex filters are defined.

6-2-1 Types of Filters: LP, HP, BP, Notch

We will start with the definition and specification of the four basic filter types.

1. *Low-pass (LP)*. A low-pass filter passes frequencies below $\dot{\omega}_p$ and stops frequencies above ω_s (see Fig. 6-7a).
2. *High-pass (HP)*. A high-pass filter passes frequencies above ω_p and stops frequencies below ω_s (see Fig. 6-7b).
3. *Band-pass (BP)*. A band-pass filter passes frequencies between ω_{p1} and ω_{p2} and stops frequencies above ω_{s2} and below ω_{s1} (see Fig. 6-7c).
4. *Notch*. A notch filter passes frequencies below ω_{p1} and above ω_{p2} and stops frequencies between ω_{s1} and ω_{s2} (see Fig. 6-7d).

The graphs in Fig. 6-7 are shown with the magnitude normalized because L_{max} and L_{min} are always normalized loss values. The frequency axis, however, was not normalized. The values of ω_p and ω_s may also appear as normalized frequencies, Ω_p and Ω_s.

There are many applications possible for each filter type. We can also combine the filters. An application for combining the filters is an equalizer for music, as shown in Fig. 6-8. The music is input to several band-pass filters designed to pass a different band of frequencies. The output is applied to a summing amplifier to control the gain of each filter.

Band-pass and notch filters may be specified unsymmetrical. For this book we will force these to be symmetrical. This will be done by increasing the requirements of the fil-

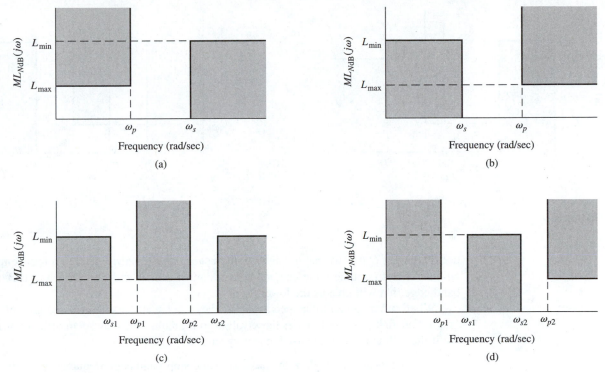

Figure 6-7

ter. The object is to revise the requirements to meet or exceed the original specification so that the filter is symmetrical.

Figure 6-9a shows a band-pass filter that is unsymmetrical. First we will increase L_{min2}'s value to L_{min1}'s value, making both stop bands equal. Next we compare the number of decades between ω_{s1} and ω_{p1} to the number of decades between ω_{p2} and ω_{s2}. If they are not equal, we make them both equal to the smaller value. In this case we need to decrease ω_{p1}, which will lower the center frequency. We could have increased ω_{s1} instead of decreasing ω_{p1}, but this would also increase the Q of the circuit. Q stands for "Quality factor" and it is the ratio of the center frequency to the bandwidth. Normally, we prefer to

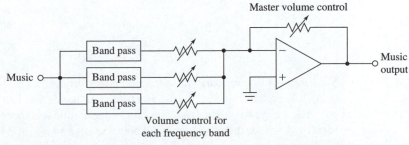

Figure 6-8

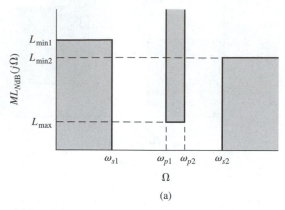

(a)

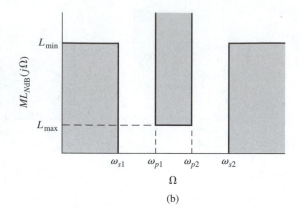

(b)

Figure 6-9

use the lowest Q that we can get away with because it will usually lead to a less complex circuit. Therefore, unless the center frequency must remain the same, we will move the band edge that will give us the lower Q.

When we are given filter specifications, they are usually just values and are not on a graph. Our first job is to transfer these values to a graph. This may seem difficult at first, but if the following steps are used, it is very simple:

1. Locate and mark all of the pass-band and stop-band edge frequencies.
2. Identify the transition bands, which always appear between a stop band and a pass band.
3. Identify the pass bands and the stop bands. An ω_p frequency is the dividing frequency between a transition band and a pass band, and an ω_s frequency is the dividing frequency between a transition band and a stop band. Because in the previous step the transition bands were identified on one side of ω_p and ω_s, the other side of ω_p and ω_s can now be identified as a pass band and stop band region, respectively.
4. Draw the blocks.

EXAMPLE 6-4

Draw the following specification, and identify the type of filter.

$$\omega_{s1} = 50 \text{ rad/sec} \qquad \omega_{s2} = 200 \text{ rad/sec}$$

$$\omega_{p1} = 10 \text{ rad/sec} \qquad \omega_{p2} = 2 \text{ krad/sec}$$

$$L_{min} = 40 \text{ dB} \qquad L_{max1} = 2 \text{ dB} \qquad L_{max2} = 5 \text{ dB}$$

Solution: First we mark the stop-band edge frequencies and the pass-band edge frequencies, as shown in Fig. 6-10a. The transition bands are then identified between each pass-band edge frequency and stop-band edge frequency. Once the transition bands are identified, we see there is a stop band between ω_{s1} and ω_{s2}, and there are two pass bands.

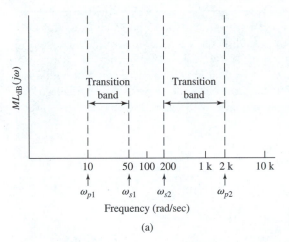

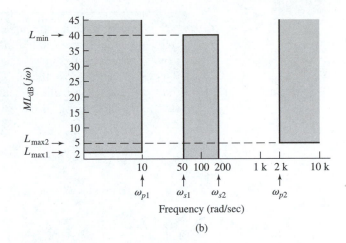

Figure 6-10

Next, the magnitude levels of L_{max} and L_{min} are drawn in the pass bands and stop bands. Keep in mind that when L_{max} or L_{min} also have a numeric subscript, they are associated with the respective ω_p and ω_s having the same numeric subscript. However, if an L_{max} or L_{min} does not have a numeric subscript, it will be associated with all the respective ω_p or ω_s whether these frequencies have a numeric subscript or not.

With everything located, the blocks can be marked and completed, as shown in Fig. 6-10b.

EXAMPLE 6-5

Revise the specifications shown in Fig. 6-10b to make the notch filter symmetrical without changing the center frequency.

Solution: First we move L_{max2} so that it is equal to L_{max1}, because if we meet the L_{max1} specification, then, we will exceed the specification of L_{max2}.

Next we calculate the width, in decades, of the two transition bands. Using Eq. (D-2) from Appendix D, we calculate the lower transition band as

$$\text{ND} = \log\left(\frac{f_2}{f_1}\right) = \log\left(\frac{\omega_2}{\omega_1}\right)$$

$$= \log\left(\frac{50}{10}\right) = 0.69897 \text{ decade}$$

and the upper transition band as

$$= \log\left(\frac{2\,\text{k}}{200}\right) = 1 \text{ decade}$$

We need to reduce the higher-frequency transition band to be equal to the lower-frequency transition band. Normally, we would increase ω_{s2} but this would change the center frequency. We must, therefore, shift ω_{p2} to a lower frequency so that it is 0.69897 decades above the ω_{s2}. This will give us a transition band equal in width to the lower transition band. We can calculate the location to move ω_{p2} to by using Eq. (D-2) and substituting in 0.69897 for the number of decades and 200 for the lower frequency.

$$\text{ND} = \log\left(\frac{f_2}{f_1}\right) = \log\left(\frac{\omega_{p2}}{\omega_{s2}}\right)$$

$$0.69897 = \log\left(\frac{\omega_{p2}}{200}\right)$$

$$\omega_{p2} = 200 \times 10^{0.69897}$$

$$= 1 \text{ krad/sec}$$

The revised filter specifications are shown in Fig. 6-11.

6-2-2 Passive and Active Filters

Filters can be put into two categories: passive and active. Active filters require external power, and passive filters do not. When the term *passive filter* is used, typically a filter using only R, L, and C is meant, but there are other types of passive filters available, such as mechanical and crystal filters.

A mechanical filter uses a transducer to convert an electrical signal to a mechanical vibration. The signal vibrates a material that is often a piece of metal cut in a particular shape. This material vibrates well at some frequencies, but will vibrate poorly at other frequencies. This allows only selected frequencies to pass. The vibration, after passing through the element, is then converted back to an electrical signal.

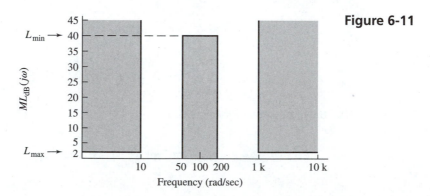

Figure 6-11

A crystal filter works very much like a mechanical filter except that it uses the piezo-electric effect. Crystals are well known for the ability to be resonant at a particular frequency, making them good as a very selective band-pass filter. Frequently, several crystals are used to obtain different filtering characteristics.

There are two types of active filters: digital and analog. Both types require external power supplies, and both can amplify and filter simultaneously. The digital filter converts the signal to digital numbers that can be manipulated by a computer. The computer can perform calculations on these numbers to filter the signal. The resulting numbers are then converted back to an analog signal. The analog filter uses an analog amplifier such as an op-amp. These analog op-amp filters are used at audio frequencies (<30 kHz) due to the frequency limit of the op-amp. Because we are at audio frequencies, only R and C elements are used. Inductors designed for these frequencies usually are large, expensive, and of poor quality (low Q). Because we have an active element, the op-amp, it is relatively easy to avoid using an inductor.

The rest of this book will use the analog op-amp active filter. When the term *active filter* is used, we will specifically be speaking of the analog op-amp active filter.

6-3 OP-AMPS

Currently, the upper frequency limit of active filters is due to the op-amp's frequency response and to a lesser degree to the capacitors. With improvements in op-amps and capacitors, this upper limit will continue to increase. We would like an analytical method that will remain the same as technology improves. To do this, we will analyze our circuits using a perfect op-amp. The op-amp will look perfect up to some limit, and as this limit increases, our analysis will continue to be appropriate.

6-3-1 Approximate Op-Amp

We must first define a current-technology op-amp. Then we will approximate the op-amp as a perfect op-amp with the limits required to make this assumption.

Using our knowledge of Bode plots and Laplace transforms, we can make an ac model of an op-amp. Let's first review the main characteristics of a common op-amp.

1. The input impedance, r_i, is in the MΩ range.
2. The output impedance, r_o, is in the range 50 to 100 Ω.
3. The open-loop voltage gain below the lowest break frequency is greater than 100,000 (100 dB).
4. The lowest break frequency is about 10 Hz. There is a second break frequency at about 1 MHz, which is close to the unity-gain point and is sometimes ignored.

We can simulate the op-amp's break frequencies by using either a series RC network driven with a voltage source or a parallel RC network driven with a current source. Because an op-amp is primarily a voltage amplifier and the gain is expressed as voltage gain, we will use the voltage source driving a series network. Figure 6-12 shows this series network.

The transfer function for the series network shown in Fig. 6-12 in a proper form to draw the Bode plot is

$$\frac{V_o}{V_{\text{in}}} = \frac{1}{RCs + 1}$$

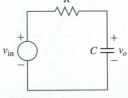

Figure 6-12

The break frequency, as shown in Chapter 5, is calculated by

$$\omega = \frac{1}{RC}$$

Knowing the break frequency required, we may choose a value for either R or C and solve for the other component. We are using this circuit to simulate a break frequency, so it is unimportant what value is initially chosen for R or C. A convenient value to choose would be $R = 1$. This makes for an easy source conversion, using the resistor and the voltage source, if a current source is required.

To make another break frequency, the output of this circuit can be used to drive another circuit using a VCVS. We can cascade several stages; however, we typically need only two break frequencies.

To complete our model, we need to simulate the input and the output circuit. The input is a very high resistance, r_i, between the inverting and noninverting terminal, as shown in Fig. 6-13. The voltage across this impedance is then amplified by the open-loop gain and passed through the break-frequency circuits. (Because this is an equivalent circuit, it does not matter where the actual open-loop gain is taken into account.) This is then passed to the output circuit, which is a voltage source with a series resistance, r_o.

It may be of some interest to examine the transfer function of this model of the op-amp. When we use a computer to evaluate a design based on an approximate model of the op-amp, we should use this accurate model to check the validity of our approximations. The transfer function, in this case, is a series of voltage divider equations multiplied together. Assuming an output load Z_L, we obtain

$$\frac{V_{\text{out}}}{V_{\text{in}}} = \left(\frac{\frac{1}{r'C'}}{s + \frac{1}{r'C'}} \right) \left(\frac{\frac{1}{r''C''}}{s + \frac{1}{r''C''}} \right) \left(\frac{-AZ_L}{r_o + Z_L} \right)$$

If we use this model for the op-amp, we will have an accurate representation of an active filter. The analysis using this model will generate some very complex equations where a number of terms will have little effect on the filter's transfer function.

If we design the circuit gain such that its gain is much less than the op-amp's open-loop gain up to the highest frequency of use, we can ignore the op-amp's break frequencies. The highest frequency of use for a low-pass filter is at the pass-band edge frequency. After we are past the break frequency, the roll-off of the op-amp can only help us. However, a

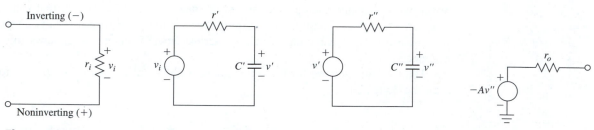

Figure 6-13

high-pass filter ideally passes frequencies from the pass-band edge frequency to infinity. In reality, there is a maximum frequency above which we are no longer interested in passing signals.

Figure 6-14a shows a low-pass filter response compared to the open-loop gain of an op-amp. For a low-pass filter, the highest frequency of use, f_H, coincides with the break frequency, as indicated in Fig. 6-14a. In order to prevent the op-amp's roll from affecting our desired response, the open-loop gain of the op-amp must be larger than the pass-band gain of our circuit at f_H. A similar situation occurs for the high-pass filter.

Figure 6-14b shows a high-pass filter response compared to the open-loop gain of an op-amp. Here we need to be given the highest frequency that our input is expected to have. At this given highest frequency of use, f_H, the op-amp's gain must be larger than the pass-band gain of our circuit so that the op-amp's roll-off does not interfere with the desired response. Therefore, we need to determine an appropriate amount of difference between the circuit's pass-band gain and the op-amp's open-loop gain for both a high-pass and a low-pass circuit.

We will recall that a typical assumption is that the open-loop gain is infinity. Often in electronics a 10% error would be acceptable. For active filters this is often not enough. In this book we will use a 5% error. A 5% error will correspond to the open-loop gain being 20 times the circuit gain at the highest frequency of use. In equation form,

$$A_{\text{op-amp}} \geqslant 20A_{\text{ckt}} \tag{6-10}$$

By meeting the condition of this equation, we can approximate the op-amp, as shown in Fig. 6-15. This simplifies the analysis tremendously, and the analytical results are almost the same as when using the full model. With this model we do the analysis assuming that the op-amp's open-loop gain is infinity for all frequencies of interest.

We can further simplify the op-amp by assuming that the input impedance, r_i, is infinit and that the output impedance, r_o, is zero. This can be assumed if we keep our resistive values in the circuit in the kΩ range.

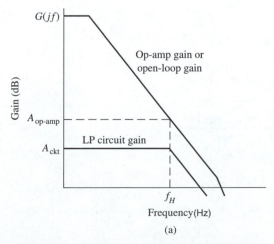

(a)

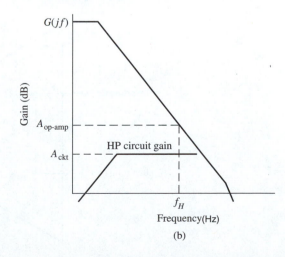

(b)

Figure 6-14

Figure 6-15

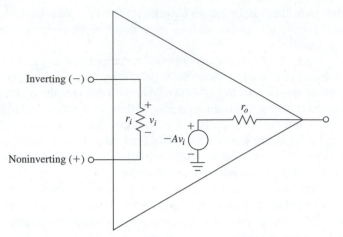

The approximate op-amp model we will use is shown in Fig. 6-16. There are some tricks to using this model successfully. Because this is an op-amp, the inverting and non-inverting terminals will have the same voltage value. When writing node equations, do not write an equation at the output node because this will generate an indeterminate equation. Even though the input terminals have the same voltage value, they cannot be considered connected since the input impedance must be infinity; consequently, both terminals will not have a current entering or leaving.

EXAMPLE 6-6

Using the approximate op-amp model, write the transfer function for the circuit in Fig. 6-17.

Solution: We do not need to write an equation at the noninverting terminal since it is fixed at v_{in} volts, and we cannot write an equation at v_o with this model. The only nodal equation we can write will be at the inverting terminal.

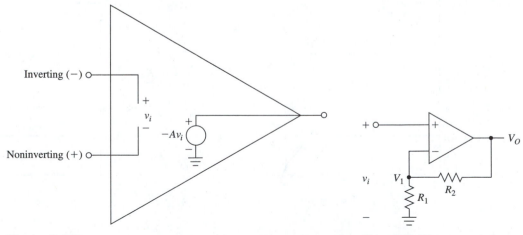

Figure 6-16

Figure 6-17

$$\left(\frac{1}{R_1} + \frac{1}{R_2}\right)V_1 - \left(\frac{1}{R_2}\right)V_o = 0$$

The inverting terminal is effectively an open circuit, so there is no current entering that terminal, and consequently no term appears in the equation for this path. However, we know from basic op-amp analysis that the inverting terminal voltage is always equal to the noninverting terminal voltage. Therefore, the voltage V_1 at the inverting terminal is equal to the voltage at the noninverting terminal, which is the input voltage, v_{in}. From this we may substitute v_{in} in for V_1, and the equation can be written as

$$\left(\frac{1}{R_1} + \frac{1}{R_2}\right)v_{in} - \left(\frac{1}{R_2}\right)V_o = 0$$

Solving for V_o/v_{in}, the transfer function is

$$\frac{V_o}{v_{in}} = 1 + \frac{R_2}{R_1}$$

We must also consider the slew rate of the op-amp. The slew rate is a specification from the manufacturer's data sheet that tells us how fast the output voltage can change. If we try to change the output voltage faster than the capability of the op-amp, then the output will become distorted. Because the slope of a sine function is steeper at higher frequencies, we will be concerned with the maximum slope of a sine function at the highest frequency of use. To find this maximum slope, we must first find the slope, or rate of change, for all values of time by taking the derivative.

$$\frac{d}{dt} V_o \sin(\omega t) = V_o \omega \cos(\omega t)$$

From this we see that the maximum slope is when $\cos(\omega t) = 1$. Therefore, the slew rate of the op-amp must be

$$SR \geq V_o \omega \qquad\qquad (6\text{-}11)$$

where SR = the slew rate of the op-amp
V_o = the peak value of the output
ω = the maximum frequency of use

This equation shows that for a fixed slew rate, as we increase the frequency of use, we must reduce the peak output voltage.

As with all our analysis, we must remember that we are working with limits based on the sinusoidal function. This means that if we want the slew-rate limit for a square wave, we must convert the square wave to its sinusoidal components, determine the highest sinusoidal frequency required, and then apply Eq. (6-11).

In Eq. (6-11) there is a limit to the peak value due to the power supply. The output voltage will only be able to get to within a few volts of the power supply. This saturation voltage can be found in the data sheets. If this data value is unavailable, we will have to make an assumption. We will use a 2-V margin because typically we will be able to get within 2 V of the power supply voltage on most any op-amp.

Because there is a limit for the output, there is a limit for the maximum input voltage. Expressed in equation form,

$$V_{\text{in max}} = \frac{V_o}{A_{\text{ckt}}} \tag{6-12}$$

Once the maximum output voltage is determined, we can calculate the maximum limit for the applied input voltage.

Summarizing the approximate op-amp shown in Fig. 6-16:

1. The inverting and noninverting terminals are open circuits, but the voltage of these terminals will be equal.
2. An equation cannot be written at the op-amp's output voltage node.
3. The highest frequency of use must have an open-loop gain defined by Eq. (6-10):

$$A_{\text{op-amp}} \geq 20A_{\text{ckt}}$$

4. The slew rate must have a value defined by Eq. (6-11):

$$\text{SR} \geq V_o\omega$$

5. V_o must be less than the saturation level of the op-amp with the particular power supply voltage being used.

EXAMPLE 6-7

An active filter is to be designed using an op-amp with a slew rate of 0.5 V/μsec. The highest frequency of interest is 10 kHz, and a gain of 5 dB is required here. (a) Determine what the op-amp's open-loop gain must be at the highest frequency of use. (b) Calculate the maximum possible input signal voltage. (c) Determine the power supply voltage, to the nearest volt, that is required for the op-amp.

Solution:

a. First we must find the voltage gain of the circuit.

$$A_{\text{ckt dB}} = 20 \log (A_{\text{ckt}})$$

$$A_{\text{ckt}} = 10^{A_{\text{ckt dB}}/20}$$

$$= 10^{5/20}$$

$$= 1.7783$$

From Eq. (6-10)

$$A_{\text{op-amp}} \geq 20A_{\text{ckt}}$$

$$\geq 20(1.7783)$$

$$\geq 35.566$$

b. From Eq. (6-11),

$$SR \geq V_o\omega$$

$$V_o \leq \frac{SR}{\omega}$$

$$\leq \frac{0.5 \text{ V}/\mu\text{sec}}{2\pi \times 10 \text{ k}}$$

$$\leq 7.9577 \text{ V peak}$$

(Don't forget the 10^{-6} from the μ in the slew rate.)
The maximum input voltage using Eq. (6-12) is

$$V_{\text{in max}} = \frac{V_o}{A_{\text{ckt}}}$$

$$= \frac{7.9577}{1.7783}$$

$$= 4.4750 \text{ V peak}$$

c. The power supply voltage should be about 2 V greater than the peak output voltage.

$$2 + 7.9577 = 9.9577 \text{ V}$$

And to the nearest volt, the supply should be

$$\pm 10$$

6-3-2 Filter Component Considerations

There are two basic component considerations. One is the component itself, and the other is the component's effect on the overall performance of the filter.

When a component ages and when it is exposed to temperature and humidity variations, its value will change. For the filter to maintain its design characteristics, we must use components that have small variations to temperature, humidity, and aging. The cost usually increases with an increase in temperature and humidity stability. We must choose the most stable components that we can afford.

A fixed-value component typically has a nomimal value and a tolerance. For a capacitor, the range of this tolerance can be astonishingly large. Therefore, ceramic or electrolytic capacitors should be avoided, and Mylar or polystyrene capacitors should be used. The tolerance for the capacitor should be ±5% or less for good results.

Resistor tolerances of ±1% are commonly available. However, it may be better to use precision potentiometers for the resistors. These are only slightly larger than a resistor and may be directly soldered onto a PC board. Precision potentiometers are also very practical for initial design testing because you can readjust the resistance value instead of swapping components. When used in a finished design, they may be very valuable in peaking a filter to compensate for aging or off-value capacitors.

When a component is not the exact calculated value or when it drifts due to aging or environment, the overall response of the filter is affected. Some components will affect the transfer function greatly, but others only slightly. The degree to which the components affect the transfer function is called *sensitivity*. The study of sensitivity is beyond the scope of this book, but the basic concept should be understood.

The sensitivity of a component is not due to the component. It is due to how the component is used in the transfer function and its magnitude relationship to the other components. With experience, we will notice that in some designs, when a component deviates from its calculated value only slightly, the filter's response deviates considerably. Sometimes a relatively large change in a component, in a different design, will change the response slightly. The difference is the sensitivity.

PROBLEMS

Section 6-1

1. (a) When the loss function is −5 dB, is the signal amplified or attenuated?
   (b) Find this value expressed in terms of a gain function.

2. Write the transfer function shown in the form of a loss function.

$$\frac{(s + 3)[(s + 2)^2 + 15]}{s^2(s + 4)^2(s + 7)}$$

3. For Fig. 6-18a:
   (a) Draw the loss function with the frequency axis normalized to 3 krad/sec and the magnitude of 6 dB normalized to 0 dB.
   (b) Find the normalized magnitude and normalized frequency of the 10 krad/sec location.
   (c) Denormalize the graph drawn in part (a) to a frequency of 15 rad/sec and the 0-dB level to −2 dB.

4. For Fig. 6-18b:
   (a) Draw the loss function with the frequency axis normalized to 50 rad/sec (notice that this is not at the break frequency) and the magnitude of −5 dB normalized to 0 dB.

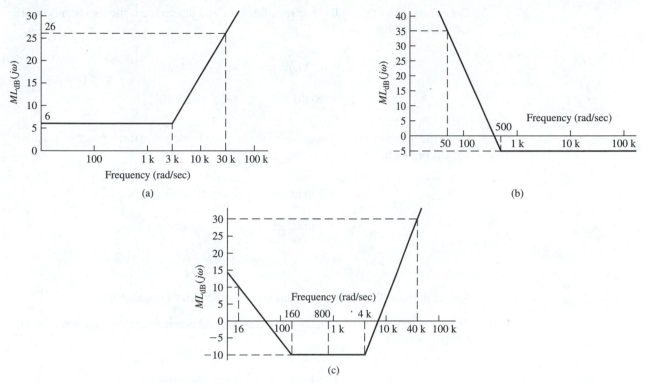

Figure 6-18

(b) What are the normalized magnitude and normalized frequency of the 5 krad/sec location?

(c) Denormalize the graph drawn in part (a) to a frequency of 1 krad/sec and the 0-dB level to 4 dB.

5. For Fig. 6-18c:

(a) Draw the loss function with the frequency axis normalized to 800 rad/sec and the magnitude of −10 dB normalized to 0 dB.

(b) Find the normalized magnitude and normalized frequency of the 5 krad/sec location.

(c) Denormalize the graph drawn in part (a) to a frequency of 1 krad/sec and the 0-dB level to +5 dB.

Section 6-2

6. Draw the graph of the following specifications and determine if the function is an HP, LP, BP, or notch.

$$\omega_s = 25 \text{ rad/sec} \qquad \omega_p = 200 \text{ rad/sec}$$

$$L_{min} = 20 \text{ dB} \qquad L_{max} = 3 \text{ dB}$$

7. Draw the graph of the following specifications and determine if the function is an HP, LP, BP, or notch.

$$\omega_s = 10 \text{ krad/sec} \qquad \omega_p = 2 \text{ krad/sec}$$

$$L_{min} = 40 \text{ dB} \qquad L_{max} = 0.5 \text{ dB}$$

8. Draw the graph of the following specifications and determine if the function is an HP, LP, BP, or notch.

$$\omega_{s1} = 2 \text{ krad/sec} \qquad \omega_{s2} = 12 \text{ krad/sec}$$

$$\omega_{p1} = 5 \text{ krad/sec} \qquad \omega_{p2} = 10 \text{ krad/sec}$$

$$L_{min1} = 40 \text{ dB} \qquad L_{min2} = 20 \text{ dB} \qquad L_{max} = 3 \text{ dB}$$

9. Revise the specifications in Problem 8 so that the filter is symmetrical.

10. Draw the graph of the following specifications and determine if the function is an HP, LP, BP, or notch.

$$\omega_{s1} = 5 \text{ krad/sec} \qquad \omega_{s2} = 10 \text{ krad/sec}$$

$$\omega_{p1} = 2 \text{ krad/sec} \qquad \omega_{p2} = 12 \text{ krad/sec}$$

$$L_{min} = 40 \text{ dB} \qquad L_{max1} = 2 \text{ dB} \qquad L_{max2} = 5 \text{ dB}$$

11. Revise the specifications in Problem 10 so that the filter is symmetrical when:
    (a) The center frequency is not changed.
    (b) The center frequency is changed.

Section 6-3

12. Using the op-amp model shown in Fig. 6-13, draw the equivalent circuit of an op-amp that has an input impedance of 1.5 MΩ, an output impedance of 70 Ω, an open-loop gain of 150,000, a lower break frequency at 8 Hz, and an upper break frequency of 1 MHz.

13. Write the transfer function for the op-amp described in Problem 12 assuming that a load of 10 kΩ is attached to the output.

14. In the pass band of an op-amp filter, V_o must be 9.5 V peak with an input voltage of 1 V peak. The highest frequency of use is 4 kHz.
    (a) What must the open-loop gain be at 4 kHz?
    (b) What must the op-amp's slew rate be in volts/μsec?
    (c) Calculate power supply voltage to the nearest volt.

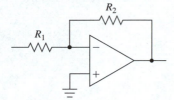

Figure 6-19

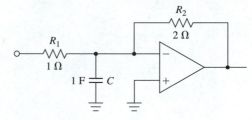

Figure 6-20

15. Using the approximate op-amp model shown in Fig. 6-16, find the transfer function of the circuit in Fig. 6-19.

16. Using the approximate op-amp model shown in Fig. 6-16, find the transfer function of the circuit in Fig. 6-20 using the component values.

Normalized Low-Pass Filter

OBJECTIVES

Upon successful completion of this chapter, you should be able to:

- Define what a circuit topology is.
- Calculate the loss function of a topology in terms of admittances.
- Select an appropriate topology to generate a desired transfer function described by a Bode plot.
- Use coefficient matching on an appropriate topology to determine the type of components and the values of those components required to design a given filter.
- Separate multiple-order filters into biquads, and determine the order in which to cascade them based on each biquad's *Q*-factor.
- Determine mathematically and graphically the filter order required for a low-pass Butterworth and Chebyshev filter, and determine the filter order for the low-pass elliptic filter by table.
- Design the component values of a normalized low-pass Butterworth, Chebyshev, and elliptic filter.
- Compare and contrast the differences in the Butterworth, Chebyshev, and elliptic filter's frequency response shape, filter order, and number of components used.

7-0 INTRODUCTION

In this chapter we will learn how to use generic active filter loss functions to make normalized low-pass filters. In Chapter 8 we will use these loss functions to make practical low-pass, high-pass, band-pass, and notch active filters.

A normalized filter will have unrealistic values of resistance and capacitance, usually a few ohms or farads. Generic loss functions are low-pass by nature and have their break or center frequency normalized. Therefore, an active filter may be normalized in both component value and frequency.

The generic loss functions are actually found in the last section of this chapter. In the first few sections we define how to apply any filter loss function to a circuit. These first few sections apply to low-pass, high-pass, band-pass, and notch filters, but we specifically apply them to a normalized low-pass filter in this chapter.

From this chapter we will have a good working knowledge of three basic normalized low-pass filters. There are a lot of mathematical ways to look at these filters, but we will look at them in the context of previous chapters. Most active-filter texts examine the mathematical wonders of these filters. We approach them from a how-to-design approach.

7-1 TOPOLOGY

Topology is the form of the circuit. It is a map of the locations of components. We will be unable to look at all the possible topologies, but we will focus our attention to five basic topologies: noninverting single feedback, noninverting dual feedback, inverting single feedback, inverting dual feedback, and twin-T.

We use nodal analysis to determine the loss function of a topology. It will be easier if we use admittance, $Y(s)$, for each component. We will also drop the s in the function notation. Therefore, we define

$$Y = \frac{1}{Z(s)} \tag{7-1}$$

Later, when we have the loss function for our topology expressed in terms of admittances, we can determine the type of components R or C to use for the admittances. By selecting which admittances are R's and which are C's, we can control what type of filter is possible. Using this loss function with R's and C's and using a numerical loss function with a desired frequency response, we can determine a circuit with numerical values for the electrical components. This process is called *coefficient matching* and will be shown in the next section. First, we need to generate the loss functions from our topologies.

Figure 7-1a shows the noninverting single-feedback topology. Because components R_1 and R_2 are almost always resistors, we will leave them as resistors. Writing the equations at voltage nodes V_1 and V_2, we have

$$(Y_A + Y_B) \quad V_1 - Y_A \quad V_{in} = 0$$

$$\left(\frac{1}{R_1} + \frac{1}{R_2}\right) V_2 - \left(\frac{1}{R_1}\right) V_o = 0$$

Multiplying through the second equation by R_1, we can combine all the resistors in the coefficient of V_2.

$$\left(1 + \frac{R_1}{R_2}\right) V_2 - V_o = 0$$

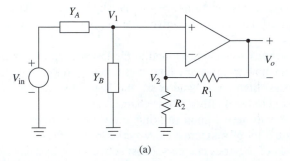

(a)

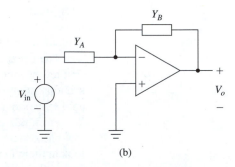

(b)

Figure 7-1

Notice that this coefficient is the same as the gain for a noninverting op-amp amplifier. We will therefore define this coefficient as K:

$$K = 1 + \frac{R_1}{R_2} \tag{7-2}$$

The second equation then becomes

$$KV_2 - V_o = 0$$

Because V_2 is the inverting terminal, it must be equal to V_1. Substituting V_1 into the second equation for V_2 and solving for V_1, we have

$$V_1 = \left(\frac{1}{K}\right) V_o$$

This is substituted back into the first equation for the node voltage V_1:

$$(Y_A + Y_B)\left(\frac{1}{K}\right) V_o - Y_A V_{\text{in}} = 0$$

Solving the equation for V_{in}/V_o, we have the loss function for Fig. 7-1a.

$$\frac{V_{\text{in}}}{V_o} = \frac{Y_A + Y_B}{KY_A} \tag{7-3}$$

By a similar procedure we can find the loss function for the inverting single-feedback topology shown in Fig. 7-1b.

$$\frac{V_{\text{in}}}{V_o} = \frac{-Y_B}{Y_A} \tag{7-4}$$

Notice that this loss function does not have a gain constant, and the noninverting single-feedback topology does. This is a characteristic difference between the inverting and non-inverting topologies.

The difference between these single-feedback topologies and dual-feedback topologies is in the filter order that is possible for them to implement. The single-feedback topologies are used for first-order filters, and the dual-feedback topologies are used for second-order filters.

Figure 7-2 shows the dual-feedback topologies. Deriving the loss function for these topologies is the same, but using algebraic substitution for the simultaneous equations is not practical. Instead, we will use determinants to solve the equations.

Let's derive the noninverting dual-feedback topology in Fig. 7-2a. We write the equations at the three nodes, substituting in V_2 and V_3 and using K as defined in Eq. (7-2).

$$(Y_A + Y_B + Y_C + Y_E)V_1 - \quad Y_C \quad V_2 - Y_B V_0 = Y_A V_{\text{in}}$$

$$-Y_C \quad V_1 + (Y_C + Y_D)V_2 + \quad 0 V_0 = 0$$

$$0 \quad V_1 + \quad K \quad V_2 - \quad V_0 = 0$$

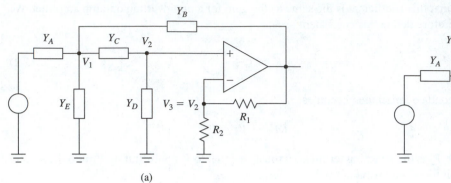

(a)   (b)

Figure 7-2

The voltage V_o is

$$
V_o = \cfrac{\begin{vmatrix} Y_A + Y_B + Y_C + Y_E & -Y_C & Y_A V_{in} \\ -Y_C & Y_C + Y_D & 0 \\ 0 & K & 0 \end{vmatrix}}{\begin{vmatrix} Y_A + Y_B + Y_C + Y_E & -Y_C & -Y_B \\ -Y_C & Y_C + Y_D & 0 \\ 0 & K & -1 \end{vmatrix}}
$$

Solving this for V_{in}/V_o yields the loss function for Fig. 7-2a.

$$
\frac{V_{in}}{V_o} = \frac{(Y_A + Y_B + Y_E)(Y_C + Y_D) + Y_C(Y_D - Y_B K)}{Y_A Y_C K} \tag{7-5}
$$

Using the same procedure, the inverting dual-feedback topology in Fig. (7-2b) can be found.

$$
\frac{V_{in}}{V_o} = \frac{Y_D(Y_A + Y_B + Y_C + Y_E) + Y_B Y_C}{-Y_A Y_C} \tag{7-6}
$$

The last topology we will use is called twin-T and is shown in Fig. 7-3. This is called twin-T because the component group Y_B-Y_C-Y_F and Y_A-Y_D-Y_E each form the shape of a T. The loss function for this topology is

$$
\frac{V_{in}}{V_o} = \frac{Y_G Y_1 Y_2 + Y_F Y_1 (Y_B + Y_C - K Y_C) + Y_E Y_2 (Y_A + Y_D)}{K(Y_A Y_E Y_2 + Y_B Y_F Y_1)} \tag{7-7}
$$

where $K = 1 + R_1/R_2$ [defined in Eq. (7-2)]
$\quad\ Y_1 = Y_A + Y_D + Y_E$
$\quad\ Y_2 = Y_B + Y_C + Y_F$

Figure 7-3

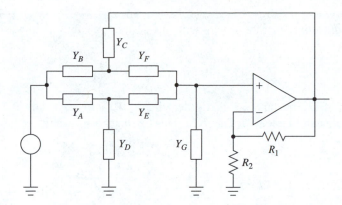

The twin-T topology is sometimes modified so that Y_D is tied to the op-amp output instead of reference. When this is done, the loss function changes only slightly. In this case, the last term of the numerator becomes

$$Y_E Y_2 (Y_A + Y_D - K Y_D)$$

Appendix E contains a summary of the topologies we will use.

7-2 COEFFICIENT MATCHING

When designing a filter, we normally start with a desired loss function, choose an appropriate topology, determine what components to use for the admittances, and match the desired loss function's coefficients to the topology loss function's coefficients. This last step is called *coefficient matching*. The desired loss function comes either from a Bode plot (as shown in Chapter 5) or from a defined function (which we will look at in Section 7-4).

The coefficient matching procedure is as follows:

1. From the topology loss function, determine which admittances should be resistors and which should be capacitors. Choose the admittances such that the desired loss function and the topology loss function have the same number of constants and powers of s present in both numerator and denominator.
2. Make the coefficient of the highest power of s in the topology loss function equal to 1.
3. Form a set of simultaneous equations by equating the coefficients of the desired loss function to the coefficients of the topology's loss function.

EXAMPLE 7-1

From a Bode plot of a desired frequency response we found the transfer function shown. Use coefficient matching to determine a suitable filter.

$$\frac{V_o}{V_{in}} = \frac{80}{s + 10}$$

Solution: First we need to write the desired transfer function as a loss function.

$$L(s) = \frac{s + 10}{80}$$

Because this is a first-order circuit, we will start with the noninverting single-feedback topology in Fig. 7-1a. This topology also has the same sign (positive). From Eq. (7-3),

$$L(s) = \frac{Y_A + Y_B}{KY_A}$$

where $K = 1 + R_1/R_2$ [defined in Eq. (7-2)].

We must choose what type of components to use for Y_A and Y_B. Because the desired loss function has an s in it, one of our admittances must be a capacitor. If we choose Y_A as a capacitor and Y_B as a resistor, the resulting loss function is

$$L(s) = \frac{sC_A + \dfrac{1}{R_B}}{KsC_A}$$

This will not work, because the desired loss function does not have an s in the denominator. If we choose Y_A as a resistor and Y_B as a capacitor, the resulting loss function is

$$L(s) = \frac{\dfrac{1}{R_A} + sC_B}{K\dfrac{1}{R_A}}$$

This will work, because we have the same powers of s in the correct locations.

We then make the coefficient of the highest powers of s equal to 1.

$$L(s) = \frac{s + \dfrac{1}{R_A C_B}}{K\left(\dfrac{1}{R_A C_B}\right)}$$

Finally, we write the simultaneous equations by coefficient matching.

$$\frac{1}{R_A C_B} = 10$$

$$\left(1 + \frac{R_1}{R_2}\right)\frac{1}{R_A C_B} = 80$$

Here we have two equations and four unknowns.

When we choose a value for C_B we will still have one more equation than unknowns. We could select a value for R_1 or R_2, but we may not choose a value for R_A because we have chosen a value of C_B in the first equation. To have the most components of the same value, it would be better to let R_1 or R_2 equal R_A.

Let's select $C_B = 1 \ \mu F$ and $R_1 = R_A$. Solving the first equation,

$$R_A = \frac{1}{10 C_B} = \frac{1}{10(1 \ \mu F)} = 100 \ k\Omega$$

Substituting the known values into the second equation gives

$$\left(1 + \frac{100 \ k\Omega}{R_2}\right) \frac{1}{(100 \ k\Omega)(1 \ \mu F)} = 80$$

$$R_2 = 14.286 \ k\Omega$$

Figure 7-4 shows the final circuit.

Frequently, the denominator constants of the two loss functions are not equated. This prevents the occurrence of negative resistance values but shifts the function's loss up or down, maintaining the loss difference between the stop band and pass band. The overall loss of the function is typically unimportant, but the differences in dB between the pass bands and the stop bands are important. If the overall loss of the function is important, we may add an amplifier to adjust the overall loss to the desired level.

Often there are more coefficients than equations, which means that there is more than one solution. We can reduce the number of unknowns by allowing some of the components to equal each other. By proper choice this will simplify the equation. This will, also, reduce production cost because we can buy in larger quantities with fewer different values.

Even when we equate some components, we may not have the same number of equations and unknowns. We will frequently have to select some values and use the equation to solve for the rest. Because resistors come in a larger variety of values and potentiometers are smaller and less expensive than variable capacitors, we desire to select capacitor values and solve for resistor values.

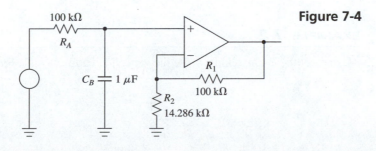

Figure 7-4

Most of the time we are not concerned with the phase of the response because we are interested only in the voltage magnitude. When this is the case, we may ignore the sign of the desired loss function and the sign of the topology's loss function. We therefore will have two choices for the design. Even when we must maintain the sign of the desired loss function, we still have two choices by adding an inverting circuit to the output of the filter to correct the sign.

In Example 7-1 we could have used the inverting single-feedback topology. The loss function for this from Eq. (7-4) was

$$\frac{V_{in}}{V_o} = \frac{-Y_B}{Y_A}$$

which does not appear to match the example's loss function. However, if we make Y_B a resistor in parallel with a capacitor and Y_A a resistor, we will be in the correct form.

$$L(s) = \frac{\dfrac{1}{R_B} + sC_B}{-\left(\dfrac{1}{R_A}\right)}$$

In this case we would either ignore the minus sign or add an inverting amplifier to the output of the filter.

This leads us to conclude that admittances do not have to be single components, but could be complex networks. We will use single-component values for admittances on our designs as much as possible so that the process of the designs will be easy to follow.

In Example 7-1 we selected a practical capacitor value, but we could have selected any value. In Chapter 8 we will learn how to change the values of the components after the design. We may therefore select any convenient value. For calculation purposes it will be easier to select values around 1. When this is done, we call the circuit a *normalized circuit*. If we had chosen $C_B = 1$ F, the resulting resistor values would have been calculated: $R_A = R_1 = 0.1 \, \Omega, R_2 = 0.014286 \, \Omega$.

We can presolve the coefficient-matching equations by using variables instead of numbers in the desired loss function. There are a number of ways to solve these, but let's define some general guidelines.

Because resistors have a larger assortment of values, we desire to choose capacitor values. We would also like to have as many component values equal as practical. We may ignore the desired function's multiplying factor, because it may be too restrictive, as shown in the following example.

EXAMPLE 7-2

Using coefficient matching, write the general equations to solve the loss function:

$$\frac{s^2 + as + b}{-h}$$

Solution: This requires an inverting dual-feedback topology if we wish to match the minus sign. By the process shown in Example 7-1, we can select the components as shown in Fig. 7-5. Making the coefficient of the highest power of s equal to 1, we therefore have

$$\dfrac{s^2 + \dfrac{1}{C_E}\left(\dfrac{1}{R_A} + \dfrac{1}{R_B} + \dfrac{1}{R_C}\right)s + \dfrac{1}{C_D C_E R_B R_C}}{-\dfrac{1}{C_D C_E R_B R_C}}$$

For this topology, we can coefficient match the h value only if h and b happen to be equal to each other. If they are not equal, we cannot match both of these coefficients. The h value is a multiplying factor that controls the overall amplification or attenuation of all frequencies, but the b helps to determine which individual frequencies are stopped and which ones are passed. Therefore, in this case we match b and just accept whatever h turns out to be. If the h is important, we will add another stage that has the amplification or attenuation we require.

We will start by allowing $R_B = R_C = R$. The loss function reduces to

$$\dfrac{s^2 + \dfrac{1}{C_E}\left(\dfrac{1}{R_A} + \dfrac{2}{R}\right)s + \dfrac{1}{C_D C_E R^2}}{-\dfrac{1}{C_D C_E R^2}}$$

Matching the coefficients gives

$$\dfrac{1}{C_E}\left(\dfrac{1}{R_A} + \dfrac{2}{R}\right) = a$$

$$\dfrac{1}{C_D C_E R^2} = b$$

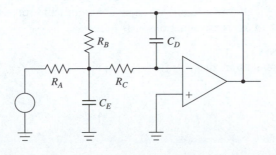

Figure 7-5

We want to select capacitor values and the second equation has only one resistor in it, so we will solve the second equation for R.

$$R = \frac{1}{\sqrt{bC_D C_E}}$$

Once we have a value for R we need to determine a value for R_A. Therefore, we will solve the first equation for R_A.

$$R_A = \frac{R}{RC_E a - 2}$$

Notice that in this equation, for R_A to be positive, $(RC_E a)$ must be greater than 2. Therefore,

$$RC_E a > 2$$

$$R > \frac{2}{C_E a}$$

This restricts the value of R previously calculated and, therefore, affects the values of C_D and C_E that we can choose because if

$$\frac{2}{C_E a} < R \quad \text{and} \quad R = \frac{1}{\sqrt{bC_D C_E}}$$

it implies that

$$\frac{2}{C_E a} < \frac{1}{\sqrt{bC_D C_E}}$$

This expression gives us a restriction on selecting the capacitors' values. If we solve it for C_D, then it will show the values of C_E that will make R_A positive values. Solving this expression for C_D, we find

$$C_D < \frac{C_E a^2}{4b}$$

We could have assumed that we choose C_D first, and then solve this equation for C_E. Either way it allows us to choose capacitor values and solve for resistor values. The order in which to use these equations is as follows:

1. Choose C_E.
2. Choose C_D such that

$$C_D < \frac{C_E a^2}{4b}$$

3. Solve for R_B and R_C.

$$R = \frac{1}{\sqrt{bC_DC_E}}$$

4. Solve for R_A.

$$R_A = \frac{R}{RC_Ea - 2}$$

5. Calculate the resulting multiplying factor.

$$h = \frac{1}{C_DC_ER^2}$$

The topologies shown in Appendix E have a list of presolved equations that were found by the process shown in Example 7-2.

7-3 BIQUADS

From Bode plots in Chapter 5 and filter specifications in Chapter 6, we can easily determine some general loss functions for low-pass, high-pass, band-pass, and notch filters.

$$\text{First-order low-pass} = \frac{s + b}{\pm h}$$

$$\text{First-order high-pass} = \frac{s + b}{\pm hs}$$

$$\text{Second-order low-pass} = \frac{s^2 + as + b}{\pm h} \qquad \zeta < 1$$

$$\text{Second-order high-pass} = \frac{s^2 + as + b}{\pm hs^2} \qquad \zeta < 1$$

$$\text{Band-pass} = \frac{s^2 + as + b}{\pm hs} \qquad \zeta > 1$$

$$\text{Notch} = \frac{s^2 + as + b}{\pm h(s^2 + e)} \qquad \zeta > 1$$

These are not the only forms, but they are a subset of a general form called a biquad. A biquad is the ratio of two quadratic functions. The loss function, expressed in biquads, is

$$L(s) = \prod_{i=1}^{int[(N+1)/2]} \frac{m_is^2 + a_is + b_i}{\pm h(n_is^2 + d_is + e_i)} \qquad (7\text{-}8)$$

where int means the integer part

N is the filter order

m_i and n_i may be either 1 or 0

h, a_i, b_i, and e_i are any real number

d_i will typically be 0 for the filters we will use

A complex filter may be composed of a loss function with a high-order numerator polynomial over a high-order denominator polynomial. This requires a high-order circuit, but all of our topologies are only capable of either first- or second-order. One solution would be to develop a new topology for each higher-order filter required, but this could be very time-consuming. A common solution to this problem is to cascade first- and second-order circuits to generate higher-order circuits, and that is how we will proceed. In order to do this, our first step will be to split up the high-order loss function into biquads.

Splitting up a high-order loss function into biquads is simply a matter of pairing each numerator quadratic factor with a denominator quadratic factor and adding a multiplying factor called h to the denominator. The only exception to this is when one of the factors in the given loss function is a first-order polynomial instead of a quadratic. Even though it is not the ratio of quadratics, for convenience, we will still refer to it as a biquad. Let's do an example of splitting up a loss function into biquads.

EXAMPLE 7-3

Split the loss function into biquads.

$$\frac{(s + 0.47558)(s^2 + 0.22958s + 1.0863)(s^2 + 0.69854s + 0.59202)}{6.6305 \times 10^{-3}(s^2 + 10.568)(s^2 + 4.3650)}$$

Solution:

$$\frac{s + 0.47558}{h_1}$$

$$\frac{s^2 + 0.22958s + 1.0863}{h_2(s^2 + 10.568)}$$

$$\frac{s^2 + 0.69854s + 0.59202}{h_3(s^2 + 4.3650)}$$

There will be a resulting multiplying constant equal to

$$h = h_1 h_2 h_3$$

From Example 7-3 we would continue by using coefficient matching or the presolved equations in Appendix E. The first equation could be a noninverting single-feedback topology, and the last two stages could be noninverting dual-feedback topologies. We could also use two inverting dual-feedback topologies for the last two stages because the net outcome would still be a positive function. If we were not required to match the sign of the multiplying constant, there are several more possibilities.

We notice there is an option of which denominator factor to associate with a numerator factor. Different pairing will yield different component values, but the overall loss function will be the same. We will pair the lowest numerator quadratic constant with the lowest denominator quadratic constant, the next-to-lowest numerator quadratic constant with the next-to-lowest denominator quadratic constant, and so on. This will tend to keep the resistors of a stage close in value to each other.

The order of which stage to cascade first, second, and so on, is determined by the Q of the numerator quadratic. Remember that Q stands for "Quality factor," and it refers to the ratio of a center frequency to the bandwidth. Therefore, when a quadratic has peaking, we may use Q to give us a relative idea of the amount of peaking. A high Q means a tall, narrow-width peak, and a low Q means a short, broad-width peak.

From Chapter 5, we found that as ζ decreased below $\sqrt{2}/2$, the peaking at the break frequency became taller and more pronounced, which corresponds to a higher Q. This relationship between ζ and the Q is expressed mathematically by

$$Q = \frac{1}{2\zeta} \qquad (7\text{-}9)$$

We can substitute this quantity into the general form of the quadratic.

$$s^2 + 2\zeta\omega_n s + \omega_n^2 \qquad (7\text{-}10)$$

so that we can express the quadratic in terms of Q as

$$s^2 + \frac{\omega_n}{Q}s + \omega_n^2 \qquad (7\text{-}11)$$

Because we may express the quadratic as powers of s with numerical coefficients a and b,

$$s^2 + as + b$$

we can express the Q as

$$Q = \frac{\sqrt{b}}{a} \qquad (7\text{-}12)$$

Using this equation, we can determine the Q of the numerator of each biquad. We will then put the biquads in order of the lowest-Q stage first to highest-Q stage last. This is done because a high-Q circuit may amplify frequencies that are close to the peak frequency so much that they can overdrive the following stage. The stage following the high-Q stage may be a circuit that was designed to reduce or remove these frequencies, but when overdriven, the output will become distorted. Therefore, if we arrange our circuits so that the lowest-Q circuits are first, then we may be able to remove or reduce these frequencies before they are amplified by the high-Q circuits.

If the loss function is an odd-order function, it will contain a first-order factor that has no possibility of containing any peak frequencies. Therefore, if the loss function has a first-order factor, the resulting circuit from this factor will be placed first.

EXAMPLE 7-4

The biquads from Example 7-3 are listed below. What is the numerator Q of each biquad, and in what order should they be put?

$$\frac{s + 0.47558}{h_1}$$

$$\frac{s^2 + 0.22958s + 1.0863}{h_2(s^2 + 10.568)}$$

$$\frac{s^2 + 0.69854s + 0.59202}{h_1(s^2 + 4.3650)}$$

Solution: The first term has a first-order numerator, so it will be the first stage in the filter. The other two biquads' Q can be calculated by Eq. (7-12).

$$Q = \frac{\sqrt{b}}{a}$$

$$Q = \frac{\sqrt{1.0863}}{0.22958} = 4.5398$$

$$Q = \frac{\sqrt{0.59202}}{0.69854} = 1.1015$$

Therefore, the first term will be the first stage, the second term will be the last stage, and the third term will be the middle stage.

7-4 LOW-PASS FILTER APPROXIMATIONS

More often than not, we are given the filter specifications as shown in Chapter 6. We cannot design a filter that is able to form blocks like a filter specification. Instead, we approximate the filter specification by staying within the limits of L_{max} and L_{min}.

There are a number of defined low-pass functions that will approximate a filter specification. We will look at three approximations: Butterworth, Chebyshev, and elliptic. These approximations are based on the magnitude of the loss. The phase is not restricted because a large number of filter applications are not concerned with the phase.

The filters will be designed using tabulated values of low-pass loss functions. These loss functions can be transformed from low-pass functions to high-pass, band-pass, and notch functions. We therefore, often call these *generic approximations* or *generic loss functions*. Appendix F contains the derivations of the equations used to define these approximations and the procedures for using the equations to generate the tables for the generic loss functions.

In the rest of this chapter, we will restrict ourselves to the normalized low-pass filter. This means that we will find the low-pass filter with the pass-band edge frequency at 1 and choose our capacitor values at 1 (when possible). Then, in Chapter 8, we will learn how to convert the normalized circuit to practical values of resistance and capacitance, to change the values so that the circuit performs at different frequencies, and to design high-pass, band-pass, and notch filters.

7-4-1 General Form

Each of the three functions are based on the loss function form

$$L(s) = \sqrt{1 + \varepsilon^2 F^2(s)} \tag{7-13}$$

It is beyond the scope of the text to delve into the mathematical reasoning of why this expression was selected for the loss function. However, a few general statements may be made about how this expression is used to determine the filter approximations. Typically, we solve Eq. (7-13) for normalized low-pass loss functions. To generate a specific loss function from this general loss function, a numerator polynomial over a denominator polynomial in terms of s is used for $F(s)$ and a numerical value is used for ε.

To determine the particular function with numerical coefficients for $F(s)$, the s in $F(s)$ is set equal to $j\Omega$, where $\Omega = \omega/\omega_p$. Next, a criterion is established for how the loss function's curve will approximate the desired normalized filter specification. This criterion determines whether the loss function will be a Butterworth, Chebyshev, or elliptic approximation of the normalized filter specification. Mathematical methods are then employed to determine the numerical values for the coefficients of the $F(j\Omega)$, which will establish the coefficients in the $F(s)$ function.

The value of ε is determined from the value of L_{max}. One of the aspects of the criterion established for $F(j\Omega)$ requires that it be equal to 1 at the normalized pass-band edge frequency. At this location, the value of the loss function will be L_{max}. Knowing these values and adjusting Eq. (7-13) to a dB form, because L_{max} is a dB quantity, we can write Eq. (7-13) as

$$L_{max} = 20 \log\left(\sqrt{1 + \varepsilon^2 F^2(j\Omega)}\right)$$

$$= 20 \log\left(\sqrt{(1 + \varepsilon^2(1)^2)}\right)$$

$$= 10 \log(1 + \varepsilon^2)$$

Solving this for ε, we have

$$\varepsilon = \sqrt{10^{0.1L_{max}} - 1} \tag{7-14a}$$

Most of the time we require the value of ε^2 instead ε, so we typically use the form

$$\varepsilon^2 = 10^{0.1L_{max}} - 1 \qquad \text{(7-14b)}$$

The particular function for $F(s)$ along with the particular value of ε are substituted into Eq. (7-13), and then the resulting function is algebraically manipulated into a factored numerator over a factored denominator. It should be noted that while the elliptic approximation will have a polynomial for the denominator, the Butterworth and Chebyshev approximations will only have a constant in the denominator.

Depending on the filter specification, different-order loss functions will result from this procedure. In the next few subsections, the loss functions for up to the fifth order have been calculated for three different L_{max} values for the Butterworth, Chebyshev, and elliptic approximations.

7-4-2 Butterworth

The response of a Butterworth filter is shown in Figs. 7-6, 7-7, and 7-8. The corresponding Butterworth loss functions are shown in Tables 7-1, 7-2, and 7-3. Notice that the graphs shown in the figures are split at the frequency 1. The graph to the left of 1 has a different vertical scale than it does on the right, but the horizontal axis is the same for both sections. This allows us to see more clearly the shape of the function before the break frequency because we have magnified it.

Each table with accompanying figure has a different L_{max} value. In each figure we see that the shape of the curve in the pass band is dead flat until the pass-band edge frequency, where the curve turns up and rolls off at a constant rate. Because of this shape in the pass band, the Butterworth approximation is often called the *maximally flat approximation*.

The order of filter required for a particular filter specification can be found in two different ways. One method is graphic and the other is by calculation. In both cases we find the normalized stop-band frequency first by using the equation

$$\Omega_s = \frac{\omega_s}{\omega_p} \qquad \text{(7-15)}$$

Using this equation will make our resulting filter have a break frequency at 1 rad/sec, not at the originally desired break frequency. In Chapter 8 we will learn how to shift this filter back up to the desired break frequency.

In the graphic method, we determine which figure of Figs. 7-6, 7-7, and 7-8 to use by our required L_{max}. We then locate our L_{min} at our Ω_s on this graph. The filter order that just touches or is above this point will be the filter order to use. The disadvantage of this method is that we must have a graph with our particular L_{max}. The calculation method does not require a graph.

In the calculation method we find the filter order by

$$n \geq \frac{\log\left(\dfrac{10^{0.1L_{min}} - 1}{\varepsilon^2}\right)}{2 \log(\Omega_s)} \qquad \text{(7-16)}$$

Table 7-1 *Butterworth:* L$_{max}$ = 3.0103 dB

N	Denominator Constant	Numerator
1	1.0000	$s + 1.0000$
2	1.0000	$s^2 + 1.4142s + 1.0000$
3	1.0000	$(s + 1.0000)(s^2 + 1.0000s + 1.0000)$
4	1.0000	$(s^2 + 1.8478s + 1.0000)(s^2 + 0.76537s + 1.0000)$
5	1.0000	$(s + 1.0000)(s^2 + 1.6180s + 1.0000)(s^2 + 0.61803s + 1.0000)$

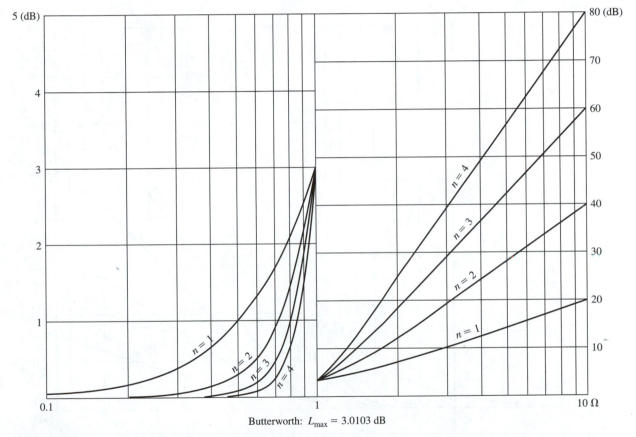

Butterworth: L_{max} = 3.0103 dB

Figure 7-6

where n is an integer
ε is defined in Eq. (7–14)

Because n must be an integer, we always round up to the next integer. Therefore, if we calculated $n = 1.03$, we would use a second-order system.

Once the filter order is determined by the graphic method or by calculation, we find the loss function in the Butterworth tables for our particular L_{max} and filter order.

Table 7-2 *Butterworth:* $L_{max} = 1.5000$ dB

N	Denominator Constant	Numerator
1	1.5569	$s + 1.5569$
2	1.5569	$s^2 + 1.7646s + 1.5569$
3	1.5569	$(s + 1.1590)(s^2 + 1.1590s + 1.3433)$
4	1.5569	$(s^2 + 2.0640s + 1.2478)(s^2 + 0.85494s + 1.2478)$
5	1.5569	$(s + 1.0926)(s^2 + 1.7678s + 1.1937)(s^2 + 0.67525s + 1.1937)$

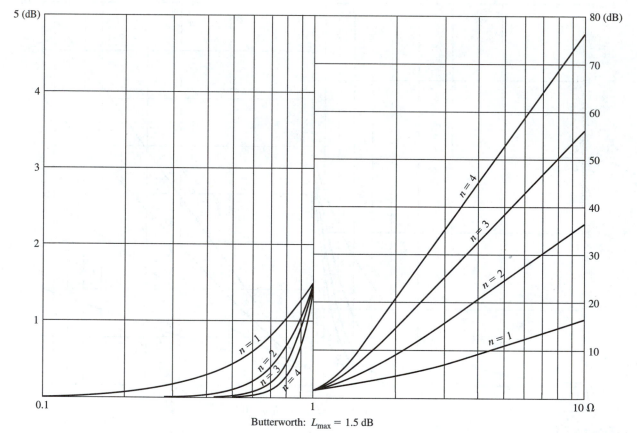

Butterworth: $L_{max} = 1.5$ dB

Figure 7-7

Table 7-3 *Butterworth:* $L_{max} = 0.2500\ dB$

N	Denominator Constant	Numerator
1	4.1081	$s + 4.1081$
2	4.1081	$s^2 + 2.8664s + 4.1081$
3	4.1081	$(s + 1.6016)(s^2 + 1.6016s + 2.5650)$
4	4.1081	$(s^2 + 2.6306s + 2.0268)(s^2 + 1.0896s + 2.0268)$
5	4.1081	$(s + 1.3266)(s^2 + 2.1464s + 1.7598)(s^2 + 0.81986s + 1.7598)$

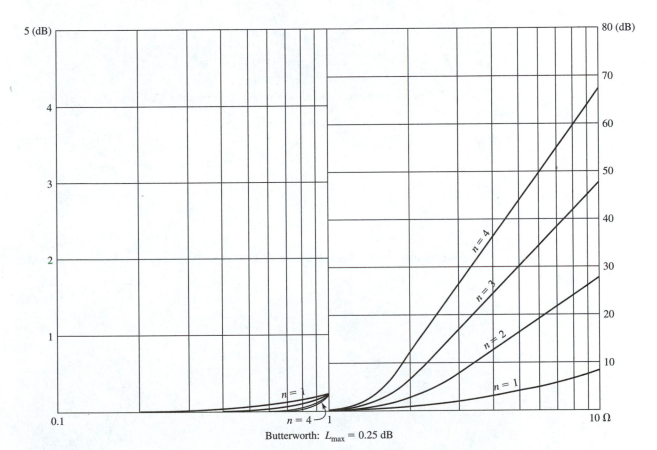

Butterworth: $L_{max} = 0.25$ dB

Figure 7-8

EXAMPLE 7-5

Design a normalized Butterworth filter with the following specifications:

$$f_s = 16 \text{ kHz} \qquad f_p = 2 \text{ kHz}$$

$$L_{min} = 32 \text{ dB} \qquad L_{max} = 1.5 \text{ dB}$$

Solution: The normalized frequencies are found by using Eq. (6-6b) and Eq. (7-15).

$$\Omega_p = \frac{f_p}{f_p} = \frac{2 \text{ kHz}}{2 \text{ kHz}} = 1$$

$$\frac{\omega_s}{\omega_p} = \frac{2\Omega (16 \text{ kHz})}{2\Omega (2 \text{ kHz})} = 8$$

Although it is not required to do so, the graph of the normalized specification is drawn in Fig. 7-9.

Using Eq. (7-14b), we find ε^2.

$$\varepsilon^2 = 10^{0.1 L_{max}} - 1$$

$$= 10^{(0.1)(1.5)} - 1$$

$$= 0.41254$$

Substituting into Eq. (7-16), we find the order to use.

$$n \geq \frac{\log\left(\dfrac{10^{0.1 L_{min}} - 1}{\varepsilon^2}\right)}{2 \log(\Omega_s)}$$

$$\geq \frac{\log\left[\dfrac{10^{(0.1)(32)} - 1}{0.41254}\right]}{2 \log(8)}$$

$$\geq 1.9844$$

Therefore, we will use a second-order loss function.

L_{max} is 1.5 dB, so we must use the Table 7-2. We then find the second entry in the table because we calculated a second-order system, and, therefore, our desired loss function is

$$\frac{s^2 + 1.7646s + 1.5569}{1.5569}$$

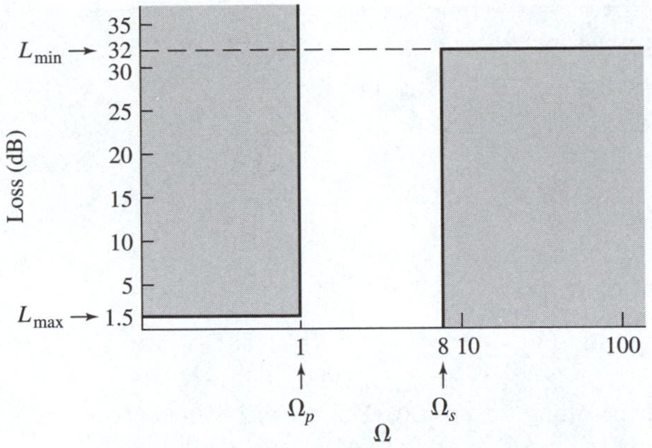

Figure 7-9

To design the circuit we use the topologies in Appendix E. Either the inverting or the noninverting low-pass dual-feedback topology will work. Let's use the noninverting topology since we will match the sign of the denominator. Using the general biquad equation (E-1) in Appendix E, we can determine

$$m = 1 \qquad a = 1.7646 \qquad b = 1.5569$$

$$h = 1.5569 \qquad n = 0 \qquad d = 0 \qquad e = 1$$

We will first determine if it is possible to use the low-pass noninverting dual-feedback topology using Eq. (E-25a) in Appendix E, Section E-3-1.

$$0 \le 2 - \frac{a}{\sqrt{b}}$$

$$\le 2 - \frac{1.7646}{\sqrt{1.5569}}$$

$$\le 0.58578$$

This is true, so we may use this topology.

We then choose a value for $C_B = C_D = C$. Because we want a normalized circuit, we will choose $C = 1$ F. Therefore, from Eq. (E-25b), R is

$$R = \frac{1}{C\sqrt{b}}$$

$$= \frac{1}{1\sqrt{1.5569}}$$

$$R_A = R_C = R = 0.80144 \ \Omega$$

Next we solve for the K value using Eq. (E-25c).

$$K = 3 - \frac{a}{\sqrt{b}}$$

$$= 3 - \frac{1.7646}{\sqrt{1.5569}}$$

$$= 1.5858$$

The resulting h from Eq. (E-25d) will then be

$$h = \frac{K}{C_B C_D R_A R_C}$$

$$= \frac{1.5858}{1(1)(0.80144)(0.80144)}$$

$$= 2.4689$$

We then calculate R_1 and R_2 using the relationship of R_1, R_2, and K as defined in Eq. (E-20).

$$K = 1 + \frac{R_1}{R_2}$$

In this equation we can assume a value for one of the resistors and solve for the other. We will arbitrarily assume a value of R_2 and solve for R_1. For economic reasons we would like to have as many resistors equal in value; therefore, we will choose R_2 equal to 0.80144. Solving for R_1, we have

$$R_1 = R_2(K - 1)$$

$$= (0.80144)(1.5858 - 1)$$

$$= 0.46947 \ \Omega$$

The final normalized circuit is shown in Fig. 7-10.

In Example 7-5 our resulting h was not equal to our desired h. If we had to match the h, we could have added another amplifier. This would require a noninverting amplifier, or we could change our design to an inverting low-pass dual-feedback topology followed by an inverting amplifier.

Figure 7-10

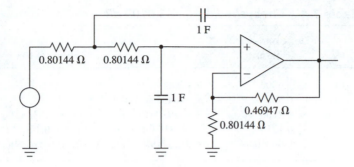

7-4-3 Chebyshev

The response of a Chebyshev filter is shown in Figs. 7-11, 7-12, and 7-13. The corresponding Chebyshev loss functions are shown in Tables 7-4, 7-5, and 7-6. The graphs in these figures are split at a normalized frequency of 1. The section of graph to the left of 1 is drawn using the vertical scale on the far left of the graph, and the section to the right of 1 is drawn using the vertical scale on the far right. The frequency axis along the bottom of the graph is continuous from left to right. This allows us to magnify the details of the graph from 0.1 up to the break frequency.

The main characteristic of this response is the equal height ripple in the pass band just below the break frequency. Because of this equal height ripple, this filter is often called the *equiripple approximation*. As long as we are below the desired L_{max} value, we will be able to meet our desired loss function's specifications.

The loss functions for the Chebyshev response are similar to the Butterworth loss functions. Notice that they even have the same first-order loss function. The only difference is that the quadratics in the Chebyshev loss functions have break frequencies lower than the quadratics in the Butterworth loss functions. Looking at the graphs, we see that to accomplish this, each quadratic has some peaking (due to ζ being less than $\sqrt{2}/2$).

Because the Chebyshev has an earlier break frequency than the Butterworth, the roll-off rate of 20 dB/dec per filter order starts sooner, and as a result, the Chebyshev will have a greater attenuation at any stop-band frequency than the Butterworth. Both types of filters eventually have the same roll-off rate, but the Chebyshev has a greater loss in the stop band.

Using the Chebyshev equations is similar to using the Butterworth equations. First we normalize the stop-band frequency using Eq. (7-15). Then we may apply the graphic method described in the Butterworth section to the Chebyshev graphs, or we may calculate the filter order by

$$n \geq \frac{\cosh^{-1}\left(\dfrac{10^{0.1L_{min}} - 1}{\varepsilon^2}\right)^{1/2}}{\cosh^{-1}(\Omega_s)} \tag{7-17}$$

where n is an integer

ε is defined in Eq. (7-14)

$$\cosh^{-1}(x) = \ln(x + \sqrt{x^2 - 1})$$

Because n must be an integer, we always round up to the next integer value.

Table 7-4   Chebyshev: $L_{max} = 3.0103$ dB

N	Denominator Constant	Numerator
1	1.0000	$s + 1.0000$
2	0.5000	$s^2 + 0.64359s + 0.70711$
3	0.2500	$(s + 0.29804)(s^2 + 0.29804s + 0.83883)$
4	0.1250	$(s^2 + 0.41044s + 0.19579)(s^2 + 0.17001s + 0.90290)$
5	0.06250	$(s + 0.17719)(s^2 + 0.28670s + 0.37689)(s^2 + 0.10951s + 0.93590)$

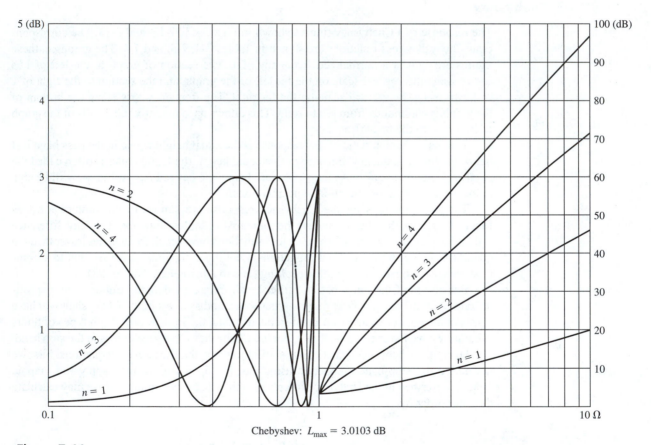

Chebyshev: $L_{max} = 3.0103$ dB

Figure 7-11

Table 7-5   Chebyshev: L$_{max}$ = 1.5000 dB

N	Denominator Constant	Numerator
1	1.5569	$s + 1.5569$
2	0.77846	$s^2 + 0.92218s + 0.92521$
3	0.38923	$(s + 0.42011)(s^2 + 0.42011s + 0.92649)$
4	0.19462	$(s^2 + 0.57521s + 0.24336)(s^2 + 0.23826s + 0.95046)$
5	0.097308	$(s + 0.24765)(s^2 + 0.40071s + 0.40682)(s^2 + 0.15306s + 0.96584)$

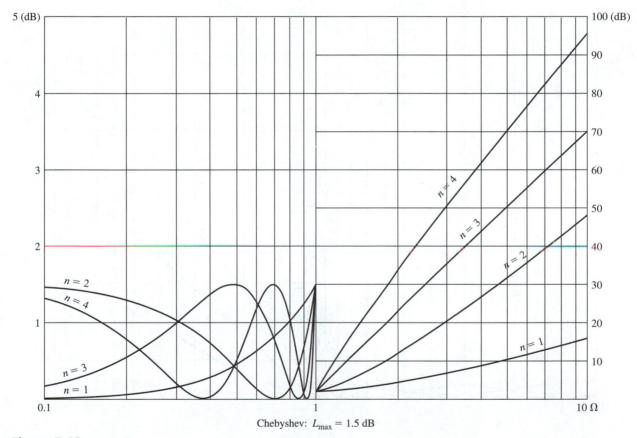

Chebyshev: L_{max} = 1.5 dB

Figure 7-12

Table 7-6   Chebyshev: L$_{max}$ = 0.2500 dB

N	Denominator Constant	Numerator
1	4.1081	$s + 4.1081$
2	2.0541	$s^2 + 1.7967s + 2.1140$
3	1.0270	$(s + 0.76722)(s^2 + 0.76722s + 1.3386)$
4	0.51351	$(s^2 + 1.0261s + 0.45485)(s^2 + 0.42504s + 1.1620)$
5	0.25676	$(s + 0.43695)(s^2 + 0.70700s + 0.53642)(s^2 + 0.27005s + 1.0954)$

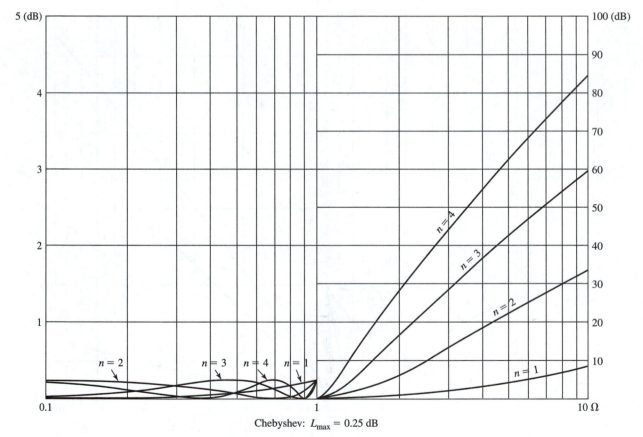

Chebyshev: $L_{max} = 0.25$ dB

Figure 7-13

Once the order is determined, we use the table containing our particular L_{max} and order.

Because the Chebyshev equations have peaking, we must be careful to cascade the lowest Q circuits first, as pointed out in Section 7-3. These tables are written with the lowest Q to highest Q when read left to right.

EXAMPLE 7-6

Design a normalized Chebyshev filter to approximate the following specifications:

$$f_p = 1 \text{ kHz} \qquad\qquad f_s = 5 \text{ kHz}$$

$$L_{max} = 3.0103 \text{ dB} \qquad L_{min} = 40 \text{ dB}$$

Solution: Using Eq. (7-15), we find the normalized stop-band frequency.

$$\Omega_s = \frac{\omega_s}{\omega_p}$$

$$= \frac{2\pi(5 \text{ kHz})}{2\pi(1 \text{ kHz})}$$

$$= 5$$

We find ε^2 from Eq. (7-14b).

$$\varepsilon^2 = 10^{0.1L_{max}} - 1$$

$$= 10^{(0.1)(3.0103)} - 1$$

$$= 1.0000$$

The correct order of filter required can then be found by Eq. (7-17).

$$n \geq \frac{\cosh^{-1}\left(\dfrac{10^{0.1L_{min}} - 1}{\varepsilon^2}\right)^{1/2}}{\cosh^{-1}(\Omega_s)}$$

$$\geq \frac{\cosh^{-1}\left[\dfrac{10^{(0.1)(40)} - 1}{1}\right]^{1/2}}{\cosh^{-1}(5)}$$

$$\geq 2.3112$$

Therefore, we need a third-order filter.

We were given $L_{max} = 3.0103$ dB, so we must use the Table 7-4. We then find the third entry in the table because we calculated a third-order system. Therefore, our desired loss function is

$$\frac{(s + 0.29804)(s^2 + 0.29804s + 0.83883)}{0.2500}$$

Here we will use a low-pass inverting single-feedback topology, followed by a low-pass inverting dual-feedback topology. In Appendix E we notice that the single-feedback topology allows us to set h to a particular value, so we may design the second-order filter first and determine the gain required for the first-order circuit to give us an overall h of 0.2500.

We will first design the second-order circuit using the biquad.

$$\frac{s^2 + 0.29804s + 0.83883}{h_1}$$

Using the equations listed in Section E-4-1 for a low-pass, inverting, dual-feedback topology, we choose $C_E = 1$ F and then select C_D based on the relationship given in Eq. (E-41a).

$$C_D < \frac{C_E a^2}{4b}$$

$$< \frac{1(0.29804)^2}{4(0.83883)}$$

$$< 0.026474$$

Because values of capacitors have the same magnitudes with different powers of 10, we will choose $C_D = 0.01$ F.

Solving for the resistors using Eq. (E-41b) and Eq. (E-41c), we have

$$R = \frac{1}{\sqrt{bC_D C_E}}$$

$$= \frac{1}{\sqrt{(0.83883)(0.01)(1)}}$$

$$R_B = R_C = R = 10.919 \ \Omega$$

$$R_A = \frac{R}{RC_E a - 2}$$

$$= \frac{10.919}{(10.919)(1)(0.29804) - 2}$$

$$= 8.7059 \ \Omega$$

The resulting h for the second-order circuit from Eq. (E-41d) is

$$h_1 = \frac{1}{C_D C_E R_B R_C}$$

$$= \frac{1}{(0.01)(1)(10.919)^2}$$

$$= 0.83883$$

We want to make the overall denominator constant equal to 0.25, so we need to solve for the h_2 required for the first-order circuit.

$$h_1 h_2 = 0.25000$$

$$h_2 = \frac{0.25000}{h_1}$$

$$= 0.29803$$

We find the procedure for the low-pass, inverting, single-feedback topology in Section E-2-1. Choose $C_B = 1$ F, and solve for R_B using Eq. (E-15a).

$$R_B = \frac{1}{b C_B}$$

$$= \frac{1}{(0.29804)(1)}$$

$$= 3.3553 \ \Omega$$

As noted above, we want a particular h so we will solve for R_A using Eq. (E-15b) with our desired h substituted in.

$$R_A = \frac{1}{h C_B}$$

$$= \frac{1}{(0.29803)(1)}$$

$$= 3.3553 \ \Omega$$

R_A and R_B do not calculate to the same exact value, but they are so close that we will use the same resistor. The resulting circuit is shown in Fig. 7-14.

Figure 7-14

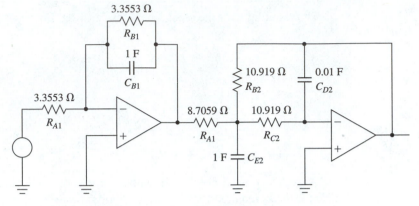

7-4-4 Elliptic

The response of the elliptic filter is shown in Figs. 7-15, 7-16, and 7-17. The corresponding elliptic loss functions are shown in Tables 7-7, 7-8, and 7-9. The graphs shown in the figures are split at 1, the same as the Butterworth and Chebyshev graphs. This type of filter is sometimes referred to as the *Cauer approximation*.

Table 7-7a   *Elliptic:* $L_{max} = 3.0103$ *dB,* $\Omega_s = 2$

N	L_{min} (dB)	Denominator Constant	Numerator	Denominator
2	22.900	7.1612×10^{-2}	$s^2 + 0.60745s + 0.75593$	$s^2 + 7.4641$
3	40.321	5.3882×10^{-2}	$s + 0.32193$	
			$s^2 + 0.26805s + 0.86251$	$s^2 + 5.1532$
4	57.775	1.2920×10^{-3}	$s^2 + 0.42709s + 0.22182$	$s^2 + 4.5933$
			$s^2 + 0.15023s + 0.91666$	$s^2 + 24.227$
5	75.229	1.6140×10^{-3}	$s + 0.19059$	
			$s^2 + 0.28496s + 0.41342$	$s^2 + 4.3650$
			$s^2 + 0.095977s + 0.94485$	$s^2 + 10.568$

Table 7-7b   *Elliptic:* $L_{max} = 3.0103$ *dB,* $\Omega_s = 5$

N	L_{min} (dB)	Denominator Constant	Numerator	Denominator
2	39.824	1.0205×10^{-2}	$s^2 + 0.63888s + 0.71429$	$s^2 + 49.495$
3	65.756	7.6540×10^{-3}	$s + 0.30136$	
			$s^2 + 0.29371s + 0.84232$	$s^2 + 33.165$
4	91.688	2.6038×10^{-5}	$s^2 + 0.41287s + 0.19936$	$s^2 + 29.203$
			$s^2 + 0.16711s + 0.90494$	$s^2 + 167.77$
5	117.62	3.2547×10^{-5}	$s + 0.17906$	
			$s^2 + 0.28654s + 0.38201$	$s^2 + 27.586$
			$s^2 + 0.10751s + 0.93723$	$s^2 + 71.405$

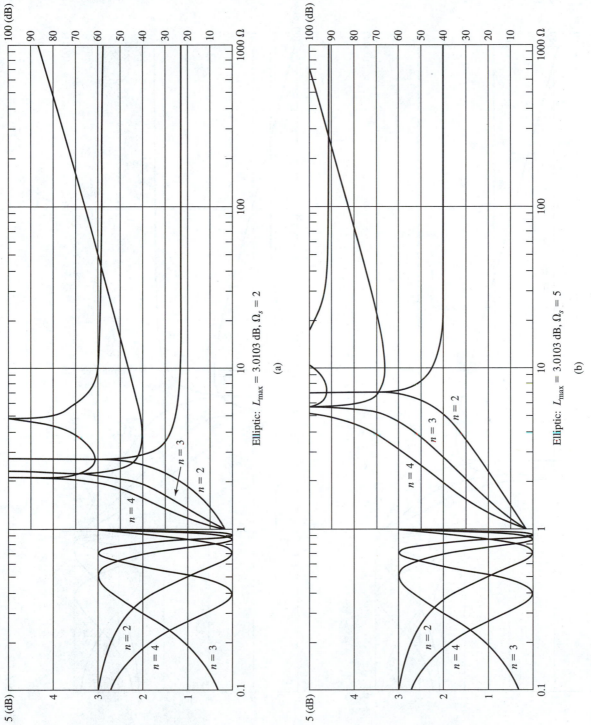

Elliptic: $L_{max} = 3.0103$ dB, $\Omega_s = 2$

(a)

Elliptic: $L_{max} = 3.0103$ dB, $\Omega_s = 5$

(b)

Figure 7-15

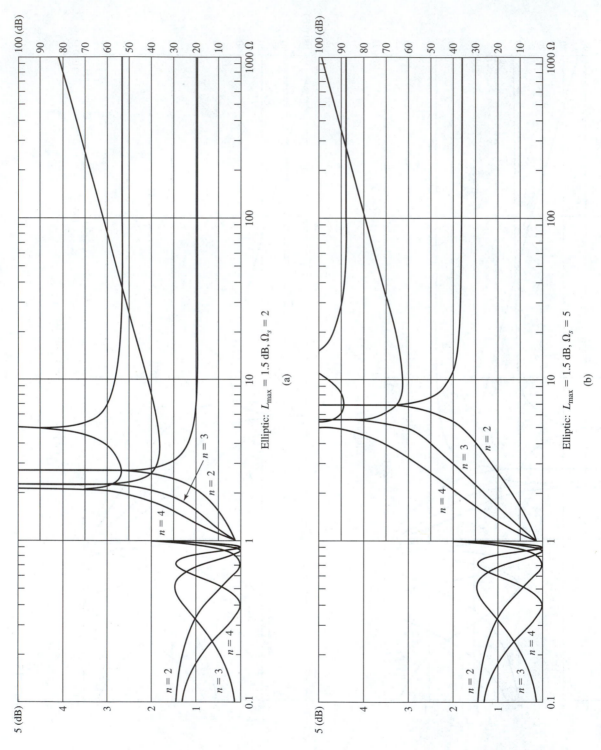

Elliptic: L_{max} = 1.5 dB, Ω_s = 2

(a)

Elliptic: L_{max} = 1.5 dB, Ω_s = 5

(b)

Figure 7-16

Table 7-8a *Elliptic:* $L_{max} = 1.5000\ dB,\ \Omega_s = 2$

N	L_{min} (dB)	Denominator Constant	Numerator	Denominator
2	19.086	1.1109×10^{-1}	$s^2 + 0.85334s + 0.98549$	$s^2 + 7.4641$
3	36.476	8.3891×10^{-2}	$s + 0.45674$	
			$s^2 + 0.37285s + 0.94651$	$s^2 + 5.1532$
4	53.929	2.0116×10^{-3}	$s^2 + 0.59927s + 0.27702$	$s^2 + 4.5933$
			$s^2 + 0.20876s + 0.96038$	$s^2 + 24.227$
5	71.383	2.5129×10^{-3}	$s + 0.26697$	
			$s^2 + 0.39784s + 0.44677$	$s^2 + 4.3650$
			$s^2 + 0.13338s + 0.97180$	$s^2 + 10.568$

Table 7-8b *Elliptic:* $L_{max} = 1.5000\ dB,\ \Omega_s = 5$

N	L_{min} (dB)	Denominator Constant	Numerator	Denominator
2	35.979	1.5887×10^{-2}	$s^2 + 0.91330s + 0.93453$	$s^2 + 49.495$
3	61.910	1.1917×10^{-2}	$s + 0.42518$	
			$s^2 + 0.41326s + 0.92952$	$s^2 + 33.165$
4	87.843	4.0539×10^{-5}	$s^2 + 0.57873s + 0.24796$	$s^2 + 29.203$
			$s^2 + 0.23392s + 0.95196$	$s^2 + 167.77$
5	113.77	5.0673×10^{-5}	$s + 0.25034$	
			$s^2 + 0.40044s + 0.41243$	$s^2 + 27.586$
			$s^2 + 0.15015s + 0.96673$	$s^2 + 71.405$

These loss functions are similar to the Chebyshev loss functions except that a denominator quadratic is added. The quadratic in the denominator has a ζ value of zero, which causes an infinite peak at a frequency in the stop band. This will pull the loss response such that the roll-off will be more than 20 dB/dec per order in the transition band. Because of this denominator quadratic, the equations must be calculated using a particular stop band, Ω_s. The denominator quadratic does have a disadvantage.

The response in the pass band will not continue to attenuate for even-order equations. For odd-order equations, the response will eventually increase in attenuation, but at a rate of only 20 dB/dec.

Because the elliptic loss functions are determined at a particular stop band, Ω_s, the most efficient use of this type of loss function will occur when the desired filter's stop band is the same as the stop band that was used to generate the elliptic loss functions. This book has tables for Ω_s of either 2 or 5. If tables of other stop bands are desired, then a computer or calculator program should be written using the information in Appendix F to generate them.

The elliptic loss functions have large peaks, so it is especially important to have the lowest Q as the first stage. The tables are organized in biquads with the lowest Q as the top entry and the highest Q as the bottom entry for each filter order. To find the order of the filter required, we use the table with the appropriate L_{max} and Ω_s. Then we look down the second column and find the first entry that has a value equal to or greater than our desired L_{min}. The following example will demonstrate this process.

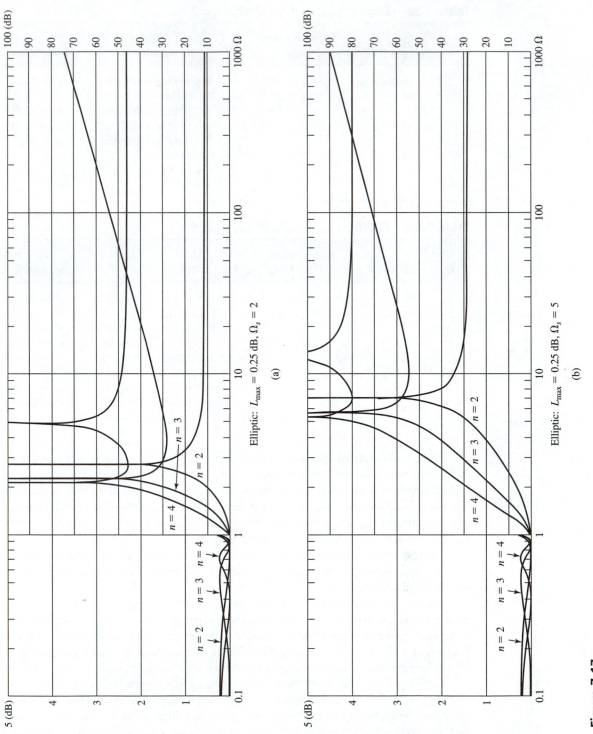

Elliptic: $L_{max} = 0.25$ dB, $\Omega_s = 2$

(a)

Elliptic: $L_{max} = 0.25$ dB, $\Omega_s = 5$

(b)

Figure 7-17

Table 7-9a *Elliptic:* $L_{max} = 0.2500$ dB, $\Omega_s = 2$

N	L_{min} (dB)	Denominator Constant	Numerator	Denominator
2	10.967	2.8290×10^{-1}	$s^2 + 1.4716s + 2.1733$	$s^2 + 7.4641$
3	28.055	2.2135×10^{-1}	$s + 0.86031$	
			$s^2 + 0.63895s + 1.3259$	$s^2 + 5.1532$
4	45.502	5.3077×10^{-3}	$s^2 + 1.0746s + 0.52880$	$s^2 + 4.5933$
			$s^2 + 0.35862s + 1.1496$	$s^2 + 24.227$
5	62.956	6.6305×10^{-3}	$s + 0.47558$	
			$s^2 + 0.69854s + 0.59202$	$s^2 + 4.3650$
			$s^2 + 0.22958s + 1.0863$	$s^2 + 10.568$

Table 7-9b *Elliptic:* $L_{max} = 0.2500$ dB, $\Omega_s = 5$

N	L_{min} (dB)	Denominator Constant	Numerator	Denominator
2	27.558	4.1887×10^{-2}	$s^2 + 1.7565s + 2.1337$	$s^2 + 49.495$
3	53.483	3.1443×10^{-2}	$s + 0.77977$	
			$s^2 + 0.74833s + 1.3373$	$s^2 + 33.165$
4	79.415	1.0697×10^{-4}	$s^2 + 1.0333s + 0.46487$	$s^2 + 29.203$
			$s^2 + 0.41512s + 1.1602$	$s^2 + 167.77$
5	105.35	1.3370×10^{-4}	$s + 0.44229$	
			$s^2 + 0.70615s + 0.54424$	$s^2 + 27.586$
			$s^2 + 0.26400s + 1.0941$	$s^2 + 71.405$

EXAMPLE 7-7

Design an elliptic filter to approximate the following filter specification:

$$\omega_p = 2 \text{ kHz} \qquad \omega_s = 10 \text{ kHz}$$

$$L_{max} = 0.2500 \text{ dB} \qquad L_{min} = 25 \text{ dB}$$

Solution: Finding the normalized stop band using Eq. (7-15), we obtain

$$\Omega_s = \frac{\omega_s}{\omega_p}$$

$$= \frac{2\Omega \, (10 \text{ kHz})}{2\Omega \, (2 \text{ kHz})}$$

$$= 5$$

Therefore, we use Table 7-9b since this is for $L_{max} = 0.25$ dB and $\Omega_s = 5$. The first entry has an $L_{min} = 27.558$ dB, which exceeds our requirement; therefore, we may use a second-order equation.

$$\frac{s^2 + 1.7565s + 2.1337}{4.1887 \times 10^{-2}(s^2 + 49.495)}$$

Referring to Appendix E, we see that the only appropriate topology is the twin-T. We first test to see if this topology works, using Eq. (E-57a).

$$0 \leq e + b - a\sqrt{e}$$

$$\leq 49.495 + 2.1337 - 1.7565\sqrt{49.495}$$

$$\leq 39.271$$

If this was not true, we would have to search for another topology that is not in Appendix E.

Next we choose $C = 1$ F because we want a normalized circuit. Therefore, $C_A = C_E = 1$ F and $C_C = 2C = 2$ F. Solving for R gives

$$R = \frac{1}{C\sqrt{e}}$$

$$= \frac{1}{1\sqrt{49.495}}$$

$$R_B = R_F = R = 0.14214 \ \Omega$$

$$R_D = \frac{R}{2} = 0.071070 \ \Omega$$

Next we choose C_G such that

$$C_G \geq \frac{C(e - b)}{2b}$$

$$\geq \frac{(1)(49.495 - 2.1337)}{(2)(2.1337)}$$

$$\geq 11.098$$

We will choose C_G equal to 20 F. Sometimes C_G needs to be $\geq$ a negative number. In this case, we may choose $C_G = 0$, but it must be at least zero.

Finding R_G, we obtain

$$R_G = \frac{2\sqrt{e}}{C(b - e) + 2C_G b}$$

$$= \frac{2\sqrt{49.495}}{1(2.1337 - 49.495) + 2(20)(2.1337)}$$

$$= 0.37041\ \Omega$$

In a case where $b = e$ in this equation and C_G is chosen as 0, R_G will be infinity. This indicates using an open circuit for both R_G and C_G.

We then calculate K.

$$K = \frac{\dfrac{2}{R_G} + C(4\sqrt{e} - a) + 2C_G(\sqrt{e} - a)}{2C\sqrt{e}}$$

$$= \frac{\dfrac{2}{0.37041} + 1(4\sqrt{49.495} - 1.7565) + 2(20)(\sqrt{49.495} - 1.7565)}{(2)(1)\sqrt{49.495}}$$

$$= 17.265$$

The resulting h is

$$h = K\frac{C}{C + 2C_G}$$

$$= 17.625\left[\frac{1}{1 + 2(20)}\right]$$

$$= 0.42111$$

Because this is not the original h desired, we could add an amplifier stage with a gain of 9.9468×10^{-2} if it was necessary.

Finally, we choose $R_2 = 0.14214\ \Omega$ and calculate R_1.

$$R_1 = R_2(K - 1)$$

$$= (0.14214)(17.265 - 1)$$

$$= 2.3120\ \Omega$$

The completed circuit is shown in Fig. 7-18.

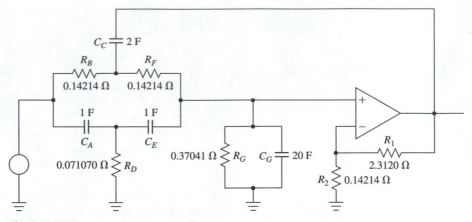

Figure 7-18

7-4-5 Comparison of Butterworth, Chebyshev, and Elliptic

The obvious difference among the three approximations is the shape of the responses. Another important difference is the filter order required to approximate a filter specification. The elliptic filter will always give us the lowest-order equation to approximate a filter specification. When we compare the elliptic to the Chebyshev or the Chebyshev to the Butterworth in a particular case, we may require the same-order filter. When there is a difference, the elliptic will be a lower order than the Chebyshev or Butterworth, and the Chebyshev will be a lower order than the Butterworth. The following example demonstrates a case where there is a difference in all cases.

EXAMPLE 7-8

Find the order of filter required for a Butterworth, Chebyshev, and elliptic filter for the following filter specification.

$$L_{max} = 3.0103 \text{ dB} \qquad L_{min} = 20 \text{ dB} \qquad \Omega_s = 2$$

Solution: We find ε^2 from Eq. (7-14).

$$\varepsilon^2 = 10^{0.1 L_{max}} - 1$$

$$= 10^{(0.1)(3.0103)} - 1$$

$$= 1$$

Substituting into Eq. (7-16), we find the Butterworth order required.

$$n \geq \frac{\log\left(\dfrac{10^{0.1 L_{min}} - 1}{\varepsilon^2}\right)}{2 \log(\Omega_s)}$$

$$\geq \frac{\log\left[\dfrac{10^{(0.1)(20)} - 1}{1}\right]}{2 \log(2)}$$

$$\geq 3.3147$$

Therefore, we will require a fourth-order Butterworth.

The order of filter required for the Chebyshev is found by using Eq. (7-17).

$$n \geq \frac{\cosh^{-1}\left(\dfrac{10^{0.1 L_{min}} - 1}{\varepsilon^2}\right)^{1/2}}{\cosh^{-1}(\Omega_s)}$$

$$\geq \frac{\cosh^{-1}\left[\dfrac{10^{(0.1)(20)} - 1}{1}\right]^{1/2}}{\cosh^{-1}(2)}$$

$$\geq 2.2690$$

Therefore, we need a third-order Chebyshev filter, which is one order less than the order required for the Butterworth.

For the elliptic we use Table 7-7a because this is for $L_{max} = 3.0103$ dB and $\Omega_s = 2$. The first entry has an $L_{min} = 22.900$ dB, which exceeds our requirement; therefore, we may use the second-order elliptic.

In this example each filter required one less order to approximate the filter specification. This is not always the case, but should be considered before deciding whether to use the Butterworth, Chebyshev, or elliptic filter.

In general, the fewer components, the less expensive the filter is to manufacture, and lower-order filters typically require fewer components. Frequently, economics are not the only factor to consider. Each type of filter has different characteristics, which can enhance or deteriorate the overall performance. To choose the best filter, we must know the exact application. If using the least op-amps is the main goal, the clear choice in Example 7-8 is the elliptic filter.

7-5 USING MATLAB

In this section we will do an example that shows how to write a MATLAB program to calculate the Rs and Cs for a normalized filter. We will also write a program that will graph the relationship between the input and output signal for a given transfer function.

EXAMPLE 7-9

Write a MATLAB program to calculate the R's and C's for an inverting, dual-feedback low-pass filter for the following low-pass function.

$$L(s) = \frac{s^2 + 1.7646s + 1.5569}{1.5569}$$

Solution: Figure 7-19 shows the completed M-file that is saved with the filename IDF.m. The first command clears the Command Window so that the disp commands following the clear command can be printed on a blank screen.

The values for a and b are requested from the user with the input command. This command prints the text within the parentheses and waits for the user to enter a value. The user can use the up or down arrow to select previously entered values for this requested input. Once the user presses the Enter key, the value last entered on the Command Window is stored in the variable to the left of the equal sign in the command. The value of Ce is entered using the same type of command.

The equations for calculating the R's and C's of the inverting, dual-feedback low-pass filter are found in Appendix E, Section E-4-1. The implementation of these equations is relatively easy except for when the Cd value is requested from the user. The Cd value must be chosen less than or equal to a value that is based on a, b, and Ce. This quantity is calculated from the equations in the appendix and stored in the variable $Cdmax$.

In order to make sure that the user enters a valid value, a while loop is constructed. Before this loop is entered, a constant, itest, is initially set to zero. Then, in the while loop, the user is asked to enter a value that is less than or equal to the $Cdmax$ value. In order to display the value of $Cdmax$, it is converted to a string using the num2str function and then concatenated to the rest of the text string using the strcat function. This string is printed in the Command Window, informing the user of the maximum value that maybe entered for Cd. If the user selects a Cd value that meets this condition, then the if command inside the while loop is executed and itest is set to 1. This causes the loop to end. If the condition is not met, then the question is asked again.

Once a, b, Ce, and Cd are selected, the values for R, Ra, and h are calculated. From these results the values for all the R's and C's are printed to the user in the Command Window using the disp command.

With the M-file written and saved, we can run the program by typing `IDF` in the Command Window. The program will then request the values of a, b, Ce, and Cd. Selecting the values of $Ce = 1$ and $Cd = 0.1$ along with the a and b that were given, the results are calculated by the program and displayed in the Command Window. The run of this program is shown in Fig. 7-20.

```
%IDF
%Inverting Dual Feedback design.
clc
disp('This finds the values for an Inverting Dual-feedback
  low pass filter')
disp(' ')
disp('(s^2 + a s + b)')
disp('--------------')
disp('        - h        ')
disp(' ')
disp('enter the values for a and b')
disp(' ')

a = input ('a = ');
b = input ('b = ');

disp(' ')
disp('Choose the Ce capacitor value for the filter');
Ce = input ('Ce = ');

Cdmax = Ce*a^2/(4*b);
disp(' ')
itest = 0; %set up loop test
while itest == 0
    disp(' ')
    disp(strcat('Select Cd such that it is < or = ',
      num2str(Cdmax),' '))
    Cd = input('Cd = ');
    if Cd <= Cdmax
        itest = 1 ; %the user's choice is valid
    end
end

R  = 1./(sqrt(b*Cd*Ce));
Ra = R/(R*Ce*a - 2.);
h  = 1./(Cd*Ce*R*R);
disp(' ')
disp('=====================================')
disp(strcat('      Ra = ',num2str(Ra)))
disp(strcat('Rb = Rc = ',num2str(R)))
disp(strcat('      Cd = ',num2str(Cd)))
disp(strcat('      Ce = ',num2str(Ce)))
disp(strcat('      h  = ',num2str(h)))
```

Figure 7-19

MATLAB M-file for Example 7-9

```
>> Sig
This finds the values for an Inverting Dual-feedback low pass filter

(s^2 + a s + b)
───────────────
       - h

enter the values for a and b

a = 1.7646
b = 1.5569

Choose the Ce capacitor value for the filter
Ce = 1

Select Cd such that it is < or = 0.5
Cd = .1


========================================
      Ra =1.0252
Rb = Rc =2.5344
      Cd =0.1
      Ce =1
      h  =1.5569
>>
```

Figure 7-20

MATLAB Command Window for Example 7-9

EXAMPLE 7-10

Write a MATLAB program to display a graph of the input and output signals for the following loss function when the input is a unity sine wave with a frequency of 0.5 Hz.

$$L(s) = \frac{s^2 + 1.7646s + 1.5569}{1.5569}$$

Solution: Figure 7-21 shows the completed M-file that is saved with the filename Sig.m. In order to make this a more versatile program, it is written to accept any Laplace transfer function. Therefore, we will need to take the reciprocal of the loss function before entering it into this program. This is very straightforward and most all of the commands have been discussed in previous sections. The only commands that require discussion are the linspace command and the plot command.

The linspace command is set up to automatically create a range of x values based on the frequency of the input signal. Instead of a fixed number for the beginning x value, $-1*T$ is used, and instead of a fixed number for the ending x value, $2*T$ is used. Since T is the value of the time period of the input sine wave, the graph will always show one cycle of the two waveforms before $t = 0$ and two cycles after $t = 0$.

```
function Sig(f,fHz)
%This shows the input and output time-domain steady state response
%  for a given transfer function having a unity sine wave input.
%Sig(f,fHz)
%        where f is a the Laplace transfer function of the circuit
%               fHz is frequency of the input sine function in Hertz
%
%
%Example:
%        s = sym('s');
%        f = 10 /(s + 10);
%        Frq(f, 100)
%


echo off
s = sym('s');
w = 2 * pi * fHz; %calculate the frequency in radians/sec
T = 1 / fHz; %calculate the time period of the signal
fx = subs(f,s, j .* w); %get the magnitude and phase at
this frequency
M = abs(fx); %get just the magnitude
P = angle(fx); %get just the phase angle (radians)
x = linspace(-1*T,2*T,100);%setup to graph -1 cycle to +3 cycles
y1 = sin(w .* x); %setup the input signal
y2 = M.*sin(w .* x + P); % setup the output signal
plot(x, y1, x, y2, x, y2, 'X');grid on; figure(gcf);
```

Figure 7-21

MATLAB M-file for Example 7-10

The plot command is set up to plot two functions on a single set of axes. This is accomplished by listing one function's x-axis matrix with its corresponding y-axis matrix followed by the other function's x-axis matrix with its corresponding y-axis matrix. Notice that after the last x–y group, there is a text character X. This causes each plotted point of the last x–y group to have an X mark at each plot location.

To run this program for the given loss function, enter the following in the Command Window.

```
s = sym('s');
f = (1.5569)/(s^2 + 1.7646 * s + 1.5569);
Sig(f,0.5)
```

Notice that f is the reciprocal of the given Loss function. Figure 7-22 shows the resulting graph.

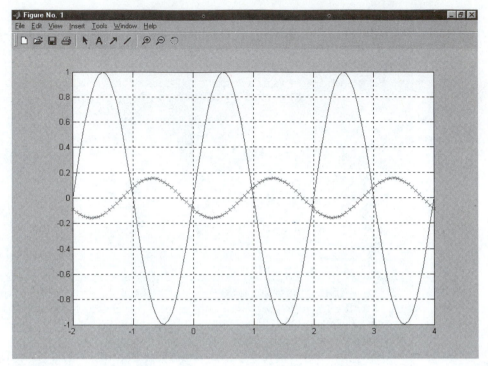

Figure 7-22

The concept in Example 7-9 could be extended to include all the filter topologies. A program for coefficient matching could also be written. Once the commands for transferring information to and from the user are known, many useful programs can be written in MATLAB.

It is significant to note that the program in the last example could be applied to circuits in Chapter 5. Using this program, we could determine the expected outputs of these passive circuits due to a sine wave input.

7-6 USING THE TI-89

In this section we will do an example that shows how to write a program for the TI-89 to calculate the R's and C's for a normalized filter. We will also write a program that will graph the relationship between the input and output signal for a given transfer function.

EXAMPLE 7-11

Write a program for the TI-89 that will calculate the R's and C's for an inverting, dual-feedback low-pass filter for the following low-pass function.

$$L(s) = \frac{s^2 + 1.7646s + 1.5569}{1.5569}$$

Solution: Figure 7-23 shows the completed program, which is named IDF. Because this is a program and not a function, the output must appear on the IO screen of the calculator. Therefore, the first instruction in this program is to clear the IO screen. After the IO screen is cleared, the text from each Disp instruction is printed on the IO screen, informing the user of what this program does.

The next group of instructions is a dialog block that produces a complex dialog box. The dialog block is made up of a Title and two Request instructions. We can have only one Title instruction, but we may have several Request instructions. The Title instruction contains the text to print at the top of the dialog box, and the Request instructions allows the user to input the values of *a* and *b*, which are stored in the variables *as* and *bs*, respectively.

Right after this dialog block, the input values of *as* and *bs* are converted from a text string into an expression and stored in *a* and *b*, respectively. This is necessary because a Request instruction receives the information entered by the user as text strings. After *a* and *b* are stored, the value of *Ce* is requested and converted to an expression in a single-line dialog box.

The values for the *R*'s and *C*'s for the inverting, dual-feedback low-pass filter are calculated by using the equations found in Appendix E, Section E-4-1. The calculation of the *R*'s and *C*'s is relatively easy to implement except for the *Cd* value. The *Cd* value must be less than or equal to a value based on *a*, *b*, and *Ce*. This maximum value is calculated from an equation in Appendix E and stored in Cdmax.

A while loop is then used to force the user to enter a valid value for *Cd*. A constant, itest, is initialized to zero just before the while loop is entered. Then, in the while loop, a dialog block is opened. The Title instruction in the Dialog block is used to inform the user what the maximum value for *Cd* must be. The text for the Title instruction is constructed by converting the expression for Cdmax into a text string using the format instruction, and then this string is concatenated to the rest of the text in the Title instruction using the & operator. If the number entered does not meet the given requirement, then itest remains zero and the loop is executed again. When the user enters a valid number, the if instruction is executed, causing itest to become equal to 1. When itest is not 0 anymore, the condition of the while loop is false, and the program goes to the next instruction after the while loop.

In this next group of instructions, *R*, *Ra*, and *h* are calculated. Using these calculated values, the results are displayed to the user on the IO screen with the next group of instructions. At the end of the program is the pause instruction that waits for the user to make note of the results, and then the last instruction causes the calculator to return to the Home screen.

After this program has been entered into the calculator, run it from the Home screen by entering IDF( ). Notice that there are no arguments passed. Figure 7-24a shows the initial display describing what the program does. Press the Enter key to cause the dialog box for entering *a* and *b* to be displayed. After entering *a* and *b*, the display will look like Fig. 7-24b. Press the Enter key twice and another Dialog box will appear for entering the value of *Ce*. Any value could be used, but enter a value of 1. The calculator screen should now look like Fig. 7-24c. When the Enter key is pressed twice after entering *Ce*, the display will indicate the maximum allowable value for *Cd* and request

```
()
Prgm
©IDF()
©No Local variables are
©defined so that we can
©access them from the Entry Line

 ClrIO ©Clear the IO screen

 Disp "This finds the values for"
 Disp "an Inverting Dual-feedback"
 Disp "low pass filter"
 Disp "(Press Enter to continue)"
 Pause

 Dialog
  Title "s^2+a*s+b/(-h)"
  Request "a",as
  Request "b",bs
 EndDlog

©convert the input strings into expressions
 expr(as)→a
 expr(bs)→b

Request "Ce",Ces
expr(Ces)→Ce

©Calculate the maximum value for Cd and ask for the Ce value
 Ce*a^2/(4*b)→Cdmax

0→itest ©set up loop test
While itest = 0
 Dialog
  Title "Select Cd <="& format(Cdmax,"e3")
  Request "Cd",Cds
 EndDlog
 if expr(Cds) < Cdmax
  1→itest
EndWhile
 expr(Cds)→Cd

©Calculate the results
1./(√ (b*Cd*Ce))→R
R/(R*Ce*a - 2.)→Ra
1./(Cd*Ce*R*R)→h
```

Figure 7-23

TI-89 program for Example 7-11

```
©Display the results
ClrIO ©Clear the IO screen
Disp"      Ra =" & format(Ra,"e5")
Disp"Rb = Rc =" & format(R,"e5")
Disp"      Cd =" & format(Cd,"e5")
Disp"      Ce =" & format(Ce,"e5")
Disp:       h =" & format(h,"e5")


pause
DispHome
EndPrgm
```

Figure 7-23

(Continued)

a value to be entered. Choosing and entering a value of 0.1, the display will now look like Fig. 7-24d. When the Enter key is pressed twice, the final results will be displayed, and it will appear as shown in Fig. 7-25. After noting the values, the Enter key is pressed once to return to the Home screen.

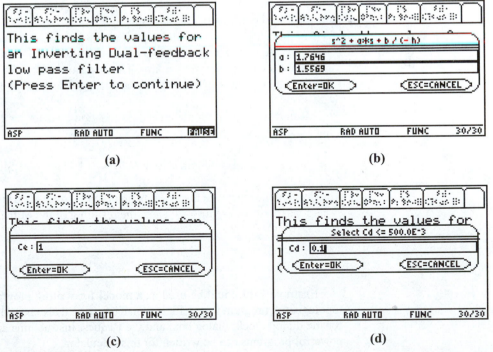

Figure 7-24

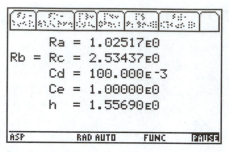

Figure 7-25

EXAMPLE 7-12

Write a program for the TI-89 that will display a graph of the input and output signals for the following loss function when the input is a unity sine wave with a frequency of 0.5 Hz.

$$L(s) = \frac{s^2 + 1.7646s + 1.5569}{1.5569}$$

Solution: Figure 7-26 shows the completed program that is stored under the name Sig. This program is written so that it will work with any Laplace transfer function. It is written for a typical transfer function, so we will have to enter the inverse of the loss function when running the program. Most all of the instructions have been discussed in previous examples, but some previously unused instructions appear in the section where the input and output functions are graphed.

In this graphing section, the Graph screen is cleared with the ClrGraph instruction. Then, by using the Define instruction, the equations for the input signal and the output signal are copied to the Y = Editor. The next instruction, Style, sets the second equation to display a thick line so that we can distinguish its graph from the graph of the first equation.

Run this program for the given loss function by entering the following on the Entry line.

$$1.5569 / (s^\wedge 2 + 1.7646*s + 1.5569) \rightarrow f:Sig(f,0.5)$$

Notice that the inverse of the loss function is entered for f because of the way that the program was written. Figure 7-27 show the resulting graph of the input and output signal.

Example 7-11 could be used as a model for writing programs for all the other filter topologies. A program for coefficient matching could also be written. Once the instructions for the dialog block, dialog box, and the Request instructions are understood, many other powerful programs can be written for this calculator.

```
(f,fHz)
Prgm
©Sig(f,fHz)
©f is the Laplace transfer function
©fHz is the frequency of the input sine function in Hz.

Local md, T, fx
©save the current mode and change mode for graphing
  getMode("All")→md
© setMode ({"Angle","Radian","Graph","Function"})

©Calculate the frequency information
 2 * π * fHz→w
 1/fHz→T ©calculate the period of sine wave

©Calculate the magnitude and phase of the sine wave for
the transfer function
 f|s=i*w→fx
 abs(fx)→M
 angle(fx)→P

©Set the x and y limits for the graph
If M > 1 Then
 M→ymax : -1*M→ymin
Else
 1→ymax : -1→ymin
EndIf

-1 * T→xmin : 2 * T→xmax


©Graph the two signals
 ClrGraph
 Define y1(x) = sin(w*x)
 Define y2(x) = M*sin(w*x+P)
 Style 2,"thick" ©Make output line thick
 DispG ©Display the graph

 Pause ©Let the user see the graph

 setMode(md) ©restore mode

EndPrgm
```

Figure 7-26

TI-89 program for Example 7-12

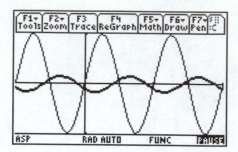

Figure 7-27

The program shown in the last example could be used for the circuits in Chapter 5. Using this program, we can predict what the output signal should look like due to a sine wave input.

PROBLEMS

Section 7-1

1. Derive the loss function for the inverting single-feedback topology shown in Fig. 7-1b.

2. Derive the loss function for the inverting dual-feedback topology shown in Fig. 7-2b.

3. Derive the loss function for the twin-T topology shown in Fig. 7-3. Assign $K = 1 + R_1/R_2$.

Section 7-2

4. Using the noninverting single-feedback topology in Fig. 7-1a with its loss function shown in Eq. (7-3), state what type of components (R, C, or combination) that Y_A and Y_B must be in order to use coefficient matching on:

   (a) $\dfrac{s + b}{h}$

   (b) $\dfrac{s + b}{hs}$

5. State the type of components (R, C, or combination) that should be used to obtain the loss function form

$$\frac{s^2 + bs + d}{hs^2}$$

   and write the resulting loss in terms of using these components when using:

   (a) The noninverting dual-feedback topology in Fig. 7-2a with its loss function shown in Eq. (7-5).
   (b) The inverting dual-feedback topology in Fig. 7-2b with its loss function shown in Eq. (7-6).

6. Design a filter for the loss function

$$\frac{s + 5}{10}$$

with the capacitor set to 1 μF. It may not be possible to match the denominator constant. If not, what is the resulting denominator constant?

(a) Use the noninverting single-feedback topology in Fig. 7-1a and loss function shown in Eq. (7-3). (Assume that $R_2 = R_A$)

(b) Use the inverting single-feedback topology in Fig. 7-1b and loss function shown in Eq. (7-4).

7. Design a filter for the loss function

$$\frac{s^2 + 1.4142s + 1}{1}$$

with the capacitors set to 1 F (normalized circuit). It may not be possible to match the denominator constant. If not, what is the resulting denominator constant? Use the noninverting dual-feedback topology in Fig. 7-2a with Y_B and Y_D as capacitors and all resistors equal in value. (Assume that $R_2 = 1\ \Omega$.)

8. Using coefficient matching and assuming that C will be chosen first, write the general equations to solve the loss function

$$\frac{s^2 + as + b}{-hs^2}$$

(a) Use the inverting dual-feedback topology in Fig. 7-2b. Assume that $Y_A = Y_B = Y_C = sC$, $Y_D = 1/R_D$, and $Y_E = 1/R_E$.

(b) Use the noninverting dual-feedback topology in Fig. 7-2a. Assume that $Y_A = Y_C = sC$, $Y_B = Y_D = 1/R$, and $Y_E = 0$. (Remember that when an admittance is 0, it means an open circuit.)

Section 7-3

9. Identify the order and type (low pass, high pass, band pass, or notch) for the loss functions shown.

(a) $\dfrac{s^2 + 5s + 25}{5}$

(b) $\dfrac{(s + 10)(s + 1000)}{s}$

(c) $\dfrac{s + 1}{s}$

10. List the biquads in a form suitable to design a cascaded filter. List the biquad in order of the first stage through the last stage.

(a) $\dfrac{(s^2 + 1.6016s + 2.5650)(s + 1.6016)}{4.1081}$

(b) $$\frac{(s^2 + 0.15306s + 0.96584)(s + 0.24765)(s^2 + 0.40071s + 0.40682)}{0.097308}$$

(c) $$\frac{(s^2 + 0.10751s + 0.93723)(s^2 + 0.28654s + 0.38201)(s + 0.17906)}{3.2547 \times 10^{-5}(s^2 + 27.586)(s^2 + 71.405)}$$

Section 7-4

Section 7-4-2

11. Using Eq. (7-14b) and Eq. (7-16), find L_{min} for a third-order Butterworth filter that has an $\Omega_s = 3$ and a $L_{max} = 0.25$ dB.

12. Design a normalized Butterworth filter to approximate the following specifications.

$$\omega_p = 1 \text{ krad/sec} \qquad \omega_s = 20 \text{ krad/sec}$$

$$L_{max} = 3.0103 \text{ dB} \qquad L_{min} = 25 \text{ dB}$$

   (a) Use only inverting topologies that will match the overall h.
   (b) Use only noninverting topologies that will match the overall h.

13. Design a normalized Butterworth filter to approximate the following specifications.

$$\Omega_s = 2 \qquad L_{max} = 1.5 \text{ dB} \qquad L_{min} = 18 \text{ dB}$$

   (a) Use the inverting topologies that match the overall h.
   (b) Use the noninverting topologies and one inverting amplifier to match the overall h. (Assume that $R_2 = 1 \Omega$.)

14. Design a third-order normalized Butterworth filter with $L_{max} = 0.25$ dB (matching the magnitude of h in the overall equation).

   (a) Use a noninverting first-order circuit (assume that $R_2 = 1 \Omega$), an inverting second-order circuit, and an inverting amplifier.
   (b) Use only two inverting stages.

Section 7-4-3

15. Design a normalized Chebyshev filter to approximate the following specifications.

$$\omega_p = 2 \text{ kHz} \qquad \omega_s = 20 \text{ kHz}$$

$$L_{max} = 3.0103 \text{ dB} \qquad L_{min} = 25 \text{ dB}$$

   (a) Use inverting topologies only. Calculate the resulting h.
   (b) Use noninverting topologies only. Calculate the resulting h. (Assume that $R_2 = 1 \Omega$.)

16. Design a normalized Chebyshev filter to approximate the specifications shown in Problem 13.

$$\Omega_s = 2 \qquad L_{max} = 1.5 \text{ dB} \qquad L_{min} = 18 \text{ dB}$$

(a) Use only inverting topologies matching the overall h.

(b) Use only noninverting topologies. Calculate the resulting h. (Assume that $R_2 = 1 \, \Omega$.)

17. Using a fourth-order Chebyshev equation with $L_{max} = 3.0103$ dB, calculate the exact magnitude and phase of the loss function at a frequency of:

(a) 0.9 rad/sec

(b) 1.2 rad/sec

Section 7-4-4

18. Design a normalized elliptic filter to approximate the specifications shown. (Assume that $R_2 = 1 \, \Omega$.)

$$\Omega_s = 2 \qquad L_{max} = 1.5 \text{ dB} \qquad L_{min} = 18 \text{ dB}$$

19. Design a normalized elliptic filter to approximate the following specifications. Use an inverting first-order circuit for the first stage.

$$\omega_p = 10 \text{ rad/sec} \qquad \omega_s = 50 \text{ rad/sec}$$

$$L_{max} = 3.0103 \text{ dB} \qquad L_{min} = 60 \text{ dB}$$

8

Practical Filters
from the Generic Loss Functions

OBJECTIVES

Upon successful completion of this chapter, you should be able to:

- Frequency shift a normalized filter to a desired frequency location.
- Impedance shift a normalized filter to realistic values of R and C.
- Gain shift an active filter to a desired gain magnitude in the pass band.
- Design high-pass filters by converting the low-pass Butterworth, Chebyshev, and elliptic loss functions into high-pass loss functions.
- Design band-pass active filters by converting the low-pass Butterworth, Chebyshev, and elliptic loss functions into band-pass loss functions.
- Define the difference between a wide-band-pass filter and a narrow-band-pass filter using Q.
- Design a wide-band-pass active filter by cascading a low-pass and a high-pass filter.
- Design narrow-band-pass Butterworth, Chebyshev, and elliptic active filters.
- Design a notch filter using a band-pass filter and a summing amplifier.
- Design notch filters using the twin-T topology.

8-0 INTRODUCTION

In this chapter we will learn how to design practical active filters. First we will learn how to change the normalized low-pass filters designed in Chapter 7 into filters with practical break frequencies and appropriate component values. Then we will learn how to design high-pass, band-pass, and notch filters from the low-pass loss function.

8-1 FREQUENCY SHIFTING

So far we have designed our circuits at a pass-band frequency of 1 rad/sec. Frequency shifting allows us to move the pass-band frequency to any frequency we desire. The technique shown in this section will work on any RC active filter, including the high-pass, band-pass, and notch. This process will change the filter loss function, but it will be unnecessary to express this shifted loss function.

Looking at the transfer function in Appendix E, we notice that the coefficients occur as an RC product. We also notice that the break frequencies always occur as the reciprocal of an RC product. Therefore, the frequency is proportional to the reciprocal of the RC combination.

$$f \propto \frac{1}{RC} \qquad (8\text{-}1)$$

If we increase the RC product by some factor, F_f, we will decrease the frequency by a proportional factor, and if we decrease the RC product by some factor, we will increase the frequency by a proportional factor.

We typically know the frequency at which the circuit was designed and the frequency to which we want to shift. We need to know the factor that relates the new frequency, f_{new} or ω_{new}, to the old frequency, f_{old} or ω_{old}.

$$f_{new} = F_f f_{old} \qquad (8\text{-}2a)$$

$$\omega_{new} = F_f \omega_{old} \qquad (8\text{-}2b)$$

From this we can solve for the frequency shifting factor.

$$F_f = \frac{f_{new}}{f_{old}} \qquad (8\text{-}3a)$$

$$F_f = \frac{\omega_{new}}{\omega_{old}} \qquad (8\text{-}3b)$$

Once this factor is calculated, we divide either all the frequency-dependent R's or all the frequency-dependent C's. (The R's in the constant K are not frequency dependent.) Therefore, we use either

$$R_{new} = \frac{R_{old}}{F_f} \qquad (8\text{-}4a)$$

or

$$C_{new} = \frac{C_{old}}{F_f} \qquad (8\text{-}4b)$$

It does not matter whether we use the R's or the C's since when we use impedance shifting in the next section, we will arrive at the same final values.

Some of the transfer functions in Appendix E have a gain constant K. Because K is a ratio of resistors, it will not be affected by frequency shifting. The resistors associated with the K factor do not have to be changed with the other R's of the circuit, but they may be changed if it is convenient. We may say that the R's associated with the K factor are not frequency dependent.

EXAMPLE 8-1

A filter is designed at a frequency of 1 rad/sec. Find the values of the components when shifted to 500 Hz.

$$R_1 = 0.8\ \Omega \qquad R_2 = 1.2\ \Omega \qquad C_1 = 1\ F \qquad C_2 = 0.1\ F$$

Solution: First we find the frequency shifting factor, F_f, from Eq. (8-3a).

$$F_f = \frac{f_{new}}{f_{old}}$$

$$= \frac{500\ Hz}{1/2\pi}$$

$$= 3141.6$$

At this point there are two possible solutions to this problem. One would be to change the R's, and the other would be to change the C's.

If we change the R's using Eq. (8-4a), the solution is

$$R_{1\ new} = \frac{R_{1\ old}}{F_f} = \frac{0.8}{3141.6} = 2.5465 \times 10^{-4}\ \Omega$$

$$R_{2\ new} = \frac{R_{2\ old}}{F_f} = \frac{1.2}{3141.6} = 3.8197 \times 10^{-4}\ \Omega$$

The final values for this approach are

$$R_1 = 2.5465 \times 10^{-4}\ \Omega \qquad R_2 = 3.8197 \times 10^{-4}\ \Omega$$

$$C_1 = 1\ F \qquad\qquad\qquad C_2 = 0.1\ F$$

If we change the C's using Eq. (8-4b), the solution is

$$C_{1\ new} = \frac{C_{1\ old}}{F_f} = \frac{1}{3141.6} = 3.1831 \times 10^{-4}\ F$$

$$C_{2\ new} = \frac{C_{2\ old}}{F_f} = \frac{0.1}{3141.6} = 3.1831 \times 10^{-5}\ F$$

The final values for this approach are

$$R_1 = 0.8\ \Omega \qquad\qquad\qquad R_2 = 1.2\ \Omega$$

$$C_1 = 3.1831 \times 10^{-4}\ F \qquad C_2 = 3.1831 \times 10^{-5}\ F$$

8-2 IMPEDANCE SHIFTING

Using frequency shifting, as shown in the last section, we can design a filter to operate at a desired break frequency, but the filter has unrealistic component values. In this section we will learn how to impedance shift these components into realistic values. Therefore, this section is the final step in designing a filter that can be physically built and used.

Impedance shifting is based on Eq. (8-1) just as frequency shifting was. If we multiply the capacitors by a factor, F_z, and divide the resistors by the same factor, the frequency will remain the same.

$$f \propto \frac{1}{\left(\dfrac{R}{F_z}\right) F_z C} \tag{8-5}$$

In this way we may change the component values, or shift their impedance, without affecting the frequency of the filter.

When impedance shifting, we are more concerned with selecting a capacitor value and calculating a resulting resistor value because there is a wider selection of resistor values manufactured than capacitor values manufactured. From Eq. (8-5) we see that the new capacitor value after impedance shifting will be

$$C_{\text{new}} = F_z C_{\text{old}} \tag{8-6a}$$

Normally, we will select the new capacitor value that we want for the final design and we will know the old capacitor value calculated previously from frequency shifting. Therefore, we can calculate the impedance shifting factor by solving for F_z in Eq. (8-6a).

$$F_z = \frac{C_{\text{new}}}{C_{\text{old}}} \tag{8-6b}$$

We then calculate the new resistor value that will keep the filter frequency the same.

$$R_{\text{new}} = \frac{R_{\text{old}}}{F_z} \tag{8-7}$$

When impedance shifting, we want to end up with capacitor values in the 0.001 μF to 1 μF range and resistor values in the 1-kΩ to 100-kΩ range because most op-amps work well in these limits. Keep in mind that the capacitors and resistors must be of good quality, as discussed in Section 6-3-2.

When fixed-value components that are not the exact values calculated are used in the final circuit, the filter usually works very well. However, if the filter specifications are critical, we may want to check the accuracy of the filter design before the circuit is built. This is done by determining the loss function using fixed-value resistors and comparing it to the loss function with exact values.

In Appendix E, for each op-amp configuration, step 2 shows the loss function in terms of the component values used in the configuration. The values of the fixed components may be plugged into this expression to determine the exact loss function. This loss function may then be plotted or the magnitude and phase calculated at the frequencies of interest to find out if the design using fixed values is close enough to the intended design.

In the following examples we will calculate the exact values and leave the choice of whether to use precision potentiometers or select fixed-value resistors to the designer.

EXAMPLE 8-2

Using the two solutions calculated in Example 8-1, determine the component values when C_1 is chosen to be 1 μF.

Solution: Using the first solution,

$$R_1 = 2.5465 \times 10^{-4} \, \Omega \qquad R_2 = 3.8197 \times 10^{-4} \, \Omega$$

$$C_1 = 1 \, F \qquad\qquad C_2 = 0.1 \, F$$

We calculate the impedance shifting factor F_z using Eq. (8-6b).

$$F_z = \frac{C_{new}}{C_{old}} = \frac{1 \, \mu F}{1 \, F} = 1 \times 10^{-6}$$

Substituting into Eqs. (8-6a) and (8-7), we obtain

$$C_{1 \, new} = F_z C_{1 \, old} = (1 \times 10^{-6})(1 \, F) = 1 \, \mu F$$

$$C_{2 \, new} = F_z C_{2 \, old} = (1 \times 10^{-6})(0.1 \, F) = 0.1 \, \mu F$$

$$R_{1 \, new} = \frac{R_{1 \, old}}{F_z} = \frac{2.5465 \times 10^{-4}}{1 \times 10^{-6}} = 254.65 \, \Omega$$

$$R_{2 \, new} = \frac{R_{2 \, old}}{F_z} = \frac{3.8197 \times 10^{-4}}{1 \times 10^{-6}} = 381.97 \, \Omega$$

Using the second solution,

$$R_1 = 0.8 \, \Omega \qquad\qquad R_2 = 1.2 \, \Omega$$

$$C_1 = 3.1831 \times 10^{-4} \, F \qquad C_2 = 3.1831 \times 10^{-5} \, F$$

Again we calculate the impedance-shifting factor F_z using Eq. (8-6b).

$$F_z = \frac{C_{new}}{C_{old}} = \frac{1 \, \mu F}{3.1831 \times 10^{-4} \, F} = 3.1416 \times 10^{-3}$$

Substituting into Eqs. (8-6a) and (8-7) gives

$$C_{1 \text{ new}} = F_z C_{1 \text{ old}} = (3.1416 \times 10^{-3})(3.1831 \times 10^{-4} \text{ F}) = 1 \ \mu\text{F}$$

$$C_{2 \text{ new}} = F_z C_{2 \text{ old}} = (3.1416 \times 10^{-3})(3.1831 \times 10^{-5} \text{ F}) = 0.1 \ \mu\text{F}$$

$$R_{1 \text{ new}} = \frac{R_{1 \text{ old}}}{F_z} = \frac{0.8}{3.1416 \times 10^{-3}} = 254.65 \ \Omega$$

$$R_{2 \text{ new}} = \frac{R_{2 \text{ old}}}{F_z} = \frac{1.2}{3.1416 \times 10^{-3}} = 381.97 \ \Omega$$

It is very significant to notice in Example 8-1 and Example 8-2 that impedance shifting and frequency shifting do not depend on whether the filter is low-pass, high-pass, or band-pass, and these two processes also do not depend on whether the circuit is from a Butterworth, Chebyshev, or elliptic loss function. Another important fact demonstrated in Example 8-1 and Example 8-2 is that no matter what component, R or C, that we frequency shifted with in the beginning, when we impedance shift to a particular value for one of the components, we always arrive at the same values for the R's and C's.

The following example demonstrates the design of a low-pass filter from beginning to end by completing an example from Chapter 7.

EXAMPLE 8-3

The normalized low-pass filter of the second-order Butterworth filter from Example 7-5 is shown in Fig. 8-1. Complete the design by frequency shifting and impedance shifting the components so that the circuit has a break frequency of 2 kHz and the 1 F capacitors are 0.01 μF.

Solution: First we need to shift the filters's frequency from 1 rad/sec to 2 kHz. Using Eq. (8-3a), we obtain

$$F_f = \frac{f_{\text{new}}}{f_{\text{old}}} = \frac{2 \text{ kHz}}{1/2\pi} = 1.2566 \times 10^4$$

Because in frequency shifting it does not matter whether we change the resistors or capacitors to shift the frequency to 2 kHz, we will arbitrarily choose to change the capacitors. In this case, both capacitors have the same value, and they will be changed by using Eq. (8-4b).

$$C_{\text{new}} = \frac{C_{\text{old}}}{F_f} = \frac{1}{1.2566 \times 10^4} = 7.9577 \times 10^{-5} \text{ F}$$

Figure 8-1

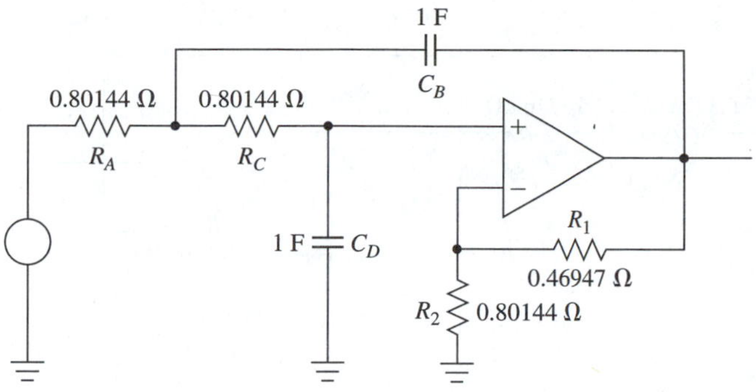

Next we need to calculate the impedance shifting factor F_z. In this case we want to make the originally given 1 F capacitor equal to 0.01 µF, but the current value of the capacitor is 7.9577×10^{-5} F. Therefore, when we substitute into Eq. (8-6b), we will use 7.9577×10^{-5} F for C_{old} and 0.01 µF for C_{new}.

$$F_z = \frac{C_{new}}{C_{old}} = \frac{0.01 \ \mu F}{7.9577 \times 10^{-5}} = 1.2566 \times 10^{-4}$$

Now we can calculate the final values for the resistors and capacitors using Eqs. (8-6a) and (8-7).

$$C_{new} = F_z \, C_{old} = (1.2566 \times 10^{-4})(7.9577 \times 10^{-5}) = 0.01 \ \mu F$$

$$R_{new} = \frac{R_{old}}{F_z} = \frac{0.80144}{1.2566 \times 10^{-4}} = 6.3776 \ k\Omega$$

The gain resistors R_1 and R_2 do not need to be included in the frequency shifting and the impedance shifting because they are not frequency-determining components. Only the ratio between them must remain the same. However, they are usually included in the frequency and impedance shifting so that at least one of the two gain resistors will have the same value as at least one of the frequency-dependent resistors. This is done for economics. With this in mind, we impedance shift the gain resistors the same as we did the other resistors, using Eq. (8-7).

$$R_{1 \ new} = \frac{R_{1 \ old}}{F_z} = \frac{0.46947}{1.2566 \times 10^{-4}} = 3.7359 \ k\Omega$$

$$R_{2 \ new} = \frac{R_{2 \ old}}{F_z} = \frac{0.80144}{1.2566 \times 10^{-4}} = 6.3776 \ k\Omega$$

The finished circuit is shown in Fig. 8-2.

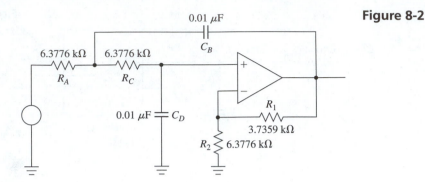

Figure 8-2

8-3 GAIN SHIFTING

The loss function tables in Chapter 7 are calculated for a pass-band loss of 0 dB. Do not get this value confused with the L_{max} value, which is the maximum loss that we can tolerate and still consider the pass-band loss to be 0 dB. Most filters are designed with a pass-band loss of 0 dB so that they can be added to an existing circuit to remove undesired frequencies in the stop band without changing the desired frequencies in the pass band. However, sometimes we want to modify the pass-band loss to something other than 0 dB.

The process of changing the pass-band loss is called *gain shifting*. This is because we are effectively changing the denominator constant of the loss function, which is the same as the gain constant when the loss function is expressed as a gain function.

We will look at two gain-shifting techniques. One technique is shifting the "gain" (actually the loss) of a loss function by controlling its denominator constant, and the other technique is shifting the gain of a circuit by using a voltage divider network. Let's look at gain shifting the loss function first.

The pass-band loss is controlled by the denominator constant h. We can define an equation that shows the new h value that we desire as the old h value multiplied by a factor M.

$$h_{new} = Mh_{old} \tag{8-8a}$$

The h_{new} and h_{old} must be denominator constants that are from loss functions having the same break frequency, and that break frequency is typically at the normalized value of 1.

When we solve Eq. (8-8a) for M we have

$$M = \frac{h_{new}}{h_{old}} \tag{8-8b}$$

Because M is a ratio, it can be calculated using normalized h values and then be used to determine the multiplying factor needed on a frequency-shifted circuit. M can also be used as a multiplying factor to modify the loss functions in Chapter 7 from their pass-band value of 0 dB to any pass-band value desired.

The amount of pass-band loss in dB that M will change a loss function can be expressed as

$$L_{shift} = -20 \log M \tag{8-9a}$$

If we know the amount we need to shift the loss, we may solve this for the multiplying factor.

$$M = 10^{-0.05 L_{shift}} \qquad \text{(8-9b)}$$

We may either multiply this factor times the normalized equation denominator before we start designing, or we may add a stage that has a loss of $-20 \log M$ (or gain of $+20 \log M$).

EXAMPLE 8-4

Modify a third-order elliptic loss function with $L_{max} = 1.5$ dB and $\Omega_s = 2$ so that it has a pass-band loss of -4 dB.

Solution: From Table 7-8a,

$$\frac{(s + 0.45674)(s^2 + 0.37285s + 0.94651)}{8.3891 \times 10^{-2}(s^2 + 5.1532)}$$

Using Eq. (8-9b) gives

$$M = 10^{-0.05 L_{shift}} = 10^{(-0.05)(-4)} = 1.5849$$

Since the loss function from Table 7-8a has a pass-band loss of 0 dB, we simply multiply the denominator by M.

$$\frac{(s + 0.45674)(s^2 + 0.37285s + 0.94651)}{(1.5849)(8.3891 \times 10^{-2})(s^2 + 5.1532)}$$

$$\frac{(s + 0.45674)(s^2 + 0.37285s + 0.94651)}{(0.13296)(s^2 + 5.1532)}$$

EXAMPLE 8-5

Find the pass-band loss of a third-order Chebyshev filter with $L_{max} = 3.0103$ if the resulting h, before frequency shifting, was 0.105.

Solution: From Table 7-4 we see that the denominator constant should have been 0.25. From Eq. (8-8b) we can find the multiplying factor that was effectively used.

$$M = \frac{h_{new}}{h_{old}}$$

$$= \frac{0.105}{0.25}$$

$$= 0.42$$

From Eq. (8-9a) we can find the amount the function was shifted.

$$L_{shift} = -20 \log M$$

$$= -20 \log (0.42)$$

$$= 7.5350 \text{ dB}$$

Therefore, the circuit has a pass-band loss of 7.5350 dB, because it was shifted by 7.5350 dB from the table loss value of 0 dB. We may say that the circuit attenuates signals in the pass band by 7.5350 dB, or another way of stating this is to say the circuit has a gain of -7.5350 dB.

In some cases, when we change the h value in the tables for a desired pass-band loss, we can design the circuit from a topology without any problems. However, with some topologies we can only calculate the resulting value of h. In Chapter 7 it was suggested that another stage could be added to either increase or decrease the overall h to the desired value, but we typically want to use as few op-amps as possible. An alternative to adding another stage is to gain shift the filter circuit. We may still have to use a second stage because this method is limited to moderate levels of not less than 0.1 to not greater than 10, but this frequently solves the problem.

By using a voltage divider circuit on the output of an op-amp, we can either attenuate or amplify the output signal. Figure 8-3 shows how the voltage divider network can be used to attenuate the overall gain or to increase the overall loss for inverting and noninverting circuits. The overall loss will be shifted by

$$L_{shift} = -20 \log \left(\frac{R_\alpha}{R_\alpha + R_\beta} \right) \tag{8-10a}$$

Knowing the shift required, we choose a value for either R_α or R_β and solve for the other resistor. Solving the equation for R_β, we have

$$R_\beta = R_\alpha (10^{0.05 L_{shift}} - 1) \tag{8-10b}$$

Figure 8-4 shows how to amplify the overall gain or decrease the overall loss for inverting and noninverting circuits. In this case we have reduced the feedback by the voltage divider of two series resistors. The op-amp must therefore increase its output in order to balance the reduced feedback voltage.

The overall loss will be shifted by

$$L_{shift} = -20 \log \left(\frac{R_\alpha + R_\beta}{R_\alpha} \right) \tag{8-11a}$$

Here again, if we know the shift required, we assume a value for one of the resistors and solve for the other. Solving the equation for R_β yields

$$R_\beta = R_\alpha (10^{-0.05 L_{shift}} - 1) \tag{8-11b}$$

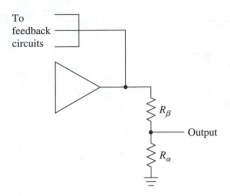

Figure 8-3

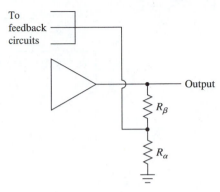

Figure 8-4

This method of gain shifting is very useful in circuit configurations where we cannot match the h in the biquad stage. We design the circuit, determine what multiplying factor to use, and add the voltage divider circuit.

EXAMPLE 8-6

Figure 8-5 is a second-order Butterworth filter with a pass-band gain of 0 dB. Find R_β if we choose R_α to be 100 kΩ and we want to change the pass-band gain to -2 dB.

Solution: If we want to shift the gain to be -2 dB, we want to shift the loss by $+2$ dB. Because we are attenuating the gain or increasing the loss, we use Eq. (8-10b).

$$R_\beta = R_\alpha(10^{0.05L_{\text{shift}}} - 1)$$

$$= (100\text{ k})[10^{(0.05)(2)} - 1]$$

$$= 25.893\text{ k}\Omega$$

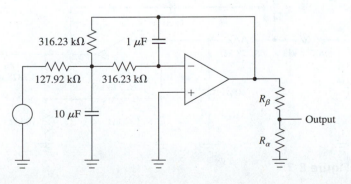

Figure 8-5

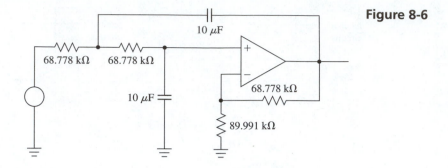

Figure 8-6

EXAMPLE 8-7

Figure 8-6 shows a Chebyshev filter. Design a voltage divider network for the output to increase the gain by 5 dB using $R_\alpha = 10 \text{ k}\Omega$.

Solution: Increasing the gain by 5 dB is the same as shifting the loss by -5 dB. Using Eq. (8-11b), we have

$$R_\beta = R_\alpha(10^{-0.05L_{\text{shift}}} - 1)$$

$$= (10 \text{ k})[10^{(-0.05)(-5)} - 1]$$

$$= 7.7828 \text{ k}\Omega$$

Figure 8-7 shows the resulting circuit. Notice that all of the feedback paths are moved to between R_α and R_β.

8-4 HIGH-PASS FILTER

So far, the loss function tables in Chapter 7 for the Butterworth, Chebyshev, and elliptic functions have been used only for low-pass filters. These loss functions are often referred to as *generic loss functions* because they can be transformed into high-pass, band-pass, and

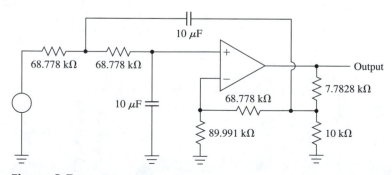

Figure 8-7

notch loss functions. In this section we will learn how to transform these generic loss functions into high-pass loss functions.

To transform a low-pass generic loss function into a high-pass loss function, we make the following substitution for s

$$s = \frac{1}{s}$$
(8-12)

This causes the stop-band frequency to occur below 1 rad/sec instead of above 1 rad/sec, while maintaining the pass-band frequency at 1 rad/sec. The low-pass loss function is flipped over the 1-rad/sec line. This procedure can be applied to the Butterworth, Chebyshev, and elliptic loss functions equally well.

For example, Fig. 8-8a shows the second-order Chebyshev filter

$$\frac{s^2 + 0.64359s + 0.70711}{0.5}$$

If we substitute $1/s$ into this, we have the loss function transformed into a high-pass loss function.

$$\frac{\left(\frac{1}{s}\right)^2 + (0.64359)\left(\frac{1}{s}\right) + 0.70711}{0.5}$$

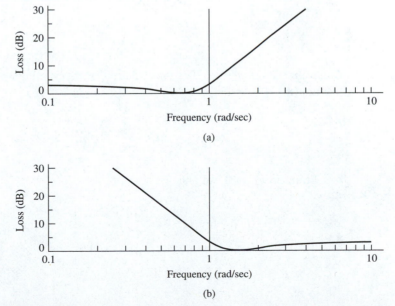

(a)

(b)

Figure 8-8

This is then expressed in the standard form as

$$\frac{s^2 + 0.91017s + 1.4142}{0.70710s^2}$$

This loss function is plotted in Figure 8-8b.

Finding the appropriate low-pass loss function to transform is very simple. It is done by using the same equations that are used for determining a low-pass function except that we calculate the Ω_s using the following equation.

$$\Omega_s = \frac{\omega_p}{\omega_s} \tag{8-13}$$

Notice that this is the reciprocal of how Ω_s is found for a low-pass filter. It is important to note that this Ω_s is not the normalized stop-band frequency for the high-pass filter. It is effectively a transformation of given high-pass specifications into the equivalent low-pass value to use for the graphs and equations requiring an Ω_s value. We will substitute this value into the low-pass equations from Chapter 7 to determine the appropriate low-pass loss function to use. The following example will demonstrate this process.

EXAMPLE 8-8

For the filter specification, design a Butterworth filter using an inverting amplifier with $C = 0.01 \, \mu F$ and the pass-band loss shifted by -10 dB.

$$\omega_p = 8 \text{ krad/sec} \qquad \omega_s = 1 \text{ krad/sec}$$

$$L_{max} = 3.0103 \text{ dB} \qquad L_{min} = 15 \text{ dB}$$

Solution: From the specifications we see that this is a high-pass filter; therefore, we use Eq. (8-13) to find Ω_s.

$$\Omega_s = \frac{\omega_p}{\omega_s} = \frac{8 \text{ k}}{1 \text{ k}} = 8$$

From Eq. (7-14b), we find

$$\varepsilon^2 = 10^{0.1 L_{max}} - 1$$

$$= 10^{(0.1)(3.0103)} - 1$$

$$= 1$$

and from Eq. (7-16),

$$n \geq \frac{\log\left(\dfrac{10^{0.1L_{\min}} - 1}{\varepsilon^2}\right)}{2\log(\Omega_s)}$$

$$\geq \frac{\log\left(\dfrac{10^{(0.1)(15)} - 1}{1}\right)}{2\log(8)}$$

$$\geq 0.82276$$

Therefore, use $n = 1$.

From the first entry in Table 7-1, the low-pass loss function to use is

$$\frac{s + 1}{1}$$

Substituting in $s = 1/s$, we have

$$\frac{\left(\dfrac{1}{s}\right) + 1}{1}$$

and putting it in a standard form, we have

$$\frac{s + 1}{s}$$

Going to Section E-2-2, we choose $C_B = 1$ F. Therefore,

$$R_B = \frac{1}{bC_B} = \frac{1}{(1)(1)} = 1\ \Omega$$

$$C_A = hC_B = (1)(1) = 1\ \text{F}$$

Next, we will frequency shift the filter using Eq. (8-3b).

$$F_f = \frac{\omega_{\text{new}}}{\omega_{\text{old}}}$$

$$= \frac{8\ \text{k}}{1}$$

$$= 8000$$

Dividing R by this value using Eq. (8-4a), we find that

$$R_{B\,\text{new}} = \frac{R_{B\,\text{old}}}{F_f} = \frac{1}{8000} = 1.25 \times 10^{-4}\ \Omega$$

We then calculate the frequency-shifting factor using Eq. (8-6b).

$$F_z = \frac{C_{\text{new}}}{C_{\text{old}}} = \frac{0.01\ \mu\text{F}}{1\ \text{F}} = 1 \times 10^{-8}$$

Substituting into Eqs. (8-6a) and (8-7) gives

$$C_{A\,\text{new}} = C_{B\,\text{new}} = F_z C_{1\,\text{old}} = (1 \times 10^{-8})(1\ \text{F}) = 0.01\ \mu\text{F}$$

$$R_{B\,\text{new}} = \frac{R_{B\,\text{old}}}{F_z} = \frac{1.25 \times 10^{-4}}{1 \times 10^{-8}} = 12.5\ \text{k}\Omega$$

Next we will change the overall loss by -10 dB, which is, in effect, increasing the overall gain by $+10$ dB. We could add an amplifier with a gain of 10, but in this case we will use a voltage divider network on the output to gain shift the amplifier, as described in Section 8-3. Because we want the loss to be shifted by -10 dB, our L_{shift} value must be -10 dB. Choosing $R_\alpha = 12.5$ kΩ and using $L_{\text{shift}} = -10$ dB, we substitute into Eq. (8-11) and solve for R_β.

$$R_\beta = R_\alpha\,(10^{-0.05 L_{\text{shift}}} - 1)$$

$$= (12.5\ \text{k})\,[10^{(-0.05)(-10)} - 1]$$

$$= 27.028\ \text{k}\Omega$$

The final circuit is shown in Fig. 8-9.

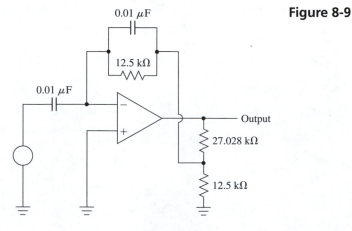

Figure 8-9

8-5 BAND-PASS FILTER

Band-pass filters can be divided into two types: wide band pass and narrow band pass. The wide-band-pass filter can be constructed by cascading low-pass and high-pass filters. The narrow band pass uses biquad band-pass stages.

8-5-1 General Band-Pass Loss Functions

Figure 8-10 shows the loss response of a normalized first-order Butterworth band-pass filter. There are two rows of frequency names along the bottom of the graph. The top row, ω_L, ω_o, and ω_H, contains the names and indicates the locations of significant frequencies used to describe any band-pass response. The bottom row contains the names and indicates the locations of frequencies used to describe band-pass filter specifications. Keep in mind that these locations could be frequencies in Hz and indicated with f instead of ω.

The pass band is normally considered to be the frequencies between the 3 dB points indicated at ω_L and ω_H in Fig. 8-10. However, we will not limit ourselves to this and will allow the use of pass bands defined between the 1.5-dB or 0.25-dB points as well as the 3.0103-dB points. When we do this, it will affect the interpretation of some of the general equations used for band-pass circuits. Let's review the general band-pass equations so we can see how they apply to band-pass filters and how they differ when using different pass-band definitions.

The center frequency, f_0 or ω_0, is the geometric mean of the frequencies of the lower point, f_L or ω_L, and the upper point, f_H or ω_H.

$$f_0 = \sqrt{f_L f_H} \tag{8-14a}$$

Expressed in radians,

$$\omega_0 = \sqrt{\omega_L \omega_H} \tag{8-14b}$$

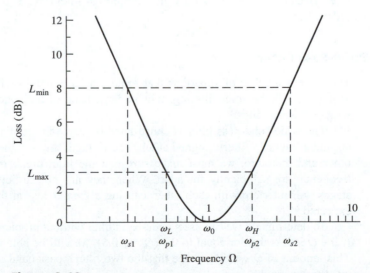

Figure 8-10

The band width, BW, is

$$BW_{Hz} = f_H - f_L \qquad \text{(8-15a)}$$

Expressed in radians,

$$BW_{R/S} = \omega_H - \omega_L \qquad \text{(8-15b)}$$

The quality factor, Q, is

$$Q = \frac{f_0}{BW_{Hz}} \qquad \text{(8-16a)}$$

In radian quantities, Q is

$$Q = \frac{\omega_0}{BW_{R/S}} \qquad \text{(8-16b)}$$

Both Eqs. (8-16a) and (8-16b) will yield the same value when applied to the same filter. Because we are not restricting ourselves to the typical endpoints, 3.0103 dB, this Q will be different from the normally defined Q when we use L_{max} different from 3.0103 dB.

The Q for a filter designed at a particular L_{max} cannot be compared to a Q found for another filter designed at another L_{max}. If we find the Q of a filter that was designed at a particular L_{max} and calculate the Q of this filter using a larger L_{max}, we will calculate a smaller Q value for this same filter. This occurs because using a larger L_{max} will increase the band width but will not change the center frequency.

It is important to remember that a second-order band-pass filter will have an upper and lower roll-off of 20 dB/dec each. Because the roll-off occurs at both ends of the response, it will require twice as many orders to have the same roll-off as a low-pass or a high-pass filter. Because we are working with symmetrical filters, we will always require an even-order filter.

8-5-2 Wide-Band-Pass Filters

There are a number of situations that require a large band of frequencies to pass. One example would be voice filtering. When a large band of frequencies is required, we use a wide-band-pass filter.

The wide-band-pass filter is constructed by cascading a high-pass filter designed at ω_L and a low-pass filter designed at ω_H. These stages are designed independently of each other and therefore, we must make sure that the high break frequency and low break frequency are sufficiently far apart. When they are not sufficiently separated, the two stages will interfere with each other, causing a greater loss at the center frequency than expected.

To determine if we can use a wide-band filter without problems, we will look at the Q. If the Q is less than or equal to 0.5 ($Q \leq 0.5$), we will be able to use a wide-band filter. This amount of Q will guarantee that the two filter's pass-band frequencies will be sufficiently separated.

Because we are cascading two stages together, the overall dB loss in the pass band will be the addition of the dB loss of each stage's pass band. For example, if the low-pass stage had a pass-band loss of 10 dB and the high-pass stage had a pass-band loss of -2 dB, then the cascaded system would have an approximately 8 dB loss at the center of the pass band. This value is approximate because the two stages do affect each other somewhat.

When cascading the two stages, we will want to put the high-pass filter first. This is because a high-pass filter has a capacitive input that will block any dc voltage present before the voltage gets to either of the op-amps. If we put the low-pass filter first, the op-amp in this stage could be overdriven by a dc voltage that would be considered small for dc values but large for ac signal values. In this book we will always put the high-pass filter first.

EXAMPLE 8-9

Design a Butterworth filter with the following specifications.

$$\omega_{p1} = 3 \text{ rad/sec} \qquad \omega_{p2} = 48 \text{ rad/sec}$$

$$\omega_{s1} = 0.25 \text{ rad/sec} \qquad \omega_{s2} = 576 \text{ rad/sec}$$

$$L_{max} = 0.25 \text{ dB} \qquad L_{min} = 30 \text{ dB}$$

Solution: We must first determine whether this is a wide-band or a narrow-band filter from Eqs. (8-14b), (8-15b), and (8-16b).

$$\omega_0 = \sqrt{\omega_L \omega_H}$$

$$= \sqrt{(3)(48)}$$

$$= 12 \text{ rad/sec}$$

$$BW_{R/S} = \omega_H - \omega_L$$

$$= 48 - 3$$

$$= 45 \text{ rad/sec}$$

$$Q = \frac{\omega_0}{BW_{R/S}}$$

$$= \frac{12}{45}$$

$$= 0.26667$$

This is ≤ 0.5 and is, therefore, a wide-band-pass filter.

The two sides of the center frequency must be symmetrical. A simple test for this is to make sure that the number of decades between ω_{s1} and ω_{p1} is the same as the number of decades between ω_{p2} and ω_{s2}

$$\mathrm{ND}_{\mathrm{low}} = \log\left(\frac{\omega_{p1}}{\omega_{s1}}\right) = \log\left(\frac{3}{0.25}\right) = 1.0792$$

$$\mathrm{ND}_{\mathrm{high}} = \log\left(\frac{\omega_{s2}}{\omega_{p2}}\right) = \log\left(\frac{576}{48}\right) = 1.0792$$

Therefore, because either side has the same distance between the stop band and the pass band and because they have the same L_{max} and L_{min}, the band-pass specification given is symmetrical.

Next we split this specification into a low-pass filter specification and a high-pass filter specification. The low-pass specification is

$$\omega_p = 48 \text{ rad/sec} \qquad \omega_s = 576 \text{ rad/sec}$$

$$L_{\mathrm{max}} = 0.25 \text{ dB} \qquad L_{\mathrm{min}} = 30 \text{ dB}$$

The high-pass specification is therefore

$$\omega_p = 3 \text{ rad/sec} \qquad \omega_s = 0.25 \text{ rad/sec}$$

$$L_{\mathrm{max}} = 0.25 \text{ dB} \qquad L_{\mathrm{min}} = 30 \text{ dB}$$

The specification is symmetrical, so we will require the same-order filter. For the low-pass filter,

$$\Omega_s = \frac{\omega_s}{\omega_p} = \frac{576}{48} = 12$$

For the high-pass filter, the corresponding low-pass stop-band frequency using Eq. (8-13) is

$$\Omega_s = \frac{\omega_p}{\omega_s} = \frac{3}{0.25} = 12$$

Because both the high-pass and the low-pass filter are symmetrical, we will find the same value for ε^2 and n. Using the equations from Chapter 7, we calculate

$$\varepsilon^2 = 5.9254 \times 10^{-2}$$

$$n \geq 1.9584 \qquad \text{use } n = 2$$

Therefore, we use a second-order Butterworth. The loss function from Table 7-3 is

$$\frac{s^2 + 2.8664s + 4.1081}{4.1081}$$

This loss function will apply to the low-pass filter and the high-pass filter. For the high-pass filter, we must transform this equation using Eq. (8-12). The result is

$$\frac{s^2 + 0.69774s + 0.24342}{s^2}$$

At this point the design follows the procedure outlined in previous sections. We must remember that the low-pass filter must be frequency shifted to 48 rad/sec and the high-pass filter must be shifted to 3 rad/sec.

Figure 8-11 shows the finished wide-band-pass filter. In each stage the gain was shifted for a pass-band loss of 0 dB. This was not requested in the problem, but filters are often designed for 0 dB in their pass band. It turned out that both stages needed to be attenuated by the same amount, and for economic reasons, both stages were given the same resistor values in their output voltage divider network.

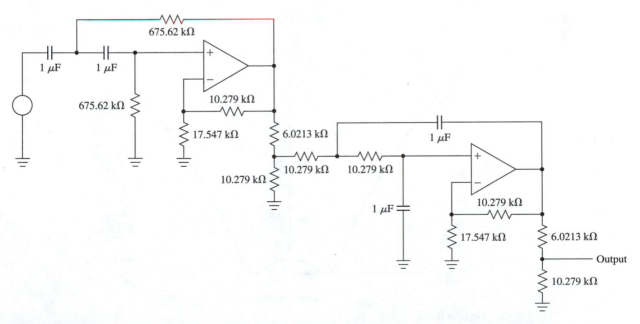

Figure 8-11

8-5-3 Narrow-Band-Pass Filters

Narrow-band-pass filters pass a smaller band width of frequencies and will have a $Q > 0.5$. This type of filter may be very useful when a small range of frequencies must be detected in a group of frequencies.

Narrow-band-pass filters may be constructed by using a biquad band-pass filter. These may, also, be cascaded to form higher-order band-pass filters.

To transform a low-pass generic loss function into a high-pass loss function with a center frequency of 1 rad/sec, we make the substitution

$$s = \frac{s^2 + 1}{\left(\dfrac{\omega_H - \omega_L}{\omega_0}\right)s} \tag{8-17}$$

where $\dfrac{\omega_H - \omega_L}{\omega_0}$ is the normalized radian band width.

This will cause the low-pass function to shift above 1 rad/sec and cause a second set of poles and zeros in a high-pass form to be generated. The high-pass poles and zeros will be shifted below 1 rad/sec and the low-pass poles and zeros will be shifted above 1 rad/sec such that the center frequency will be 1 rad/sec. An example of a Chebyshev band-pass filter response is shown in Fig. 8-12. Once this substitution is made, we design the filter from the equations in Appendix E like any other filter.

These band-pass filters are assumed to be symmetrical, but if they are not, we should apply the techniques shown in Chapter 6 to make the specifications symmetrical. Because the filters will be symmetrical, we may use either the high-pass side or the low-pass side to determine the filter order required (just as we did for the wide-band-pass filter).

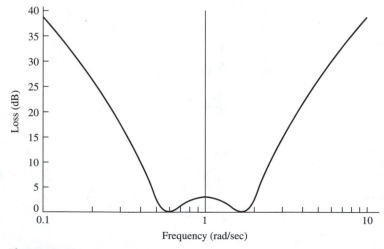

Figure 8-12

EXAMPLE 8-10

Design a Chebyshev filter with the following specifications.

$$\omega_{p1} = 400 \text{ rad/sec} \qquad \omega_{p2} = 10 \text{ krad/sec}$$

$$\omega_{s1} = 200 \text{ rad/sec} \qquad \omega_{s2} = 20 \text{ krad/sec}$$

$$L_{\text{max}} = 3.0103 \text{ dB} \qquad L_{\text{min}} = 5 \text{ dB}$$

Use $C = 0.22 \text{ } \mu\text{F}$. Calculate the resulting loss at the center frequency.

Solution: First we find the order of filter required using the specifications on the low-pass side (the upper break-frequency side). Using the equations in Chapter 7, we find that

$$\Omega_s = \frac{\omega_{s2}}{\omega_{p2}} = \frac{20 \text{ k}}{10 \text{ k}} = 2$$

$$\varepsilon^2 = 1$$

and

$$n \geq 0.71037 \qquad \text{use } n = 1$$

The loss function for a first-order Chebyshev filter from Table 7-4 is

$$\frac{s + 1}{1}$$

Next we calculate ω_0. We use the pass-band frequencies ω_{p1} and ω_{p2} because they are the upper and lower break frequencies. From Eq. (8-14b),

$$\omega_0 = \sqrt{\omega_L \omega_H}$$

$$= \sqrt{(400)(10 \text{ k})}$$

$$= 2 \text{ krad/sec}$$

The filter is symmetrical, so we could have used the stop-band frequencies and calculated the same value, but we should use the pass-band frequencies to be technically correct.

Using Eq. (8-17), we let

$$s = \frac{s^2 + 1}{\left(\dfrac{\omega_H - \omega_L}{\omega_0}\right)s}$$

$$= \frac{s^2 + 1}{\left(\dfrac{10\,k - 400}{2\,k}\right)s}$$

$$= \frac{s^2 + 1}{4.8s}$$

Substituting this into the equation form of Table 7-4 gives us

$$\frac{\dfrac{s^2 + 1}{4.8s} + 1}{1}$$

which becomes

$$\frac{s^2 + 4.8s + 1}{4.8s}$$

Notice that this has changed the equation order to second order.

We were not given directions on the type of topology to use so we will arbitrarily choose the inverting dual-feedback topology. Following the procedure in Section E-4-3, we assign

$$C_C = C_B = 1 \text{ F}$$

We then calculate

$$R_A = R_B = \frac{a}{Cb} = \frac{4.8}{(1)(1)} = 4.8 \ \Omega$$

$$R_D = \frac{2}{Ca} = \frac{2}{(1)(4.8)} = 0.41667 \ \Omega$$

$$h = \frac{1}{C_B R_A} = \frac{1}{(1)(4.8)} = 0.20833$$

Currently, the filter is designed with a center frequency of $\omega_0 = 1$ rad/sec. The desired center frequency is $\omega_0 = 2$ krad/sec. We calculate the frequency shifting factor F_f using Eq. (8-3b).

$$F_f = \frac{\omega_{\text{new}}}{\omega_{\text{old}}}$$

$$= \frac{2\text{ k}}{1}$$

$$= 2 \times 10^3$$

Shifting the capacitors using Eq. (8-4b) yields

$$C_{\text{new}} = \frac{C_{\text{old}}}{F_f} = \frac{1}{2\text{ k}} = 5 \times 10^{-4}\text{ F}$$

Finally, we impedance shift the circuit. Using Eq. (8-6b), we find the shifting factor.

$$F_z = \frac{C_{\text{new}}}{C_{\text{old}}} = \frac{0.22\ \mu\text{F}}{5 \times 10^{-4}\text{ F}} = 4.4 \times 10^{-4}$$

Substituting into Eqs. (8-6a) and (8-7) gives

$$C_{\text{new}} = F_z C_{\text{old}} = (4.4 \times 10^{-4})(5 \times 10^{-4}\text{ F}) = 0.22\ \mu\text{F}$$

$$R_A = R_E = R_{\text{new}} = \frac{R_{\text{old}}}{F_z} = \frac{4.8}{4.4 \times 10^{-4}} = 10.909\text{ k}\Omega$$

$$R_{D\text{ new}} = \frac{R_{D\text{ old}}}{F_z} = \frac{0.41667}{4.4 \times 10^{-4}} = 946.97\ \Omega$$

The final circuit is shown in Fig. 8-13.

We did not take into account the overall gain, but we were not given a gain requirement. The resulting h, before frequency shifting, for this design was 0.20833. Using Eq. (8-8b), we can determine the effective multiplying factor.

$$M = \frac{h_{\text{new}}}{h_{\text{old}}}$$

$$= \frac{0.20833}{4.8}$$

$$= 0.043403$$

Figure 8-13

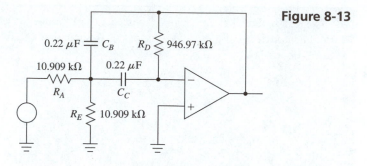

Using Eq. (8-9a), we can determine the dB shift from 0 dB or the overall loss in the pass band.

$$L_{shift} = -20 \log(M)$$

$$= -20 \log(0.043403)$$

$$= 27.250 \text{ dB}$$

Figure 8-14 shows the loss response of this filter.

8-6 NOTCH FILTER

In this section we examine two ways to design a notch filter. The first way is to use a band-pass filter and a summing amplifier. The other way is to cascade single biquad notch filters.

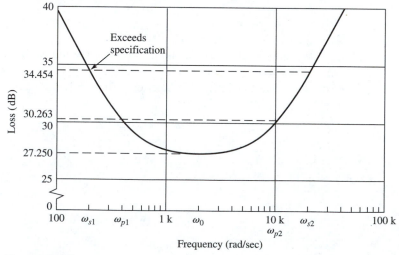

Figure 8-14

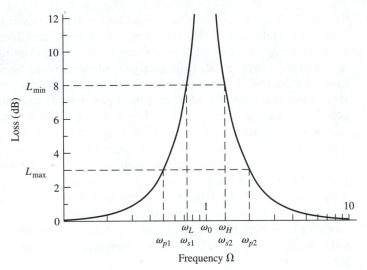

Figure 8-15

Figure 8-15 shows the loss response of a normalized first-order Butterworth notch filter. The center frequency, bandwidth, and Q are calculated with the same equations used for a band-pass filter—Eqs. (8-14), (8-15), and (8-16)—except that the stop-band frequencies, ω_{s1} and ω_{s2}, are used for ω_L and ω_H in the equations.

Because ω_L and ω_H are the stop-band frequencies, the Q of a particular notch filter is dependent on the L_{min} value instead of the L_{max} value, as was the case for the band-pass filters. Therefore, we must remember to use the same L_{min} value when comparing different notch filter Qs and their band widths.

8-6-1 Parallel Design

Figure 8-16 shows the basic parallel design. An inverting band-pass filter is designed using the center frequency and band-width requirements for the notch. With the band-pass filter as one path, two parallel paths for the signal are made using a summing amplifier.

Because the output of the band-pass filter is inverted and the path around the band-pass filter through R_2 is not inverted, the signals from the two parallel paths will subtract. Figure 8-17 shows this process, but because this process is easier to comprehend when using graphs of gain functions instead of loss functions, the graphs are drawn as gain functions.

Figure 8-17a shows the output gain function of the summing amplifier due to the inverting band-pass filter, and Fig. 8-17b shows the output gain function due to the path

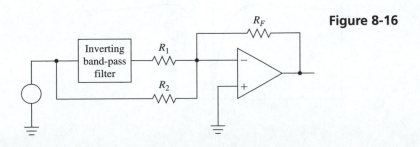

Figure 8-16

through R_2. When the input signal is much higher or much lower in frequency than the pass band, the output due to the inverting band-pass filter is insignificant. Therefore, the output of the summing amplifier is almost entirely due to the signal through the R_2 path. When the input signal has frequencies in the pass band of the inverting band-pass filter, the inverting band-pass filter will have a significant output to combine with the signal from the R_2 path. If the pass-band gain of the inverting band-pass filter is equal to the gain through the R_2 path, then the signals will be equal and opposite and will therefore cancel each other out. In other words, the frequencies from the input signal that are in the pass band will be removed from the signal. Therefore, we have notched out a band of frequencies from the signal.

The overall loss function for the circuit in Fig. 8-16 can be determined by subtracting 1 from the inverting band-pass filter's loss function. However, because the loss function is the reciprocal of the gain function, we must find a common numerator instead of a common denominator to perform this subtraction. If we assume that the loss for the inverting band-pass filter is a numerator polynomial divided by a denominator polynomial, $L_{\text{band pass}} = N(s)/D(s)$, then the resulting transfer function will be $L_{\text{notch}} = N(s)/[D(s) - N(s)]$

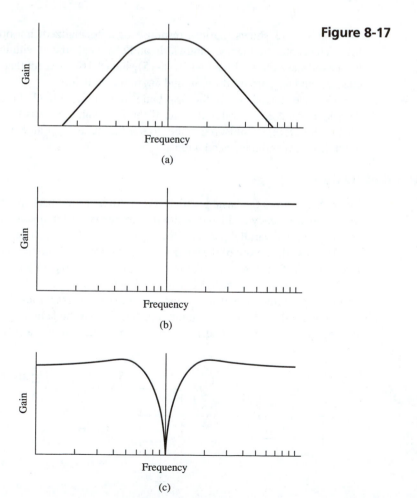

Figure 8-17

EXAMPLE 8-11

Design a notch filter using a second-order Chebyshev band-pass filter having an $L_{\max} = 3.0103$ dB, $\omega_0 = 2$ krad/sec, and a $\text{BW}_{R/S} = 9.6$ krad/sec. Design the notch for a gain of 4 dB above and below the center frequency, and no output at the center frequency.

Solution: To make a second-order band-pass filter, we start with a first-order low-pass generic loss function. Therefore, using the first entry in Table 7-4, we find

$$\frac{s + 1}{1}$$

Using Eq. (8-17), we determine the function required to transform this low-pass loss function into a band-pass function.

$$s = \frac{s^2 + 1}{\left(\dfrac{\omega_H - \omega_L}{\omega_0}\right)s} = \frac{s^2 + 1}{\left(\dfrac{9600}{2000}\right)s} = \frac{s^2 + 1}{4.8\,s}$$

Substituting this into the first-order low-pass loss function, we have

$$\frac{s^2 + 4.8s + 1}{4.8s}$$

From this equation we notice that this is the same band-pass filter designed in Example 8-10. We will refer to this example for the rest of the design of the band-pass filter part. The completed band-pass filter is shown in Fig. 8-13. This band-pass filter is then put into the configuration shown in Fig. 8-16 to make the notch filter.

We now turn our attention to setting up the gain. We were assigned to have a gain of 4 dB above and below the center frequency. A gain of 4 dB is the same as a loss of -4 dB. Because the gain function and loss function are reciprocals and because the gain function multiplier is in the numerator, we can use the loss multiplier equation, Eq. (8-9b), to find the gain multiplier needed.

$$M = 10^{-0.05L_{\text{shift}}}$$

$$= 10^{(-0.05)(-4)}$$

$$= 1.5849$$

This value could also have been found simply by using $G_{\text{dB}} = 20 \log(G)$.

We know that the gain of an inverting amplifier through the R_2 path is equal to (R_f/R_2), and we just found that this gain must be equal to 1.5849. If we let R_F be equal to 100 kΩ, then

$$R_2 = \frac{R_F}{M} = \frac{100\text{ k}}{1.5849} = 63.096\text{ k}\Omega$$

From Example 8-10 the denominator factor h, before frequency shifting, was 0.20833 for the band-pass filter. In order for the band-pass filter to have a pass-band gain of 0 dB, the denominator factor, before frequency shifting, should have been 4.8. Using Eq. (8-8b), we therefore need a multiplying factor of

$$M = \frac{h_{\text{new}}}{h_{\text{old}}}$$

$$= \frac{4.8}{0.20833}$$

$$= 23.040$$

We also need to increase this by an additional gain of 4 dB or a loss of -4 dB, which we previously calculated as a multiplying factor of 1.5849. We must have both the band-pass filter and the bypass path equal in order for the output of the notch to be nulled at the center frequency. Because we are using a summing amplifier, we can include both of these multiplying factors in R_1.

$$R_1 = \frac{R_F}{M} = \frac{100\text{ k}}{(23.040)(1.5849)} = 2.7385\text{ k}\Omega$$

Figure 8-18 shows the final circuit.

Example 8-11 could have been specified in several different ways. One way, for example, could be to specify the stop bands and pass bands as

$$\omega_{p1} = 200\text{ rad/sec} \qquad \omega_{p2} = 20\text{ krad/sec}$$

$$\omega_{s1} = 400\text{ rad/sec} \qquad \omega_{s2} = 10\text{ krad/sec}$$

$$L_{\text{max}} = 3.0103\text{ dB} \qquad L_{\text{min}} = 5\text{ dB}$$

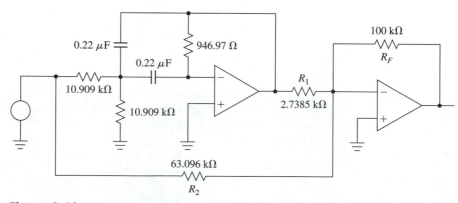

Figure 8-18

Using the parallel design, we would transform these specifications into band-pass filter specifications. This is easily done by swapping the frequencies of ω_{p1} and ω_{p2} with the frequencies of ω_{s1} and ω_{s2}, respectively. Once this is done the design proceeds the same as before.

8-6-2 Cascaded Design

Using biquads, we can cascade notch-filter stages to obtain a filter with the desired characteristics. In this design we will find, using the topologies in Appendix E, that the twin-T is the only possibility.

To convert a generic low-pass loss function from Chapter 7 to a notch loss function with a center frequency of 1 rad/sec, we make the substitution

$$s = \frac{\left(\dfrac{\omega_H - \omega_L}{\omega_0}\right)s}{s^2 + 1} \tag{8-18}$$

where $\dfrac{\omega_H - \omega_L}{\omega_0}$ is the normalized radian bandwidth.

We notice that this is the reciprocal of the band-pass filter transformation equation (8-17). This substitution will cause a second set of poles and zeros to appear in the low-pass equation similar to the band-pass filter transformation, but this substitution will also flip the function response upside down.

EXAMPLE 8-12

Design a fourth-order Chebyshev notch filter with $Q = 5$, $\omega_0 = 4$ krad/sec, and $L_{max} = 1.5$ dB.

Solution: We require a fourth order so we will use a second-order low-pass loss function to transform into a notch loss function. From Table 7-5,

$$\frac{s^2 + 0.92218s + 0.92521}{0.77846}$$

To be able to transform this loss function into a notch loss function, we need to know $(\omega_H - \omega_L)$, which is also the same as $\text{BW}_{R/S}$. We were given Q and ω_0, so we can use Eq. (8-16b),

$$Q = \frac{\omega_0}{\text{BW}_{R/S}}$$

and solve for $BW_{R/S}$.

$$BW_{R/S} = \frac{\omega_0}{Q} = \frac{4\,k}{5} = 800 \text{ rad/sec}$$

We may then state that

$$(\omega_H - \omega_L) = 800 \text{ rad/sec}$$

Therefore, substituting into Eq. (8-18), we obtain

$$s = \frac{\left(\dfrac{\omega_H - \omega_L}{\omega_0}\right)s}{s^2 + 1}$$

$$= \frac{\left(\dfrac{800}{4\,k}\right)s}{s^2 + 1}$$

$$= \frac{0.2s}{s^2 + 1}$$

Substituting this into the low-pass equation, we have

$$\frac{\left(\dfrac{0.2s}{s^2 + 1}\right)^2 + 0.92218\left(\dfrac{0.2s}{s^2 + 1}\right) + 0.92521}{0.77846}$$

Simplifying, we have

$$\frac{s^4 + 0.19935s^3 + 2.0432s^2 + 0.19935s + 1}{0.84139(s^2 + 1)^2}$$

Using a calculator or Bairstow's method shown in Appendix F, we can factor this into biquads.

$$\frac{1}{0.84139}\left(\frac{s^2 + 0.090605s + 0.83323}{s^2 + 1} + \frac{s^2 + 0.10874s + 1.2002}{s^2 + 1}\right)$$

Figure 8-19 shows the response of this equation.

Following the design equations given in Section E-5, frequency shifting to 4 krad/sec, and impedance shifting, we will find the values shown in Fig. 8-20. The combined resulting h for the two stages is 1.663.

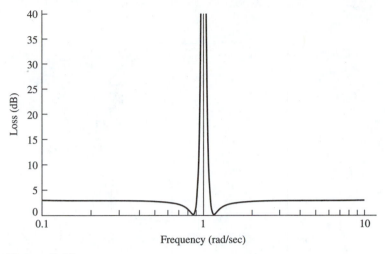

Figure 8-19

8-7 USING MATLAB

In this section we will go through an example on how to use MATLAB to convert a low-pass function into a high-pass function, and we will go through an example on drawing the Bode plot of a band-pass filter that uses fixed-value components.

EXAMPLE 8-13

Using MATLAB, transform the given elliptic low-pass function into a band-pass function where $\omega_L = 2$ krad/sec and $\omega_H = 8$ krad/sec.

$$L(s) = \frac{s^2 + 0.60745s + 0.75593}{7.1612 \times 10^{-2}(s^2 + 7.4641)}$$

Solution: Using Eq. (8-14b), we can calculate the center frequency of the band-pass function as

$$\omega_0 = \sqrt{\omega_L \omega_H} = \sqrt{2\,k\,8\,k} = 4000 \text{ krad/sec}$$

We then substitute this value and the values for ω_L and ω_H into Eq. (8-17).

$$s = \frac{s^2 + 1}{\left(\dfrac{\omega_H - \omega_L}{\omega_0}\right)s} = \frac{s^2 + 1}{\left(\dfrac{8\,k - 2\,k}{4\,k}\right)s} = \frac{s^2 + 1}{1.5s}$$

When this expression is substituted into the low-pass function for s, the low-pass function will be converted to a high-pass function. This will be easily done in MATLAB.

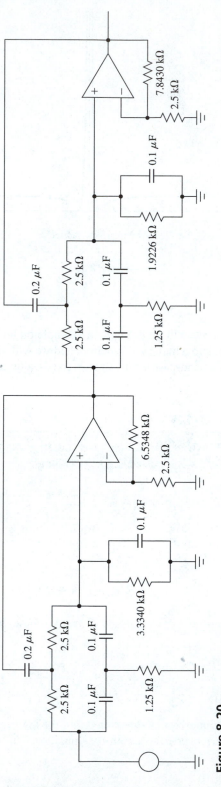

Figure 8-20

From MATLAB we define the variable *s* as a symbol and enter the low-pass loss function. Enter the following into the Command Window.

```
s  = sym('s');
fL = (s^2 + 0.60745*s + 0.75593) /
     (0.071612 * (s^2 + 7.4641));
```

The new expression for *s* will be substituted into the fL function using the subs( ) command, and the resulting expression will be simplified using the simplify( ) command. The simplify( ) command will cancel any common factor that may exist between the numerator and denominator factors. Enter into the Command Window the following.

```
fH = subs(fL, (s^2 + 1)/(1.5*s), 's');
fH = simplify (fH);
```

The function in fH contains the high-pass function, but if we display it in the Command Window, we would notice that the coefficients of the numerator and denominator contain very complex numerical expressions. In order to make sense of the answer, we must break up the fH function into a numerator and a denominator, and then find the roots of the numerator and denominator. To break up the function into a numerator and a denominator, enter the following command.

```
[n,d] = numden (fH);
```

This command stores the numerator of fH into *n* and the denominator of fH into *d*. From these variables we can find the roots of the numerator of fH and the denominator of fH.

Enter the following into the Command Window to find the roots of the numerator.

```
format short e;
nmatrix = sym2poly(n);
roots (nmatrix)
```

The format command tells MATLAB to print the answer using five significant digits for numerical values. The sym2poly( ) command converts the numerator into a matrix containing the coefficients of the polynomial. The roots( ) command calculates the roots of the nmatrix matrix.

The last two commands could have been combined in one command, thus avoiding the new variable. However, we will want to use the first element in the nmatrix variable along with the first element from a similar matrix for the denominator to calculate the new denominator multiplying factor for the band-pass function.

Enter the following to find the roots of the denominator of fH.

```
dmatrix = sym2poly(d);
roots (dmatrix)
```

The multiplying factor for the denominator can now be found by using the first element of the nmatrix and the first element of the dmatrix. Enter the following.

```
dmatrix(1) / nmatrix(1)
```

Notice that this is the denominator value divided by the numerator value, which is upside down to what would normally be done. This is because the multiplying factor of a loss function goes in the denominator.

Figure 8-21 shows the Command Window after entering the preceding commands. From the values shown in the Command Window, we can conclude that the elliptic band-pass function is

$$\frac{[(s + 0.34821)^2 + (1.7668)^2][(s + 0.10738)^2 + (0.54483)^2]}{0.071612\,[s^2 + (4.3291)^2][s^2 + (0.23100)^2]}$$

$$\frac{(s^2 + 0.6942s + 3.24283)(s^2 + 0.21476s + 0.30837)}{0.071612\,(s^2 + 18.7411)(s^2 + 0.053361)}$$

EXAMPLE 8-14

A Chebyshev band-pass filter was designed in Example 8-10, and the resulting circuit is shown in Fig. 8-13. Assume that this filter is built with standard-value resistors of 10 kΩ and 1 kΩ, and also assume that the capacitors are the standard value 0.22 μF, as shown in the figure. Draw the resulting frequency response of this circuit with these fixed-value components using the Frq( ) program from Example 5-12.

Solution: From the information given above we can conclude that the filter's components are

$$R_A = R_E = 10 \text{ k}\Omega$$

$$R_D = 1 \text{ k}\Omega$$

$$C_B = C_C = 0.22 \text{ μF}$$

Because the circuit shown in Fig. 8-13 is an inverting dual-feedback topology and because it is a band-pass configuration, the loss function is given by Eq. (E-48) found in Appendix E.

$$\frac{s^2 + \left(\dfrac{1}{C_C R_D} + \dfrac{1}{C_B R_D}\right)s + \dfrac{1}{C_B C_C}\left(\dfrac{R_A + R_E}{R_A R_D R_E}\right)}{\left(\dfrac{-1}{C_B R_A}\right)s}$$

```
>> s = sym('s');
>> fL = (s^2 + 0.60745*s + 0.75593) / (0.071612 *
        (s^2 + 7.4641));
>> fH = subs (fL, (s^2 + 1)/(1.5*s), 's');
>> fH = simplify (fH);
>> [n,d] = numden (fH);
>> format short e;
>> nmatrix = sym2poly(n);
>> roots (nmatrix)

ans =

-3.4821e-001 +1.7668e+000i
-3.4821e-001 -1.7668e+000i
-1.0738e-001 +5.4483e-001i
-1.0738e-001 -5.4483e-001i

>> dmatrix = sym2poly(d);
>> roots (dmatrix)

ans =

         0 +4.3291e+000i
         0 -4.3291e+000i
         0 +2.3100e-001i
         0 -2.3100e-001i

>> dmatrix / nmatrix

ans =

2.9523e-001
```

Figure 8-21   MATLAB Command Window for Example 8-13

Substituting in the values for the given R's and C's, the loss function is

$$\frac{s^2 + 9090.9s + 4.1322 \times 10^6}{-454.55s}$$

Because this function has a center frequency of 2 krad/sec, as seen from Example 8-10, a reasonable range of frequencies to display would be from 100 rad/sec to 100 krad/sec. To graph this band-pass loss function in this range, enter the following into the Command Window.

```
f = (s^2 +9090.9*s + 4.1322e6)/(-454.55 * s);
Frq(f,100,3)
```

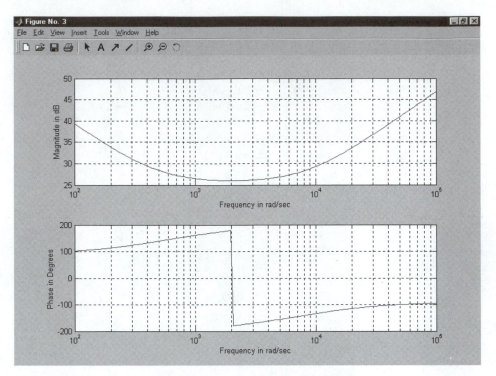

Figure 8-22

Figure 8-22 shows the resulting graph of the magnitude and phase. Keep in mind that there is not a sudden change in the phase angle. Normally, MATLAB keeps all the phase angles between $\pm 180°$, and the phase, in reality, keeps increasing smoothly on up to 270°.

8-8 USING THE TI-89

This section will cover how to use the TI-89 to convert a low-pass function into a high-pass function, and it will cover how to draw the plot of a band-pass filter that uses fixed-value components.

EXAMPLE 8-15

Using the TI-89, transform the given elliptic low-pass function into a band-pass function where $\omega_L = 2$ krad/sec and $\omega_H = 8$ krad/sec.

$$L(s) = \frac{s^2 + 0.60745s + 0.75593}{7.1612 \times 10^{-2}(s^2 + 7.4641)}$$

Solution: From Eq. (8-14b), we calculate the center frequency of the band-pass function as

$$\omega_0 = \sqrt{\omega_L\,\omega_H} = \sqrt{2\,k\,8\,k} = 4000 \text{ krad/sec}$$

Then, substituting this value and the values for ω_L and ω_H into Eq. (8-17), we have

$$s = \frac{s^2 + 1}{\left(\dfrac{\omega_H - \omega_L}{\omega_0}\right)s} = \frac{s^2 + 1}{\left(\dfrac{8\,k - 2\,k}{4\,k}\right)s} = \frac{s^2 + 1}{1.5s}$$

This expression is substituted into the low-pass function for s to convert the low-pass function into a high-pass function. Enter the following on the Entry line.

$(x\char`\^2 +0.60745*x+0.75593)/(0.071612*(x\char`\^2+7.4641))|x = (s\char`\^2+1)/(1.5*s)\rightarrow f\,H$

Figure 8-23a shows the calculator's screen after the Enter key is pressed. The expression just calculated is the high-pass function, but we need it in a different form to be able to design the op-amp circuit.

Because the loss functions have the multiplying constant in the denominator, we need to determine the denominator multiplying factor. From Fig. 8-23a we see that the numerator constant is 13.9641 and that the coefficients of the highest power of s in both the numerator and the denominator factors are 1. Therefore, the denominator multiplying constant is the reciprocal of 13.9641, which is 0.0716122.

The roots of the numerator will be found next. Enter the following on the Entry line.

cZeros(getNum(fH), s)

This is a compound instruction. The inside instruction, getNum( ), pulls out the numerator polynomial of the fH function. The outside instruction, cZeros( ), finds the roots of the resulting numerator polynomial using s as the variable. Figure 8-23b shows the result of this calculation.

The denominator roots are found in a similar manner by substituting the netDenom( ) for the getNum( ) instruction. Enter the following on the Entry line.

cZeros(getDenom(fH), s)

Figure 8-23c shows the result of this calculation.

From the preceding information, the high-pass function can be written in the following form.

$$\frac{[(s + 0.34821)^2 + (1.7668)^2][(s + 0.10738)^2 + (0.54483)^2]}{0.071612[s^2 + (4.3291)^2][s^2 + (0.23100)^2]}$$

$$\frac{(s^2 + 0.6942s + 3.24283)(s^2 + 0.21476s + 0.30837)}{0.071612(s^2 + 18.7411)(s^2 + 0.053361)}$$

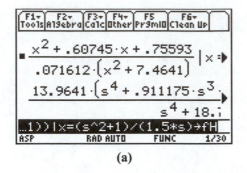

(a)

(b)

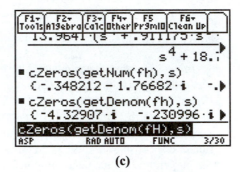

(c)

Figure 8-23

EXAMPLE 8-16

A Chebyshev band-pass filter was designed in Example 8-10, and the resulting circuit is shown in Fig. 8-13. Assume that this filter is built with standard-value resistors of 10 kΩ and 1 kΩ, and also assume that the capacitors are the standard value 0.22 μF, as shown in the figure. Draw the resulting frequency response of this circuit with these fixed-value components using the Frq( ) program from Example 5-13.

Solution: From the information given above we can conclude that the filter's components are

$$R_A = R_E = 10 \text{ k}\Omega$$

$$R_D = 1 \text{ k}\Omega$$

$$C_B = C_C = 0.22 \text{ }\mu\text{F}$$

Because the circuit shown in Fig. 8-13 is an inverting dual-feedback topology and because it is a band-pass configuration, the loss function is given by Eq. (E-48) found in Appendix E.

$$\frac{s^2 + \left(\dfrac{1}{C_C R_D} + \dfrac{1}{C_B R_D}\right)s + \dfrac{1}{C_B C_C}\left(\dfrac{R_A + R_E}{R_A R_D R_E}\right)}{\left(\dfrac{-1}{C_B R_A}\right)s}$$

Substituting in the values for the given Rs and Cs, the loss function is

$$\frac{s^2 + 9090.9s + 4.1322 \times 10^6}{-454.55s}$$

A good range for the frequency axis is 100 rad/sec to 100 krad/sec, since this function has a center frequency of 2 krad/sec. Enter the following on the Entry line.

$$\text{frq}((s^2 + 9090.9*s + 4.1322\text{E}6)/(-454.55 * s),100,3)$$

Figure 8-24 shows the graph of the magnitude and phase on the split screen. Notice that because the TI-89 keeps the phase angle values between $\pm 180°$, there is a sudden jump where it makes one of these transitions. If the calculator allowed the phase angle to continue going positive, the phase graph would go to $270°$.

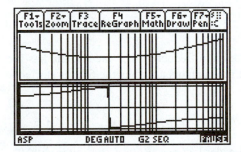

Figure 8-24

PROBLEMS

Section 8-1

1. A filter is designed at a frequency of 1 rad/sec. Changing the resistors, find the values of the components when shifted to 2 kHz.

$$R_1 = 1\ \Omega \qquad R_2 = 2\ \Omega \qquad C_1 = 2F \qquad C_2 = 0.2\ F$$

2. A filter is designed at a frequency of 10 Hz. Changing the capacitors, find the values of the components when shifted to 1 rad/sec.

$$R_1 = 1.8\ \Omega \qquad R_2 = 0.2\ \Omega \qquad C_1 = 10\ F \qquad C_2 = 1\ F$$

3. A filter is designed at a frequency of 1 kHz. Changing the capacitors, find the values of the components when shifted to 5 kHz.

$$R_1 = 3\ \Omega \qquad R_2 = 2.2\ \Omega \qquad C_1 = 0.01\ F \qquad C_2 = 0.1\ F$$

4. A filter is designed at a frequency of 1 rad/sec. Changing the resistors, find the values of the components when shifted to 2.5 krad/sec.

$$R_1 = 2\ k\Omega \qquad R_2 = 2.2\ k\Omega \qquad C_1 = 2\ \mu F \qquad C_2 = 0.1\ \mu F$$

Section 8-2

5. Impedance shift

$$R_1 = 4\ \Omega \qquad R_2 = 0.2\ \Omega \qquad C_1 = 10\ F \qquad C_2 = 1\ F$$

so that:

(a) $C_1 = 10\ \mu F$
(b) $C_2 = 10\ \mu F$

6. A filter is designed at a frequency of 1 rad/sec and has the component values listed. Find the value of the components when the filter is shifted to 10 krad/sec and the C_1 is shifted to 2.2 μF.

$$R_1 = 0.8\ \Omega \qquad R_2 = 0.1\ \Omega \qquad C_1 = 2\ F \qquad C_2 = 1\ F$$

7. Design a second-order low-pass Butterworth filter with $L_{max} = 0.25$ dB using a non-inverting dual-feedback topology. Calculate the resulting h value, and then frequency shift the filter to 4.2 kHz and impedance shift it so that $C_A = 0.001\ \mu F$. Assume $R_2 = R_A$. (This problem is continued in Problem 13.)

8. Design a fourth-order low-pass Butterworth filter with $L_{max} = 3.0103$ dB with two noninverting dual-feedback topologies. Calculate the resulting h value, and then frequency shift the filter to 2 krad/sec and impedance shift it so that $C = 0.1\ \mu F$. Assume $R_2 = R_A$. (This problem is continued in Problem 14.)

Section 8-3

9. Modify the first-order Butterworth function with $L_{max} = 1.5$ dB so that it has a pass-band loss of:

   (a) -5 dB
   (b) $+5$ dB

10. Modify the fourth-order elliptic function with $L_{max} = 0.25$ dB and $\Omega_s = 5$ so that it has a pass-band loss of:

    (a) -8 dB
    (b) $+10$ dB

11. Showing the points to connect the op-amp output and the feedback circuit, design a resistive voltage divider network (with $R_\alpha = 22$ kΩ) to shift a filter's loss by:

    (a) $+3.0103$ dB
    (b) -2.5 dB

12. Showing the points to connect the op-amp output and the feedback circuit, design a resistive voltage divider network (with $R_\beta = 22$ kΩ) to shift the filter's gain by:

    (a) $+3.0103$ dB
    (b) -2.5 dB

13. Using a resistive voltage divider network ($R_\alpha = R_B$), shift the pass-band gain of the Butterworth filter in Problem 7 to a pass-band gain of:

    (a) 20 dB
    (b) -3 dB

14. Using a resistive voltage divider network ($R_\alpha = R_A$), shift the pass-band gain of the Butterworth filter in Problem 8 to a pass-band gain of:

    (a) -2 dB
    (b) 10 dB

Section 8-4

15. Transform a third-order Chebyshev equation with $L_{max} = 1.5$ dB into a high-pass Chebyshev equation.

16. Transform a third-order elliptic equation with $L_{max} = 0.25$ dB and $\Omega_s = 2$ into a high-pass elliptic equation.

17. For the filter specification, design a Butterworth filter using a noninverting topology. Set $C = 0.22$ µF, and let $R_2 = R_B = R_\alpha$. Set the pass-band loss to -6 dB.

$$\omega_p = 440 \text{ rad/sec} \qquad \omega_s = 110 \text{ rad/sec}$$

$$L_{max} = 0.25 \text{ dB} \qquad L_{min} = 12 \text{ dB}$$

18. For the filter specification, design an elliptic filter. Use a twin-T topology. Make $C_A = 0.1\ \mu F$, and $R_2 = R_B = R_\alpha$.

$$\omega_p = 9\ krad/sec \qquad \omega_s = 4.5\ krad/sec$$

$$L_{max} = 3.0103\ dB \qquad L_{min} = 20\ dB$$

Section 8-5

Section 8-5-1

19. Find the Q, $BW_{R/S}$, BW_{Hz}, f_0, and ω_0 for the specifications shown.

$$\omega_{p1} = 2.4\ krad/sec \qquad \omega_{p2} = 3.75\ krad/sec$$

$$\omega_{s1} = 1.2\ krad/sec \qquad \omega_{s2} = 7.5\ krad/sec$$

$$L_{max} = 0.25\ dB \qquad L_{min} = 40\ dB$$

20. Find ω_L and ω_H for a $BW_{R/S}$ of 500 rad/sec at a center frequency of 2 krad/sec.

Section 8-5-2

21. For the filter specification, design a Chebyshev filter. Use inverting topologies, design the filter with a 0-dB pass-band loss. Make $C_{A1} = C_{E2} = 0.022\ \mu F$, $C_{D2} = 0.1 \times C_{E2}$, and $R_\alpha = 10\ k\Omega$.

$$f_{p1} = 720\ Hz \qquad f_{p2} = 4.5\ kHz$$

$$f_{s1} = 160\ Hz \qquad f_{s2} = 20.25\ kHz$$

$$L_{max} = 1.5\ dB \qquad L_{min} = 12\ dB$$

22. For the filter specification, design an elliptic filter using a twin-T topology with a pass-band loss of -10 dB. Make $C_A = 0.01\ \mu F$, and $R_2 = R_\alpha = 10\ k\Omega$.

$$\omega_{p1} = 1.35\ krad/sec \qquad \omega_{p2} = 9.6\ krad/sec$$

$$\omega_{s1} = 675\ rad/sec \qquad \omega_{s2} = 19.2\ krad/sec$$

$$L_{max} = 0.25\ dB \qquad L_{min} = 8\ dB$$

Section 8-5-3

23. Transform the third-order Butterworth function with $L_{max} = 1.5$ dB into a band-pass Butterworth function with $\omega_0 = 5$ krad/sec having a Q of 10.

24. Transform a fourth-order elliptic function with $L_{max} = 3.0103$ dB and $\Omega_s = 2$ into a band-pass elliptic function with $\omega_0 = 50$ rad/sec having a Q of 2.

25. For the filter specification, write the normalized Chebyshev equation.

$$f_{p1} = 2 \text{ kHz} \qquad f_{p2} = 4.5 \text{ kHz}$$

$$f_{s1} = 400 \text{ Hz} \qquad f_{s2} = 22.5 \text{ kHz}$$

$$L_{max} = 3.0103 \text{ dB} \qquad L_{min} = 30 \text{ dB}$$

26. For the filter specification, design a Butterworth filter using a noninverting topology. Make $C = 2.2 \text{ μF}$ and $R_2 = R$. Calculate the resulting pass-band gain.

$$\omega_{p1} = 37.5 \text{ rad/sec} \qquad \omega_{p2} = 54 \text{ rad/sec}$$

$$\omega_{s1} = 1 \text{ rad/sec} \qquad \omega_{s2} = 2.025 \text{ krad/sec}$$

$$L_{max} = 0.25 \text{ dB} \qquad L_{min} = 10 \text{ dB}$$

Section 8-6

Section 8-6-1

27. Using parallel design, design a notch filter using a second-order Butterworth band-pass filter having $L_{max} = 0.25 \text{ dB}$, $\omega_0 = 200 \text{ rad/sec}$, and $BW_{R/S} = 200 \text{ rad/sec}$. Design the gain outside the notch to be 5 dB, and design the notch center to have no output. Make $C_1 = 1 \text{ μF}$ and the summing amplifier with an $R_F = 100 \text{ k}\Omega$.

28. Using parallel design, design a notch filter using a second-order Chebyshev band-pass filter having $L_{max} = 0.25 \text{ dB}$, $\omega_0 = 200 \text{ rad/sec}$, and $BW_{R/S} = 200 \text{ rad/sec}$. Design the gain outside the notch to be 5 dB, and design the notch center to have no output. Make $C_1 = 1 \text{ μF}$ and the summing amplifier with an $R_F = 100 \text{ k}\Omega$. (This is the same as Problem 27 except using a Chebyshev circuit.)

Section 8-6-2

29. Transform a third-order Chebyshev equation with $L_{max} = 3.0103 \text{ dB}$ into a normalized notch Chebyshev equation with $\omega_0 = 200 \text{ rad/sec}$ having a Q of 5.

30. Design a second-order Butterworth notch having $L_{max} = 3.0103 \text{ dB}$, $Q = 2$, and $BW_{R/S} = 500 \text{ rad/sec}$. Make $C_A = 0.1 \text{ μF}$ and $R_2 = R_B$.

31. For the filter specification, design an elliptic filter using any reasonable topology with any reasonable capacitor values.

$$\omega_{p1} = 1.96 \text{ krad/sec} \qquad \omega_{p2} = 9 \text{ krad/sec}$$

$$\omega_{s1} = 3.92 \text{ krad/sec} \qquad \omega_{s2} = 4.5 \text{ krad/sec}$$

$$L_{max} = 3.0103 \text{ dB} \qquad L_{min} = 20 \text{ dB}$$

Appendix A

Transform Tables

A-1 TRANSFORM PAIRS

f(t)	F(s)	
$\delta(t)$	1	(P-1)
$u(t)$	$\dfrac{1}{s}$	(P-2)
$t\,u(t)$	$\dfrac{1}{s^2}$	(P-3)
$e^{-bt}\,u(t)$	$\dfrac{1}{s+b}$	(P-4)
$\sin(\omega t)\,u(t)$	$\dfrac{\omega}{s^2+\omega^2}$	(P-5)
$\cos(\omega t)\,u(t)$	$\dfrac{s}{s^2+\omega^2}$	(P-6)

A-2 TRANSFORM OPERATIONS

f(t)	F(s)	
$h(t-a)\,u(t-a)$	$e^{-as}\,H(s)$	(O-1)
$e^{-bt}\,h(t)\,u(t)$	$H(s+b)$	(O-2)
$th(t)\,u(t)$	$-\dfrac{d}{ds}[H(s)]$	(O-3)
$\left\{\dfrac{d}{dt}[h(t)]\right\}u(t)$	$s\,H(s)-h(0)$	(O-4a)
$\dfrac{d}{dt}[h(t)\,u(t)]$	$s\,H(s)$	(O-4b)
$\displaystyle\int_0^t h(t)\,u(t)\,dt$	$\dfrac{H(s)}{s}$	(O-5)

A-3 TRANSFORM IDENTITIES

$$\mathscr{L}[f(t)] = \mathscr{L}[f(t)\,u(t)] = F(s) \tag{I-1}$$

$$\mathscr{L}^{-1}[F(s)] = f(t)\,u(t)$$

$$f(t) \Leftrightarrow F(s) \tag{I-2}$$

$$\mathscr{L}[K f(t)] = K\mathscr{L}[f(t)] \tag{I-3}$$

$$\mathscr{L}^{-1}[K F(s)] = K\mathscr{L}^{-1}[F(s)]$$

$$\mathscr{L}[f_1(t) + f_2(t) + \cdots] = \mathscr{L}[f_1(t)] + \mathscr{L}[f_2(t)] + \cdots \tag{I-4}$$

$$\mathscr{L}^{-1}[F_1(s) + F_2(s) + \cdots] = \mathscr{L}^{-1}[F_1(s)] + \mathscr{L}^{-1}[F_2(s)] + \cdots$$

The following are often mistaken as identities but they are not.

$$\mathscr{L}[f_1(t) f_2(t)] \quad \textit{is not} \quad \mathscr{L}[f_1(t)]\mathscr{L}[f_2(t)]$$

$$f(t) \quad \textit{is not} \quad F(s)$$

A-4 EXTENDED TRANSFORM PAIRS

f(t)	F(s)	
$t^n u(t)$	$\dfrac{n!}{s^{(n+1)}}$	**(P-7a)**
$\dfrac{1}{(n-1)!} t^{(n-1)} u(t)$	$\dfrac{1}{s^n}$	**(P-7b)**
$t^n e^{-bt} u(t)$	$\dfrac{n!}{(s+b)^{(n+1)}}$	**(P-8a)**
$\dfrac{1}{(n-1)!} t^{(n-1)} e^{-bt} u(t)$	$\dfrac{1}{(s+b)^n}$	**(P-8b)**
$e^{-\alpha t} \sin(\omega t)\, u(t)$	$\dfrac{\omega}{[(s+\alpha)^2 + \omega^2]}$	**(P-9)**
$e^{-\alpha t} \cos(\omega t)\, u(t)$	$\dfrac{(s+\alpha)}{[(s+\alpha)^2 + \omega^2]}$	**(P-10)**
$\sin(\omega t + \theta)\, u(t)$	$\dfrac{s \sin\theta + \omega \cos\theta}{[(s+\omega)^2 + \omega^2]}$	**(P-11)**
$e^{-\alpha t} \sin(\omega t + \theta)\, u(t)$	$\dfrac{s \sin\theta + (\omega \cos\theta + \alpha \sin\theta)}{[(s+\alpha)^2 + \omega^2]}$	**(P-12a)**
$\sqrt{A^2 + K^2}\, e^{-\alpha t} \sin(\omega t + \theta)\, u(t)$ where $K = \dfrac{(B - A\alpha)}{\omega}$ $\theta = \tan^{-1}\left(\dfrac{A}{K}\right)$ The signs of K (x-direction) and A (y-direction) determine the quadrant of the angle.	$\dfrac{As + B}{[(s+\alpha)^2 + \omega^2]}$	**(P-12b)**

Appendix B

Laplace Derivations

This appendix derives the Laplace transform pairs, the Laplace operations, and the inverse formula for complex poles.

B-1 DERIVING LAPLACE TRANSFORM PAIRS

B-1-1 δ(*t*) Function (P-1 Pair)

$$\mathcal{L}[\delta(t)] = \int_0^\infty \delta(t)\, e^{-st}\, dt$$

For an impulse, the value is zero except at $t = 0$. Therefore, the integral will obtain a value only at $t = 0$. Since e^{-st} has a value of 1 at $t = 0$ and the impulse has an area of 1 at $t = 0$, then

$$\int_0^\infty \delta(t)\, e^{-st}\, dt = 1$$

B-1-2 *u*(*t*) Function (P-2 Pair)

$$\mathcal{L}[u(t)] = \int_0^\infty u(t)\, e^{-st}\, dt$$

After $t = 0$, $u(t)$ is equal to a constant value of 1.

$$\int_0^\infty e^{-st}\, dt$$

If we let

$$u = -st$$

then

$$du = -s\, dt$$

Rearranging the function yields

$$\frac{1}{-s} \int_0^\infty e^{-st}(-s)\, dt = \left. \frac{e^{-st}}{-s} \right|_0^\infty = \frac{1}{s}$$

B-1-3 *t* Function (P-3 Pair)

$$\mathcal{L}[t\,u(t)] = \int_0^\infty t\,u(t)\,e^{-st}\,dt$$

After $t = 0$, $u(t)$ is equal to 1.

$$\int_0^\infty te^{-st}\,dt$$

Using

$$\int u\,dv = uv - \int v\,du$$

we let

$$u = t \quad \text{and} \quad dv = e^{-st}\,dt$$

Therefore,

$$du = dt \quad \text{and} \quad v = \frac{-e^{-st}}{s}$$

Substituting into the new form gives us

$$\frac{-te^{-st}}{s}\bigg|_0^\infty - \int_0^\infty \frac{-e^{-st}}{s}\,dt$$

$$0 - \frac{1}{s^2}\int_0^\infty e^{-st}(-s)\,dt$$

$$\frac{-e^{-st}}{s^2}\bigg|_0^\infty = \frac{1}{s^2}$$

B-1-4 e^{-bt} Function (P-4 Pair)

$$\mathcal{L}[e^{-bt}\,u(t)] = \int_0^\infty e^{-bt}\,u(t)\,e^{-st}\,dt$$

After $t = 0$, $u(t)$ is equal to 1, and we may combine the two exponential functions.

$$\int_0^\infty e^{-t(s+b)}\,dt$$

If we let

$$u = -t(s + b)$$

then

$$du = -(s + b)\, dt$$

Rearranging the function yields

$$\frac{1}{-(s + b)} \int_0^\infty e^{-t(s+b)}[-(s + b)]\, dt$$

$$\left. \frac{e^{-t(s+b)}}{-(s + b)} \right|_0^\infty = \frac{1}{s + b}$$

B-1-5 sin(ωt) Function (P-5 Pair)

$$\mathcal{L}[\sin(\omega t)\, u(t)] = \int_0^\infty \sin(\omega t)\, u(t)\, e^{-st}\, dt$$

After $t = 0$, $u(t)$ is equal to 1, and using Euler's identities, we obtain

$$\int_0^\infty \frac{e^{j\omega t} - e^{-j\omega t}}{2j}\, e^{-st}\, dt$$

Combining exponentials and splitting up the integral, we have

$$\frac{1}{2j} \left[\int_0^\infty e^{-t(s-j\omega)}\, dt - \int_0^\infty e^{-t(s+j\omega)}\, dt \right]$$

Both integrals are integrated using the same process.

$$u = -t(s - j\omega) \quad \text{and} \quad u = -t(s + j\omega)$$

Therefore,

$$du = -(s - j\omega)\, dt \quad \text{and} \quad du = -(s + j\omega)\, dt$$

This results in

$$\frac{1}{2j} \left[\frac{e^{-t(s-j\omega)}}{-(s - j\omega)} - \frac{e^{-t(s+j\omega)}}{-(s + j\omega)} \right]_0^\infty$$

$$\frac{1}{2j} \left(\frac{1}{s - j\omega} - \frac{1}{s + j\omega} \right)$$

$$\frac{\omega}{s^2 + \omega^2}$$

B-1-6 cos(ωt) Function (P-6 Pair)

$$\mathcal{L}[\cos(\omega t)\, u(t)] = \int_0^\infty \cos(\omega t)\, u(t)\, e^{-st}\, dt$$

After $t = 0$, $u(t)$ is equal to 1, and using Euler's identities, we obtain

$$\int_0^\infty \frac{e^{j\omega t} + e^{-j\omega t}}{2}\, e^{-st}\, dt$$

Combining the exponentials, we have

$$\frac{1}{2}\left[\int_0^\infty e^{-t(s-j\omega)}\, dt + \int_0^\infty e^{-t(s+j\omega)}\, dt\right]$$

Both integrals are integrated using the same process.

$$u = -t(s - j\omega) \quad \text{and} \quad u = -t(s + j\omega)$$

Therefore,

$$du = -(s - j\omega)\, dt \quad \text{and} \quad du = -(s + j\omega)\, dt$$

This results in

$$\frac{1}{2}\left[\frac{e^{-t(s-j\omega)}}{-(s - j\omega)} + \frac{e^{-t(s+j\omega)}}{-(s + j\omega)}\right]_0^\infty$$

$$\frac{1}{2}\left(\frac{1}{s - j\omega} + \frac{1}{s + j\omega}\right)$$

$$\frac{s}{s^2 + \omega^2}$$

B-2 DERIVING LAPLACE TRANSFORM OPERATIONS

B-2-1 h(t−a) u(t − a) (O-1 Operation)

$$\mathcal{L}[h(t - a)\, u(t - a)] = \int_0^\infty h(t - a)\, u(t - a)\, e^{-st}\, dt$$

After $t = a$, $u(t - a)$ is equal to 1. Since $u(t - a)$ is zero before $t = a$, the integral is zero between $t = 0$ and $t = a$. We may therefore change the lower limit of the integral so that it starts at $t = a$, and drop the $u(t - a)$.

$$\int_a^\infty h(t - a)\, e^{-st}\, dt$$

We may change the variable to x by letting

$$x = t - a$$

Then taking the derivative of this, we find

$$dx = dt$$

The lower limit of the integral then becomes zero.
Rewriting the function yields

$$\int_0^\infty h(x) \, e^{-s(x+a)} \, dx$$

When we split up the exponential, the e^{-sa} part is a constant and may be factored out of the integral. The remaining part, by definition, is $H(s)$.

$$e^{-sa} \int_0^\infty h(x) \, e^{-sx} \, dx = e^{-as} H(s)$$

B-2-2 $e^{-bt}\, h(t)\, u(t)$(O-2 Operation)

$$\mathcal{L}[e^{-bt} h(t) u(t)] = \int_0^\infty e^{-bt} h(t) u(t) e^{-st} \, dt$$

Removing the $u(t)$, we may rewrite this as

$$\int_0^\infty h(t) \, e^{-(s+b)t} \, dt$$

By definition this is

$$H(s + b)$$

B-2-3 $th(t)\, u(t)$ (O-3 Operation)

$$\mathcal{L}[th(t) u(t)] = \int_0^\infty th(t) u(t) e^{-st} \, dt$$

Since $u(t)$ is 1 after $t = 0$, we may rewrite this as

$$\int_0^\infty th(t) \, e^{-st} \, dt$$

If we take the derivative of this function with respect to s, the time variable is considered to be a constant. When the derivative is performed on the $H(s)$ integral and multiplied by -1, we will see that the result is equal to the integral we require.

$$\int_0^\infty th(t)\, e^{-st}\, dt \stackrel{?}{=} \frac{-d}{ds} H(s)$$

$$\stackrel{?}{=} \frac{-d}{ds} \int_0^\infty h(t)\, e^{-st}\, dt$$

$$\stackrel{?}{=} \int_0^\infty \frac{-d}{ds} h(t)\, e^{-st}\, dt$$

$$= \int_0^\infty th(t)\, e^{-st}\, dt$$

B-2-4 $\left\{\dfrac{d}{dt}[h(t)]\right\} u(t)$ **(O-4a Operation)**

$$\mathcal{L}\left[\left\{\frac{d}{dt}[h(t)]\right\} u(t)\right] = \int_0^\infty \left\{\frac{d}{dt}[h(t)]\right\} u(t)\, e^{-st}\, dt$$

After $t = 0$, $u(t)$ is equal to 1.

$$\int_0^\infty \frac{d}{dt}[h(t)]\, e^{-st}\, dt$$

Using

$$\int u\, dv = uv - \int v\, du$$

we let

$$u = e^{-st} \quad \text{and} \quad dv = \frac{d}{dt} h(t)\, dt$$

Therefore,

$$du = -se^{-st}\, dt \quad \text{and} \quad v = h(t)$$

Substituting into the new form gives us

$$h(t)\, e^{-st} \Big|_0^\infty - \int_0^\infty -sh(t)\, e^{-st}\, dt$$

$$s\, H(s) - h(0)$$

B-2-5 $\dfrac{d}{dt}$ **[h(t) u(t)] (O-4b Operation)**

$$\mathcal{L}\left[\frac{d}{dt}\,[h(t)\,u(t)]\right] = \int_0^\infty \frac{d}{dt}\,[h(t)\,u(t)]\,e^{-st}\,dt$$

In this case we must take the derivative before integrating.

$$\int_0^\infty \left\{\left[\frac{d}{dt}\,h(t)\right]u(t) + h(0)\,\delta(t)\right\} e^{-st}\,dt$$

Separating this into two integrals, we have

$$\int_0^\infty \left[\frac{d}{dt}\,h(t)\right]u(t)\,e^{-st} + \int_0^\infty h(0)\,\delta(t)\,e^{-st}\,dt$$

The first term, as seen in Section B-2-4 is $s\,H(s) - h(0)$, and the second term, as seen in Section B-1-1, is $h(0)$ since $h(0)$ is a constant. Adding these two terms together, we have

$$[s\,H(s) - h(0)] + h(0)$$

Therefore,

$$s\,H(s)$$

B-2-6 $\displaystyle\int_0^t h(t)\,u(t)\,dt$ **(O-5 Operation)**

$$\mathcal{L}\left[\int_0^t h(t)\,u(t)\,dt\right] = \int_0^\infty \int_0^t h(t)\,u(t)\,dt\,e^{-st}\,dt$$

After $t = 0$, $u(t)$ is equal to 1.

$$\int_0^\infty \int_0^t h(t)\,dt\,e^{-st}\,dt$$

Using

$$\int u\,dv = uv - \int v\,du$$

we let

$$u = \int_0^t h(t)\,dt \quad \text{and} \quad dv = e^{-st}\,dt$$

Therefore,

$$du = h(t)\, dt \quad \text{and} \quad v = \frac{e^{-st}}{-s}$$

Substituting into the new form gives us

$$\frac{\left[\int_0^t h(t)\, dt\right] e^{-st} \Big|^\infty}{-s} \Bigg|_0 - \int_0^\infty \frac{e^{-st} h(t)}{-s}\, dt$$

$$0 - \frac{1}{-s} \int_0^\infty e^{-st} h(t)\, dt$$

This becomes

$$\frac{H(s)}{s}$$

B-3 DERIVING COMPLEX POLES FORMULA

This section derives an equation for the sinusoidal form for the general case of multiple-order complex–conjugate poles. A function $F(s)$ that contains complex–conjugate poles can be represented as

$$F(s) = \frac{k_1/(1-1)!}{(s + \alpha - j\omega)^r} + \frac{k_2/(2-1)!}{(s + \alpha - j\omega)^{r-1}} + \cdots + \frac{k_r/(r-1)!}{(s + \alpha - j\omega)} + \cdots$$

$$+ \frac{k_1^*/(1-1)!}{(s + \alpha + j\omega)^r} + \frac{k_2^*/(2-1)!}{(s + \alpha + j\omega)^{r-1}} + \cdots + \frac{k_r^*/(r-1)!}{(s + \alpha + j\omega)} + \cdots$$

$$+ \text{ rest of the function}$$

The "*" indicates the constant associated with the conjugate root. From this point we will drop the notation "+ rest of the function" and just remember that it is always present.

We may write these factors in a summation form as

$$F(s) = \sum_{n=1}^{r} \frac{k_n/(n-1)!}{(s + \alpha - j\omega)^{r-n+1}} + \sum_{n=1}^{r} \frac{k_n^*/(n-1)!}{(s + \alpha + j\omega)^{r-n+1}}$$

To find the value of k_r and k^*_r for the inverse of a single pole value, we may use Eq. (3-14), which give the residue for a single pole value.

$$R_{-a,\, n} = \frac{d^{n-1}}{ds^{n-1}} F(s)\, (s + a)^r \Big|_{s=-a}$$

For our case the residue was called k and the pole was complex. Therefore, we may write

$$k_n = \frac{d^{n-1}}{ds^{n-1}} F(s) \, (s + \alpha - j\omega)^r \Big|_{s = -\alpha + j\omega} \qquad \text{and}$$

$$k_n^* = \frac{d^{n-1}}{ds^{n-1}} F(s) \, (s + \alpha + j\omega)^r \Big|_{s = -\alpha - j\omega}$$

We want to define the variable $R_{-\alpha+j\omega}$ as the residue for finding both roots at the same time. Therefore, we define

$$R_{-\alpha+j\omega,\, n} = \frac{d^{n-1}}{ds^{n-1}} F(s) \, [(s + \alpha)^2 + \omega^2]^r \big|_{s = -\alpha + j\omega}$$

We now need to define the k_n equation in terms of $R_{-\alpha+j\omega}$. We begin by letting $n = 1$ in the last equation.

$$R_{-\alpha+j\omega,\, 1} = F(s) \, [(s + \alpha)^2 + \omega^2]^r \big|_{s = -\alpha + j\omega}$$

Then factoring the complex quadratic into two separate complex factors, we have

$$R_{-\alpha+j\omega,\, 1} = F(s) \, [(s + \alpha - j\omega)(s + \alpha + j\omega)]^r \big|_{s = -\alpha + j\omega}$$

Therefore,

$$\frac{R_{-\alpha+j\omega,\, 1}}{(s + \alpha + j\omega)^r} = F(s) \, (s + \alpha - j\omega)^r \big|_{s = -\alpha + j\omega}$$

Substituting this into the k_n equation above, we have

$$k_n = \frac{d^{n-1}}{ds^{n-1}} F(s) \, (s + \alpha - j\omega)^r \Big|_{s = -\alpha + j\omega}$$

$$k_n = \frac{d^{n-1}}{ds^{n-1}} \frac{R_{-\alpha+j\omega,\, 1}}{(s + \alpha + j\omega)^r} \Big|_{s = -\alpha + j\omega}$$

From the last equation we can find the k values for any value of n. Applying this equation, we will see a pattern developing as n goes from 1 to r.

$$k_1 = \frac{R_{-\alpha+j\omega,\, 1}}{(s + \alpha + j\omega)^r} \Big|_{s = -\alpha + j\omega}$$

$$= \frac{1}{(2j\omega)^r} R_{-\alpha+j\omega,\, 1}$$

$$k_2 = \frac{d}{ds} \frac{R_{-\alpha+j\omega,\,1}}{(s + \alpha + j\omega)^r} \bigg|_{s=-\alpha+j\omega}$$

$$= \frac{\left(\frac{d}{ds} R_{-\alpha+j\omega,\,1}\right)(s + \alpha + j\omega)^r - R_{-\alpha+j\omega,\,1}\, r(s + \alpha + j\omega)^{r-1}}{(2j\omega)^r} \bigg|_{s=-\alpha+j\omega}$$

From the definition of $R_{\alpha+j\omega,n}$

$$\frac{d}{ds} R_{-\alpha+j\omega,\,1} = R_{-\alpha+j\omega,\,2}$$

and substituting in $s = -\alpha + j\omega$, we have

$$k_2 = \frac{1}{(2j\omega)^r}\left[R_{-\alpha+j\omega,\,2} - \frac{r}{2j\omega} R_{-\alpha+j\omega,\,1}\right]$$

Continuing the procedure, k_3 and k_4 are found to be

$$k_3 = \frac{1}{(2j\omega)^r}\left[R_{-\alpha+j\omega,\,3} - \frac{2r}{2j\omega} R_{-\alpha+j\omega,\,2} + \frac{r(r+1)}{(2j\omega)^2} R_{-\alpha+j\omega,\,1}\right]$$

$$k_4 = \frac{1}{(2j\omega)^r}\left[R_{-\alpha+j\omega,\,4} - \frac{3r}{2j\omega} R_{-\alpha+j\omega,\,3} + \frac{3r(r+1)}{(2j\omega)^2} R_{-\alpha+j\omega,\,2}\right.$$
$$\left. - \frac{r(r+1)(r+2)}{(2j\omega)^3} R_{-\alpha+j\omega,\,1}\right]$$

By a similar process we can find the conjugate k values.

$$k_1^* = \frac{1}{(-2j\omega)^r} R_{-\alpha-j\omega,\,1}$$

$$k_2^* = \frac{1}{(-2j\omega)^r}\left[R_{-\alpha-j\omega,\,2} - \frac{r}{-2j\omega} R_{-\alpha-j\omega,\,1}\right]$$

$$k_3^* = \frac{1}{(-2j\omega)^r}\left[R_{-\alpha-j\omega,\,3} - \frac{2r}{-2j\omega} R_{-\alpha-j\omega,\,2} + \frac{r(r+1)}{(-2j\omega)^2} R_{-\alpha-j\omega,\,1}\right]$$

$$k_4^* = \frac{1}{(-2j\omega)^r}\left[R_{-\alpha-j\omega,\,4} - \frac{3r}{-2j\omega} R_{-\alpha-j\omega,\,3} + \frac{3r(r+1)}{(-2j\omega)^2} R_{-\alpha-j\omega,\,2}\right.$$
$$\left. - \frac{r(r+1)(r+2)}{(-2j\omega)^3} R_{-\alpha-j\omega,\,1}\right]$$

Here we see a similar pattern being established.

In both the k and $k*$, the part of the expression inside the brackets will be a magnitude, M, with a phase angle. Because k and $k*$ are complex conjugates, the magnitudes inside the brackets will be the same with the phase angles equal and opposite. We may therefore express the k's as

$$k_n = \frac{1}{(2j\omega)^r} M_n \angle \theta_n$$

$$k_n^* = \frac{1}{(-2j\omega)^r} M_n \angle -\theta_n$$

where $M_1 \angle \theta_1 = R_{-\alpha+j\omega,\,1}$

$$M_2 \angle \theta_2 = R_{-\alpha+j\omega,\,2} - \frac{r}{2j\omega} R_{-\alpha+j\omega,\,1}$$

and so on.

We may write the magnitude and phase angle in exponential form as

$$M_n \angle \pm \theta_n = M_n e^{\pm j\theta_n}$$

Substituting the k_n with the expression that is in terms of M_n into

$$F(s) = \sum_{n=1}^{r} \frac{k_n/(n-1)!}{(s+\alpha-j\omega)^{r-n+1}} + \sum_{n=1}^{r} \frac{k_n^*/(n-1)!}{(s+\alpha+j\omega)^{r-n+1}}$$

we have

$$F(s) = \sum_{n=1}^{r} \frac{M_n e^{j\theta_n}/(n-1)!}{(2j\omega)^r (s+\alpha-j\omega)^{r-n+1}} + \sum_{n=1}^{r} \frac{(-1)^r M_n e^{-j\theta_n}/(n-1)!}{(2j\omega)^r (s+\alpha+j\omega)^{r-n+1}}$$

Note that $1/(-2j\omega)^r = (-1)^r/(2j\omega)^r$. Rearranging this, we have

$$F(s) = \sum_{n=1}^{r} \left\{ \frac{M_n}{(n-1)!(2j\omega)^r} \left[\frac{e^{j\theta_n}}{(s+\alpha-j\omega)^{r-n+1}} + \frac{e^{-j\theta_n}}{(s+\alpha+j\omega)^{r-n+1}} \right] \right\}$$

We now find the inverse of this equation using the transform pair (P-7b).

$$\frac{1}{s^n} \Leftrightarrow \frac{1}{(n-1)!} t^{(n-1)} u(t)$$

From this we have

$$f(t) = \sum_{n=1}^{r} \left\{ \frac{M_n}{(n-1)!(2j\omega)^r(r-n)!} \left[e^{j\theta_n} e^{-(\alpha-j\omega)t} t^{r-n} + (-1)^r e^{-j\theta_n} e^{-(\alpha+j\omega)t} t^{r-n} \right] \right\}$$

Rearranging the exponential expressions yields

$$f(t) = \sum_{n=1}^{r} \left\{ \frac{M_n e^{-\alpha t} t^{r-n}}{(2j\omega)^r (n-1)!(r-n)!} \left[e^{j(\omega t+\theta_n)} + (-1)^r e^{-j(\omega t+\theta_n)} \right] \right\}$$

Using Euler's identities gives us

$$f(t) = \sum_{n=1}^{r} \left\{ \frac{M_n e^{-\alpha t} t^{r-n}}{(2j\omega)^r (n-1)!(r-n)!} \left[j[(1 - (-1)^r]\sin(\omega t + \theta_n) \right. \right.$$

$$\left. \left. + [(1 + (-1)^r]\sin(\omega t + \theta_n + 90°) \right] \right\}$$

We notice that for odd values of r the second sine term inside the brackets, [ ], goes to zero, and for even values of r the first sine term inside the brackets goes to zero. Therefore, we will either have a sine term with the j or with the 90°, but we will not have both at the same time. When the j term is present with odd r, the $(2j\omega)^r$ factor in the denominator will also have a j term to cancel it. Therefore, in the final answer we will never have a j term. Knowing these facts, we can condense these two terms into a single term.

The j terms can be replaced by

$$(-1)^{\text{int}(r/2)}$$

and the 90° part that appears on even values of r can be replaced by

$$\frac{1 + (-1)^r}{2}(90°)$$

Combining these into our equation, we have the final inverse equation for multiple-order complex conjugate poles.

$$f(t) = \left[\sum_{n=1}^{r} \frac{2(-1)^{\text{int}(r/2)} M_n}{(2\omega)^r (n-1)!(r-n)!} t^{r-n} \sin\left(\omega t + \theta_n + \frac{1 + (-1)^r}{2}(90°) \right) \right] e^{-\alpha t} u(t)$$

where $n = 1, 2, 3, \cdots, r$

int $(r/2)$ means the integer portion. Therefore, int$(5/2)$ = int$(4/2)$ = 2.

Appendix C

Basic DC Circuit Equations

This appendix is a review of basic dc circuit equations. Only dc sources and resistors are used, so the review can be done before or during the introduction of Laplace transforms.

C-1 IDENTIFYING SERIES CIRCUITS

An element is in series with another element if and only if both elements have the same current. The current must not simply be the same value of current, but it must be the same flow of current.

In Fig. C-1a the 10-Ω resistors have the same current value flowing through them, but not the same flow of current. In Fig. C-1b the two resistors have the same current and are therefore in series.

C-2 IDENTIFYING PARALLEL CIRCUITS

An element is in parallel with another element if and only if both elements have the same potential across them. The potential must not simply be the same value of potential, but it must be the same potential.

In Fig. C-1b the 10-Ω resistors have the same value of potential across them, but not the same potential across them. In Fig. C-1a the two resistors have the same potential across them, and are therefore in parallel.

C-3 SERIES VOLTAGE SOURCES

Voltage sources in series may be replaced by a single voltage source having the magnitude of the algebraic sum of the individual sources. The algebraic sum is the addition of potential rises in a loop and the subtraction of potential drops in that loop.

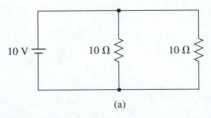
(a)

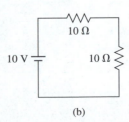
(b)

Figure C-1

C-4 PARALLEL CURRENT SOURCES

Current sources in parallel may be replaced by a single current source having the magnitude of the algebraic sum of the individual sources. The algebraic sum is the addition of currents entering a node and the subtraction of currents leaving that node.

C-5 OHM'S LAW

The voltage across a resistor is equal to the current through it times the resistor's value.

$$V = IR \qquad \text{(C-1)}$$

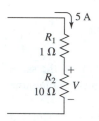

Figure C-2

EXAMPLE **C-1**

What is the voltage across R_2 in Fig. C-2?

Solution: Because R_1 is in series with R_2, the same 5 A flows in both elements.

$$V = IR_2 = (5)(10) = 50 \text{ V}$$

C-6 VOLTAGE AND CURRENT MEASUREMENTS

The $+/-$ signs in this book do *not* refer to the polarity of the voltage. They refer to the polarity of a voltmeter used to test the point. The $+$ indicates the test probe (or red test lead), and the $-$ indicates the reference (or black test lead). This means that the $+$ is measured with respect to the $-$. If the $+$ side is at a higher potential than the reference $(-)$, the value is positive. If the $+$ side is at a lower potential than the reference, the value is negative.

EXAMPLE **C-2**

Find the voltage across the resistors in Fig. C-3.

Solution: Taking a clockwise loop using Kirchhoff's voltage law in Fig. C-3a, we have

$$E - V = 0$$

$$V = E = 10 \text{ V}$$

Taking a clockwise loop using Kirchhoff's voltage law in Fig. C-3b yields

$$E + V = 0$$

$$V = -E = -10 \text{ V}$$

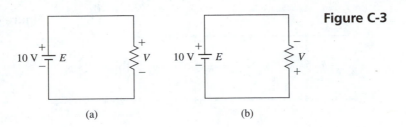

Figure C-3

(a)   (b)

A similar situation exists with current. The current arrow does *not* refer to current direction but rather to the direction to measure. If the current flows in the direction of the arrow, the value is positive. If the current flows opposite the direction of the arrow, the value is negative.

Using the +/− and arrows this way, we can determine if a component is taking energy from the circuit or supplying energy to the circuit. When we draw the current arrow between the voltage +/− signs so that the current value and voltage value have the same sign (both positive or both negative), the following observations can be made.

1. If the arrowhead is closest to the − sign, energy is being removed from the circuit.
2. If the arrowhead is closest to the + sign, energy is being supplied to the circuit.

If we measure the current and voltage as indicated in Fig. C-4, the values of the current and voltage will both be positive or both negative because resistors always remove energy from the circuit.

A voltage source, current source, capacitor, or inductor can take energy from or supply energy to the circuit. In Fig. C-5, if the voltage and current are measured and both values have the same sign, the source is being charged by removing energy from the circuit. If the signs are opposite, the voltage source is supplying energy to the circuit.

The capacitor and inductor work the same way. The only difference is that the capacitor stores energy in an electrostatic field and the inductor stores energy in an electromagnetic field.

Some circuits use voltage points rather than +/− indications. This method is often used by manufacturers on data sheets. When this is used, we will see voltage indicated as V_{ab}, V_{ba}, V_{BE}, V_{GS}, and so on. When a circuit has points marked a and b and they appear as subscripts in a variable, the first subscript is the point measured and the second subscript is the reference. If V_{ab} is to be measured, then put the + sign next to the a point on the circuit drawing and the − sign next to the b point on the circuit drawing. If V_{ba} is to be measured, then put the + sign next to the b point and the − sign next to the a point.

Figure C-4

Figure C-5

C-7 VOLTAGE DIVIDER RULE

The voltage across any element in a series network is equal to the total voltage across the series network times the element you want to know the voltage across divided by the total resistance of the series network.

For the circuit in Fig. C-6, the voltage V_x across R_x is

$$V_x = \frac{V_T R_x}{R_1 + R_2 + \cdots + R_N} \tag{C-2}$$

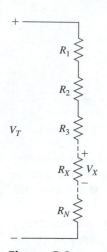

Figure C-6

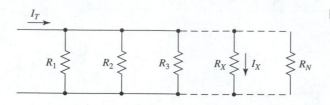

Figure C-7

For two resistors, R_1 and R_2, in series, Eq. (C-2) becomes

$$V_1 = \frac{V_T R_1}{R_1 + R_2} \tag{C-3a}$$

$$V_2 = \frac{V_T R_2}{R_1 + R_2} \tag{C-3b}$$

C-8 CURRENT DIVIDER RULE

The current through any element in a parallel network is equal to the total current entering the parallel network times the total resistance of the parallel network divided by the element you want to know the current through.

For the circuit in Fig. C-7, the current I_x through R_x is

$$I_x = \frac{I_T R_T}{R_x} \tag{C-4}$$

where R_T is the resistance the current I_T sees.

For two resistors, R_1 and R_2, in parallel, the total resistance becomes $R_T = (R_1 R_2)/(R_1 + R_2)$, and substituting into Eq. (C-4) gives

$$I_1 = \frac{I_T R_2}{R_1 + R_2} \tag{C-5a}$$

$$I_2 = \frac{I_T R_1}{R_1 + R_2} \tag{C-5b}$$

C-9 KIRCHHOFF'S VOLTAGE LAW

The sum of the voltages around a complete path is zero. The path must start at a point and return to that same point without going through a component or section of wire more than once. We determine whether to add or subtract, not by the physical polarity of the voltages, but rather by the way the voltages were measured.

Figure C-8

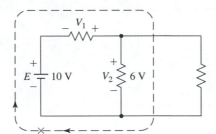

EXAMPLE C-3

Find V_1 in Fig. C-8.

Solution: The path is shown in Fig. C-8 by the dashed line. We start at the × and return to the × in a clockwise direction. This could also be done in a counterclockwise direction with the same result.

$$E + V_1 - V_2 = 0$$

$$10 + V_1 - 6 = 0$$

$$V_1 = -10 + 6$$

$$V_1 = -4 \text{ V}$$

EXAMPLE C-4

Find the voltage V_{ab} in Fig. C-9.

Solution: The a will be assigned + and the b will be assigned − due to the order of the subscript in V_{ab}. This time we will use a counterclockwise direction.

$$-V_2 - V_{ab} + V_1 - E = 0$$

$$-2 - V_{ab} + 5 - 15 = 0$$

$$V_{ab} = -2 + 5 - 15$$

$$V_{ab} = -12 \text{ V}$$

If we had been looking for V_{ba}, we would assign the +/− the opposite way and the result would be $V_{ba} = +12$ V.

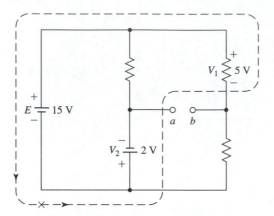

Figure C-9

C-10 KIRCHHOFF'S CURRENT LAW

The sum of the currents at a node is equal to zero. We determine whether to add or subtract by the direction the current was measured. A current measured as entering a node is added, and a current measured as leaving a node is subtracted.

EXAMPLE C-5

Find I_3 and I_5 in Fig. C-10.

Solution: At node N_1,

$$I_1 - I_2 + I_3 = 0$$

$$4 - 3 + I_3 = 0$$

$$I_3 = -4 + 3$$

$$I_3 = -1 \text{ A}$$

Notice that the value is negative. We will not change the direction we measured I_3. Instead, we will just realize that the current is flowing in the opposite direction.

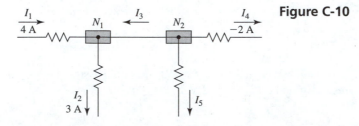

Figure C-10

Because Kirchhoff's current law is done with the direction of measurement, I_3 is considered leaving node N_2 and entering node N_1. At node N_2,

$$-I_3 - I_4 - I_5 = 0$$

$$-(-1) - (-2) - I_5 = 0$$

$$I_5 = 1 + 2$$

$$I_5 = 3 \text{ A}$$

Appendix D

Semilog Graphs

When we graph a function's response to different frequencies, we use semilog or log-log graph paper. Log-log graphs are rarely seen, so we will concentrate on semilog graphs. Semilog graphs have the horizontal axis based on a logarithmic ($\log_{10}$) scale and the vertical axis based on a linear scale. Frequently, the only information we may have about a circuit is a graph. We must therefore know how to work with semilog graphs.

D-1 HOW TO READ A LOG SCALE

Figure D-1 shows only the major divisions of a log scale. Each major increment represents a power of 10. Because it is based on powers of 10, the log scale will never start at 0. The scale will start at 1×10 to some positive or negative integer power. Mathematically, we could start at any value, except zero, with each increment 10 times the previous major division, but we exclusively use 1×10 to some integer power.

Each major division is called a *cycle* even though the divisions have nothing to do with a periodic waveform. Graph paper with four major divisions marked off is called *four-cycle paper,* which is used in Fig. D-1. It is called *five-cycle paper* when five areas are marked off. Linear values are marked off between major divisions, which will, therefore, have logarithmic spacing.

Each cycle is divided into 10 secondary steps, as shown in Fig. D-2. These steps are the linear increments from one major division to the next major division. The increments always get smaller as we increase the frequency to approach the next major division. The secondary marks are often numbered and are multiplied times the major division value to the left. For example, if the major division to the left of the secondary mark 6 is 1×10^{-3}, the value is $6 \times (1 \times 10^{-3}) = 6 \times 10^{-3}$.

The marks between the secondary divisions are marked in different ways depending on the manufacture and the number of cycles. These marks between the secondary marks are also linear increments. To find the linear value increment of each step, we count the number of spaces between the secondary marks and use the reciprocal. If these are 10 spaces, the linear increment value is 0.1, and if there are 5 spaces, the increment value is 0.2. These are multiplied by the value of the major division to the left and added to the secondary value.

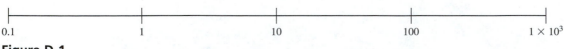

Figure D-1

Figure D-2

Figure D-3

EXAMPLE **D-1**

Figure D-3 shows one cycle. Find the value of the point marked.

Solution: We are in the secondary division 2, and there are 10 spaces between the secondary division 2 and the secondary division 3. The value is

$$(2 \times 100) + \left[\left(4 \times \frac{1}{10} \right) \times 100 \right] = 240$$

D-2 CALCULATING DISTANCES ON A LOG SCALE

Measurements on a log axis are identical to measurements on a linear axis except that we use the logarithm of the value instead of the value. This means that what we do on a linear scale we do on the log scale using the logarithm of the number. Since this is the case, we should review how we measure distance on a linear scale first.

Figure D-4 shows a linear scale. To find the distance between two points X_1 and X_2, we find the number of units between the two points, $X_2 - X_1$. For example, if $X_1 = 1.2$ and $X_2 = 3.7$, the number of units between the points is

$$NU = X_2 - X_1$$

$$= 3.7 - 1.2 \qquad \textbf{(D-1)}$$

$$= 2.5 \text{ units}$$

On a log scale, the unit is a decade because it is a change in frequency by a factor of 10. On a linear scale, to find the number of decades, ND, we solve for ND in the equation

$$10^{ND} = \frac{X_1}{X_2}$$

By using the log of the values, we can find the same quantities on a log scale. Because we are using frequency we will use f_1 and f_2 instead of X_1 and X_2, where f_1 will be less than f_2 and both must be in the same units.

Figure D-4

The number of decades between two frequencies will be

$$\log 10^{ND} = \log(f_2) - \log(f_1)$$

$$ND = \log(f_2) - \log(f_1) \quad \text{[compare with Eq. (D-1)]}$$

or

$$ND = \log\left(\frac{f_2}{f_1}\right) \quad f_1 < f_2 \tag{D-2}$$

Sometimes units of octaves are used. This term comes from the octave in music and is usually applied to electronic circuits that reproduce music. In electronics terms this means "to double the frequency." On a linear scale, to find the number of octaves we would solve for the number of octaves, NO, in the equation

$$2^{NO} = \frac{X_2}{X_1}$$

On a log scale we do the same thing except that we use the log of the values.

$$NO \log(2) = \log(f_2) - \log(f_2)$$

$$NO = \frac{\log(f_2) - \log(f_1)}{\log(2)} \tag{D-3a}$$

$$= \frac{\log\left(\dfrac{f_2}{f_1}\right)}{\log(2)} \quad f_1 < f_2$$

which we notice is in the form of the change of base identity for logarithms; therefore, an alternative form is

$$NO = \log_2\left(\frac{f_2}{f_1}\right) \quad f_1 < f_2 \tag{D-3b}$$

Each equation has been restricted to $f_1 < f_2$, but this is for convenience in working on the semilog graphs of frequency responses. Without this restriction we would calculate a negative value when f_1 is greater than f_2. This would indicate a movement to the left of f_1 just as a negative value on a linear scale would indicate. Also, each equation is a ratio of frequencies, which means that f could be in hertz or rad/sec.

EXAMPLE D-2

Find the number of decades and the number of octaves between:

a. $f_1 = 2$ rad/sec and $f_2 = 8$ rad/sec
b. $f_1 = 200$ rad/sec and $f_2 = 800$ rad/sec

Solution:
a. Using Eq. (D-2) gives

$$ND = \log\left(\frac{f_2}{f_1}\right)$$

$$= \log\left(\frac{8}{2}\right)$$

$$= 0.60206$$

The number of octaves using Eq. (D-3a) are

$$NO = \frac{\log\left(\frac{f_2}{f_1}\right)}{\log(2)}$$

$$= \frac{0.60206}{\log(2)}$$

$$= 2$$

This stands to reason because two octaves means to multiply by 2, two times.

b. These two frequencies are different by a factor of 100 from the frequencies in part a. It is important to notice that these frequencies are separated by the same number of decades or octaves as the frequencies in part a. Using Eq. (D-2), we have

$$ND = \log\left(\frac{f_2}{f_1}\right)$$

$$= \log\left(\frac{800}{200}\right)$$

$$= 0.60206$$

The number of octaves using Eq. (D-3a) is

$$NO = \frac{\log\left(\frac{f_2}{f_1}\right)}{\log(2)}$$

$$= \frac{0.60206}{\log(2)}$$

$$= 2$$

D-3 CALCULATING ROLL-OFF RATES ON SEMILOG GRAPHS

The roll-off rate, RR, is the slope, rise over run, of a straight line on a semilog graph. When the graph is showing the output gain in dB of a circuit, then the y-axis is in dB and the roll-off rate is measured in dB/decade or dB/octave. When the graph is showing the output phase shift in degrees of a circuit, then the y-axis is in degrees and the RR is measured in degrees/decade or degrees/octave.

For a graph of output gain in dB, the rise part of the roll-off rate is the difference between the dB values and is called ΔA. In equation form,

$$\Delta A = A_{2dB} - A_{1dB} \tag{D-4}$$

where A_{2dB} corresponds to the dB value at f_2
A_{1dB} corresponds to the dB value at f_1

The roll-off rate can then be expressed by

$$RR = \frac{\Delta A_{dB}}{ND} \tag{D-5a}$$

$$RR = \frac{\Delta A_{dB}}{NO} \tag{D-5b}$$

where ND is defined in Eq. (D-2)
NO is defined in Eq. (D-3)

Equation (D-5a) has the units of dB/dec, and Eq. (D-5b) has the units of dB/oct.

Using similar reasoning, a graph of the phase shift in degrees, the rise part of the roll-off rate, is the difference between the angle values and is called ΔD. In equation form,

$$\Delta D = D_2 - D_1 \tag{D-6}$$

where D_2 corresponds to the phase angle value at f_2
D_1 corresponds to the phase angle value at f_1

The roll-off rate can then be expressed by

$$RR = \frac{\Delta D}{ND} \tag{D-7a}$$

$$RR = \frac{\Delta D}{NO} \tag{D-7b}$$

where ND is defined in Eq. (D-2)
NO is defined in Eq. (D-3)

EXAMPLE D-3

For Fig. D-5, find the roll-off rate in dB/dec, and find the dB value at f_3.

Solution: There is enough information using f_1 and f_2 to find the slope of the line. To find the slope we first find the change in dB using Eq. (D-4).

$$\Delta A_{dB} = A_{2dB} - A_{1dB}$$

$$= (5.5630) - (-4.8945)$$

$$= 10.4575 \text{ dB}$$

Then using Eq. (D-2), we find the number of decades between these frequencies.

$$ND = \log\left(\frac{f_2}{f_1}\right)$$

$$= \log\left(\frac{3000}{900}\right)$$

$$= 0.52288 \text{ decade}$$

Substituting into Eq. (D-5a), we will find the roll-off rate.

$$RR = \frac{\Delta A_{dB}}{ND}$$

$$= \frac{10.4575 \text{ dB}}{0.52288 \text{ dec}}$$

$$= 20 \text{ dB/dec}$$

Because roll-off rates will be in multiples of 20, we usually round off to the nearest multiple of 20.

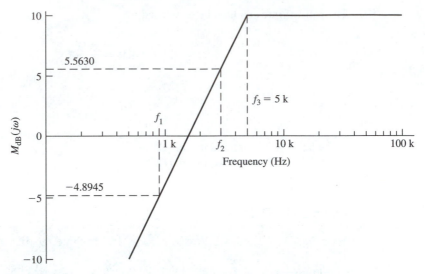

Figure D-5

Now that we know the roll-off rate we can apply Eqs. (D-2), (D-3), (D-4), and/or Eq. (D-5) to find any point on the slope, but we will have to modify our equations to accommodate f_2 and f_3.

First we find the number of decades between the two points using Eq. (D-2).

$$ND = \log\left(\frac{f_3}{f_2}\right) \qquad f_2 < f_3$$

$$= \log\left(\frac{5000}{3000}\right)$$

$$= 0.22185 \text{ decade}$$

Using Eq. (D-5a), we can find the change in dB between the two points.

$$RR = \frac{\Delta A_{dB}}{ND}$$

$$\Delta A_{dB} = (RR)(ND)$$

$$= (20 \text{ dB/dec})(0.22185 \text{ dec})$$

$$= 4.4370 \text{ dB}$$

This tells us the dB difference between the two points. To find the value of the point, we use a modified Eq. (D-4)

$$\Delta A_{dB} = A_{3dB} - A_{2dB}$$

and solve for A_{3dB}.

$$A_{3dB} = \Delta A_{dB} + A_{2dB}$$

$$= 4.4370 \text{ dB} + 5.5630 \text{ dB}$$

$$= 10 \text{ dB}$$

D-4 CONSTRUCTION OF SEMILOG GRAPH PAPER

The construction of semilog graph paper is a simple process. Knowing how to construct it will be helpful if we want to make a sketch on blank paper, write a program that displays a semilog graph, or need to find points on unfamiliar graph paper. Let's begin by summarizing some of the important points of the previous sections.

The log scale on semilog graph paper typically represents frequency, which can either be in radians or hertz. The log scale uses log base 10. The basic unit of measurement is the decade, where a decade represents a change by a factor of 10. What we can do on a linear scale we can do on a log scale except that we use the log of the value.

On a linear scale we choose a physical distance for one unit, which is the length per unit, LPU. ON a log scale we choose a physical distance for the decade, which is the length per decade, LPD. In both cases we measure everything in proportion to this physical distance. The linear scale and log scale are measured in exactly the same way except that on the log scale we use the log of the point's value. Let's first look at the linear scale and then translate it to the log scale.

On a linear scale if we have two points X_1 and X_2, where $X_1 < X_2$, the number of units between them using Eq. (D-1) will be NU = $X_2 - X_1$. To find the physical distance (PD_{lin}), we multiply by the length per unit (LPU).

$$PD_{lin} = NU \text{ LPU} \qquad \textbf{(D-8)}$$

On a log scale if we have two points f_1 and f_2 where $f_1 < f_2$, the number of decades between them using Eq. (D-2) will be ND = $\log(f_2/f_1)$. To find the physical distance (PD_{log}), we multiply by the length per decade (LPD).

$$PD_{log} = ND \text{ LPD} \qquad \textbf{(D-9)}$$

From Eqs. (D-8) and (D-9) we can find the physical distance between any two points. Let's first examine the physical distance of secondary divisions between major divisions.

On a linear scale if we want to mark off the values in tenths, we can find the physical distance of each tenth by finding the physical distance between 0 and 0.1, 0 and 0.2, 0 and

0.3, and each other interval. We then plug into Eq. (D-8) to find where to mark it off. Because the number of units between any one-tenth interval using Eq. (D-1) will give us the same value, the spacing will be equal in each one-tenth interval.

On a log scale each interval using Eq. (D-2) will give us a different value, and we must measure each one individually. We must therefore find the physical distance from the left major division to each point we want. We could calculate the distance from the last interval to the next interval, but more accuracy is obtained if we always start at the left major division.

EXAMPLE D-4

Where are the secondary divisions between major divisions of a log scale when the physical distance is 2 inches for each decade?

Solution: Using Eq. (D-2), we find the number of decades from the major division to the increment we want. We can assume any starting value for the major division, as seen in Example D-2. For convenience we will assume that we are starting at 1. Then using Eq. (D-9), we multiply the number of decades by 2 inches per decade.

$$ND = \log\left(\frac{f_2}{f_1}\right) \qquad\qquad PD_{log} = ND\ LPD$$

$$ND = \log\left(\frac{2}{1}\right) = 0.30103 \qquad PD = (0.30103)(2) = 0.60206 \text{ in.}$$

$$ND = \log\left(\frac{3}{1}\right) = 0.47712 \qquad PD = (0.47712)(2) = 0.95424 \text{ in.}$$

$$ND = \log\left(\frac{4}{1}\right) = 0.60206 \qquad PD = (0.60206)(2) = 1.2041 \text{ in.}$$

$$ND = \log\left(\frac{5}{1}\right) = 0.69897 \qquad PD = (0.69897)(2) = 1.3979 \text{ in.}$$

$$ND = \log\left(\frac{6}{1}\right) = 0.77815 \qquad PD = (0.77815)(2) = 1.5563 \text{ in.}$$

$$ND = \log\left(\frac{7}{1}\right) = 0.84510 \qquad PD = (0.84510)(2) = 1.6902 \text{ in.}$$

$$ND = \log\left(\frac{8}{1}\right) = 0.90309 \qquad PD = (0.90309)(2) = 1.8062 \text{ in.}$$

$$ND = \log\left(\frac{9}{1}\right) = 0.95424 \qquad PD = (0.95424)(2) = 1.9085 \text{ in.}$$

From this last example we can see an easy technique for approximating a log scale without measuring. If we take the log of the increment we want and multiply it by 100%, we have the percent of the total distance to the next major division. For example, if we are looking for 31.2×10^3, the increment is 3.12, and $\log(3.12) = 0.49415$. So we go about 50% of the distance to the next major division.

Appendix E

OP-AMP Topologies

This appendix shows the topologies used in this book. Each section starts with the general topology and the transfer function in terms of admittances. Each subsection shows specific cases that implement different types of filters (low pass, high pass, and band pass). The subsections use the following format:

1. The admittances are assigned as R and/or C components.
2. The resulting transfer function is written.
3. Some components are assigned to be equal in value.
4. The procedure is given to solve for component values of the circuit in terms of the loss function in the biquad form of

$$\frac{ms^2 + as + b}{\pm h(ns^2 + ds + e)} \tag{E-1}$$

For each circuit the specific biquad form will be listed.

E-1 NONINVERTING SINGLE-FEEDBACK TOPOLOGY

See Fig. E-1.

$$\frac{V_{\text{in}}}{V_o} = \frac{Y_A + Y_B}{KY_A} \tag{E-2}$$

where

$$K = 1 + \frac{R_1}{R_2}$$

E-1-1 Low Pass

1.
$$Y_A = \frac{1}{R_A} \qquad Y_B = sC_B \tag{E-3}$$

2.
$$\frac{V_{\text{in}}}{V_o} = \frac{s + \dfrac{1}{C_B R_A}}{\dfrac{K}{C_B R_A}} \tag{E-4}$$

450

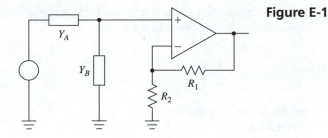

Figure E-1

3. There are no components to set equal to each other.
4. To solve for the component values in a low-pass, noninverting, single-feedback circuit from a loss function in the form of

$$\frac{s+b}{h} \qquad \text{(E-5)}$$

use the following procedure.
(a) Choose C_B.
(b) Solve for R_A

$$R_A = \frac{1}{bC_B} \qquad \text{(E-6a)}$$

(c) There are two ways to work with K and h.

$$\text{If } h \geq \frac{1}{R_A C_B} \qquad \text{then } K = hR_A C_B$$

$$\text{else choose } K \geq 1 \qquad \text{(E-6b)}$$

$$\text{and } h = \frac{K}{R_A C_B}$$

E-1-2 High Pass

1. $$Y_A = sC_A \qquad Y_B = \frac{1}{R_B} \qquad \text{(E-7)}$$

2. $$\frac{V_{in}}{V_o} = \frac{s + \dfrac{1}{C_A R_B}}{Ks} \qquad \text{(E-8)}$$

3. There are no components to set equal to each other.

4. To solve for the component values in a high-pass, noninverting, single-feedback circuit from a loss function in the form of

$$\frac{s + b}{hs} \tag{E-9}$$

use the following procedure.
(a) Choose C_A.
(b) Solve for R_B.

$$R_B = \frac{1}{bC_A} \tag{E-10a}$$

(c) There are two ways to work with K and h.

$$\text{If } h \geq 1 \qquad \text{then } K = h$$

$$\text{else choose } K \geq 1 \tag{E-10b}$$

$$\text{and } h = k$$

E-2 INVERTING SINGLE-FEEDBACK TOPOLOGY

See Fig. E-2.

$$\frac{V_{in}}{V_o} = \frac{Y_B}{-Y_A} \tag{E-11}$$

E-2-1 Low Pass

1.
$$Y_A = \frac{1}{R_A} \qquad Y_B = sC_B + \frac{1}{R_B} \tag{E-12}$$

Notice that Y_B is a capacitor in parallel with a resistor.

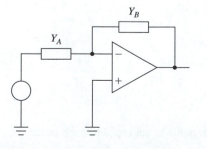

Figure E-2

2.
$$\frac{V_{in}}{V_o} = \frac{s + \dfrac{1}{C_B R_B}}{\dfrac{-1}{C_B R_A}}$$
(E-13)

3. There are no components to set equal to each other.
4. To solve for the component values in a low-pass, inverting, single-feedback circuit from a loss function in the form of

$$\frac{s + b}{-h}$$
(E-14)

use the following procedure.
(a) Choose C_B.
(b) Solve for R_B.

$$R_B = \frac{1}{b C_B}$$
(E-15a)

(c) There are two ways to work with R_A and h.
Either calculate R_A:

$$R_A = \frac{1}{h C_B}$$
(E-15b)

or choose R_A and calculate the resulting h:

$$h = \frac{1}{R_A C_B}$$

E-2-2 High Pass

1.
$$Y_A = s C_A \qquad Y_B = s C_B + \frac{1}{R_B}$$
(E-16)

Notice that Y_B is a capacitor in parallel with a resistor.

2.
$$\frac{V_{in}}{V_o} = \frac{s + \dfrac{1}{C_B R_B}}{\dfrac{-C_A}{C_B} s}$$
(E-17)

3. There are no components to set equal to each other.

4. To solve for the component values in a high-pass, inverting, single-feedback circuit from a loss function in the form of

$$\frac{s + b}{-hs} \tag{E-18}$$

use the following procedure.
(a) Choose C_B.
(b) Solve for R_B.

$$R_B = \frac{1}{bC_B} \tag{E-19a}$$

(c) There are two ways to work with C_A and h. Either calculate C_A:

$$C_A = hC_B \tag{E-19b}$$

or choose C_A and calculate the resulting h:

$$h = \frac{C_A}{C_B}$$

E-3 NONINVERTING DUAL-FEEDBACK TOPOLOGY

See Fig. E-3.

$$\frac{V_{in}}{V_o} = \frac{(Y_A + Y_B + Y_E)(Y_C + Y_D) + Y_C(Y_D - Y_BK)}{Y_AY_CK} \tag{E-20}$$

where

$$K = 1 + \frac{R_1}{R_2}$$

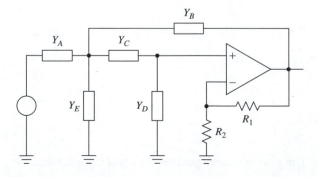

Figure E-3

E-3-1 Low Pass

1.
$$Y_A = \frac{1}{R_A} \qquad Y_B = sC_B$$

$$Y_C = \frac{1}{R_C} \qquad Y_D = sC_D \qquad \text{(E-21)}$$

$$Y_E = 0 \qquad \text{(no component)}$$

2.
$$\frac{V_{\text{in}}}{V_o} = \frac{s^2 + \left(\dfrac{1}{C_B R_A} + \dfrac{1}{C_D R_C} + \dfrac{1 - K}{C_B R_C}\right)s + \dfrac{1}{C_B C_D R_A R_C}}{\dfrac{K}{C_B C_D R_A R_C}} \qquad \text{(E-22)}$$

3.
$$C_B = C_D = C \quad \text{and} \quad R_A = R_C = R \qquad \text{(E-23)}$$

4. To solve for the component values in a low-pass, noninverting, dual-feedback circuit from a loss function in the form of

$$\frac{s^2 + as + b}{h} \qquad \zeta < 1 \qquad \text{(E-24)}$$

use the following procedure.
   (a) To use this circuit, the following must be true.

$$0 \le 2 - \frac{a}{\sqrt{b}} \qquad \text{(E-25a)}$$

   (b) Choose C.
   (c) Calculate R.

$$R = \frac{1}{C\sqrt{b}} \qquad \text{(E-25b)}$$

   (d) Calculate K.

$$K = 3 - \frac{a}{\sqrt{b}} \qquad \text{(E-25c)}$$

   (e) Calculate the resulting h.

$$h = \frac{K}{C_B C_D R_A R_C} \qquad \text{(E-25d)}$$

E-3-2 High Pass

1.
$$Y_A = sC_A \qquad Y_B = \frac{1}{R_B} \tag{E-26}$$

$$Y_C = sC_C \qquad Y_D = \frac{1}{R_D}$$

$$Y_E = 0 \qquad \text{(no component)}$$

2.
$$\frac{V_{in}}{V_o} = \frac{s^2 + \left(\dfrac{1}{C_C R_D} + \dfrac{1}{C_A R_D} + \dfrac{1-K}{C_A R_B} \right)s + \dfrac{1}{C_A C_C R_B R_D}}{Ks^2} \tag{E-27}$$

3.
$$C_A = C_C = C \quad \text{and} \quad R_B = R_D = R \tag{E-28}$$

4. To solve for the component values in a high-pass, noninverting, dual-feedback circuit from a loss function in the form of

$$\frac{s^2 + as + b}{hs^2} \qquad \zeta < 1 \tag{E-29}$$

use the following procedure.
(a) To use this circuit, the following must be true.

$$0 \leq 2 - \frac{a}{\sqrt{b}} \tag{E-30a}$$

(b) Choose C.
(c) Calculate R.

$$R = \frac{1}{C\sqrt{b}} \tag{E-30b}$$

(d) Calculate K.

$$K = 3 - \frac{a}{\sqrt{b}} \tag{E-30c}$$

(e) Calculate the resulting h.

$$h = K \tag{E-30d}$$

E-3-3 Band Pass

1.
$$Y_A = \frac{1}{R_A} \qquad Y_B = \frac{1}{R_B}$$

$$Y_C = sC_C \qquad Y_D = \frac{1}{R_D} \qquad \text{(E-31)}$$

$$Y_E = sC_E$$

2.
$$\frac{V_{in}}{V_o} = \frac{s^2 + \left(\dfrac{1}{C_E R_A} + \dfrac{1}{C_C R_D} + \dfrac{1}{C_E R_D} + \dfrac{1-K}{C_E R_B}\right)s + \dfrac{1}{C_C C_E}\left(\dfrac{1}{R_A R_D} + \dfrac{1}{R_B R_D}\right)}{\left(\dfrac{K}{C_E R_A}\right)s} \qquad \text{(E-32)}$$

3.
$$C_C = C_E = C \quad \text{and} \quad R_A = R_B = R_D = R \qquad \text{(E-33)}$$

4. To solve for the component values in a band-pass, noninverting, dual-feedback circuit from a loss function in the form of

$$\frac{s^2 + as + b}{hs} \qquad \zeta > 1 \qquad \text{(E-34)}$$

use the following procedure.
   (a) To use this circuit, the following must be true.

$$0 \le 3 - \frac{a}{\sqrt{b}} \qquad \text{(E-35a)}$$

   (b) Choose C.
   (c) Calculate R.

$$R = \frac{1}{C\sqrt{b}} \qquad \text{(E-35b)}$$

   (d) Calculate K.

$$K = 4 - \frac{a}{\sqrt{b}} \qquad \text{(E-35c)}$$

   (e) Calculate the resulting h.

$$h = \frac{K}{C_E R_A} \qquad \text{(E-35d)}$$

E-4 INVERTING DUAL-FEEDBACK TOPOLOGY

See Fig. E-4.

$$\frac{V_{in}}{V_o} = \frac{Y_D(Y_A + Y_B + Y_C + Y_E) + Y_B Y_C}{-Y_A Y_C} \tag{E-36}$$

E-4-1 Low Pass

1.
$$Y_A = \frac{1}{R_A} \qquad Y_B = \frac{1}{R_B}$$

$$Y_C = \frac{1}{R_C} \qquad Y_D = sC_D \tag{E-37}$$

$$Y_E = sC_E$$

2.
$$\frac{V_{in}}{V_o} = \frac{s^2 + \frac{1}{C_E}\left(\frac{1}{R_A} + \frac{1}{R_B} + \frac{1}{R_C}\right)s + \frac{1}{C_D C_E R_B R_C}}{\dfrac{-1}{C_D C_E R_B R_C}} \tag{E-38}$$

3.
$$R_B = R_C = R \tag{E-39}$$

4. To solve for the component values in a low-pass, inverting, dual-feedback circuit from a loss function in the form of

$$\frac{s^2 + as + b}{-h} \qquad \zeta < 1 \tag{E-40}$$

use the following procedure.

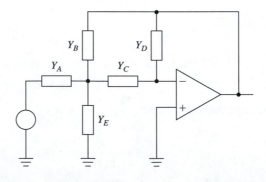

Figure E-4

(a) Choose C_E.

(b) Choose C_D such that

$$C_D \leq \frac{C_E a^2}{4b} \tag{E-41a}$$

(c) Calculate R.

$$R = \frac{1}{\sqrt{bC_D C_E}} \tag{E-41b}$$

(d) Calculate R_A.

$$R_A = \frac{R}{RC_E a - 2} \tag{E-41c}$$

(e) Calculate the resulting h.

$$h = \frac{1}{C_D C_E R_B R_C} \tag{E-41d}$$

E-4-2 High Pass

1.
$$Y_A = sC_A \qquad Y_B = sC_B$$

$$Y_C = sC_C \qquad Y_D = \frac{1}{R_D} \tag{E-42}$$

$$Y_E = \frac{1}{R_E}$$

2.
$$\frac{V_{in}}{V_o} = \frac{s^2 + \left(\dfrac{C_A}{C_B C_C R_D} + \dfrac{1}{C_C R_D} + \dfrac{1}{C_B R_D} \right)s + \dfrac{1}{C_B C_C R_D R_E}}{\left(\dfrac{-C_A}{C_B} \right)s^2} \tag{E-43}$$

3.
$$C_A = C_B = C_C = C \tag{E-44}$$

4. To solve for the component values in a high-pass, inverting, dual-feedback circuit from a loss function in the form of

$$\frac{s^2 + as + b}{-hs^2} \qquad \zeta < 1 \tag{E-45}$$

use the following procedure.

(a) Choose C.

(b) Calculate R_D.

$$R_D = \frac{3}{Ca} \tag{E-46a}$$

(c) Calculate R_E.

$$R_E = \frac{a}{3Cb} \tag{E-46b}$$

(d) Calculate the resulting h.

$$h = \frac{C_A}{C_B} = 1 \tag{E-46c}$$

E-4-3 Band Pass

1.
$$Y_A = \frac{1}{R_A} \qquad Y_B = sC_B$$

$$Y_C = sC_C \qquad Y_D = \frac{1}{R_D} \tag{E-47}$$

$$Y_E = \frac{1}{R_E}$$

2.
$$\frac{V_{in}}{V_o} = \frac{s^2 + \left(\dfrac{1}{C_C R_D} + \dfrac{1}{C_B R_D}\right)s + \dfrac{1}{C_B C_C}\left(\dfrac{R_A + R_E}{R_A R_D R_E}\right)}{\left(\dfrac{-1}{C_B R_A}\right)s} \tag{E-48}$$

3.
$$C_C = C_B = C \quad \text{and} \quad R_A = R_E = R \tag{E-49}$$

4. To solve for the component values in a band-pass, inverting, dual-feedback circuit from a loss function in the form of

$$\frac{s^2 + as + b}{-hs} \qquad \zeta > 1 \tag{E-50}$$

use the following procedure.

(a) Choose C.

(b) Calculate R.

$$R = \frac{a}{Cb} \tag{E-51a}$$

(c) Calculate R_D.

$$R_D = \frac{2}{Ca} \tag{E-51b}$$

(d) Calculate the resulting h.

$$h = \frac{1}{C_B R_A} \tag{E-51c}$$

E-5 TWIN-T TOPOLOGY

See Figure E-5.

$$\frac{V_{\text{in}}}{V_o} = \frac{Y_G Y_1 Y_2 + Y_F Y_1 (Y_B + Y_C - KY_C) + Y_E Y_2 (Y_A + Y_D)}{K(Y_A Y_E Y_2 + Y_B Y_F Y_1)} \tag{E-52}$$

where

$$Y_1 = Y_A + Y_D + Y_E$$

$$Y_2 = Y_B + Y_C + Y_F$$

$$K = 1 + \frac{R_1}{R_2}$$

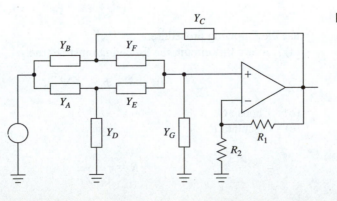

Figure E-5

E-5-1 Low Pass, High Pass, and Band Pass

1.
$$Y_A = sC_A \qquad Y_B = \frac{1}{R_B}$$

$$Y_C = sC_C \qquad Y_D = \frac{1}{R_D}$$

$$Y_E = sC_E \qquad Y_F = \frac{1}{R_F}$$
(E-53)

$$Y_G = sC_G + \frac{1}{R_G}$$

3. Due to the complexity of this circuit's transfer function, we will do step 3 first, and then step 2 will be the transfer function with these assumptions.

$$C_A = C_E = C \qquad R_B = R_F = R$$
(E-54)

$$C_C = 2C \qquad R_D = \frac{R}{2}$$

2.
$$\frac{V_{in}}{V_o} = \frac{s^2 + C_X\left(\dfrac{2}{CR_G} + \dfrac{4 - 2K}{CR} + \dfrac{2C_G}{C^2R}\right)s + C_X\left(\dfrac{2}{C^2R_GR} + \dfrac{1}{C^2R^2}\right)}{KC_X\left(s^2 + \dfrac{1}{C^2R^2}\right)}$$
(E-55)

where

$$C_X = \frac{C}{C + 2C_G}$$

4. To solve for the component values in a low-pass, high-pass, and band-pass twin-T circuit from a loss function in the form of

$$\frac{s^2 + as + b}{h(s^2 + e)}$$
(E-56)

use the following procedure.
(a) To use this circuit, the following must be true.

$$0 \leq e + b - a\sqrt{e}$$
(E-57a)

(b) Choose C.
(c) Calculate R.

$$R = \frac{1}{C\sqrt{e}}$$
(E-57b)

(d) Choose C_G such that

$$C_G \geq \frac{C(e - b)}{2b} \qquad \text{(E-57c)}$$

Note: C_G may be zero, which is an open circuit, or a positive value.

(e) Calculate R_G.

$$R_G = \frac{2\sqrt{e}}{C(b - e) + 2C_G b} \qquad \text{(E-57d)}$$

(f) Calculate K.

$$K = \frac{\dfrac{2}{R_G} + C(4\sqrt{e} - a) + 2C_G(\sqrt{e} - a)}{2C\sqrt{e}} \qquad \text{(E-57e)}$$

(g) Calculate the resulting h.

$$h = KC_X \qquad \text{(E-57f)}$$

Appendix F

Filter Table Calculations

This appendix shows how to make tables for the Butterworth, Chebyshev, and elliptic loss functions and how the equations for generating these tables are derived. From this information a computer program could be written to make the loss function tables for all three filter types. First we will look at the general derivation for these filters, and then we will look at each one specifically.

F-1 GENERAL PROCEDURE

The Butterworth, Chebyshev, and elliptic filter equations are defined by the magnitude of the loss function squared.

$$ML^2(j\Omega) = 1 + \varepsilon^2 F^2(j\Omega) \tag{F-1}$$

where ε = a numerical constant
$F(j\Omega)$ = the particular function that determines whether the loss function will be Butterworth, Chebyshev, or elliptic

In all of these filters, ε will be determined in the same way. When $F(j\Omega)$ is equal to 1, we are at a maximum in the pass band. Therefore, 20 log $(ML(j\Omega))$ will be equal to L_{max}. We know what L_{max} is required and we know that $F(j\Omega)$ is 1 at a maximum, so we can solve Eq. (F-1) for ε.

$$L_{max} = 20 \log \sqrt{1 + \varepsilon^2}$$

$$\varepsilon = \sqrt{10^{0.1 L_{max}} - 1} \tag{F-2}$$

When the function $F(j\Omega)$ is defined in Eq. (F-1), we replace $j\Omega$ with s, making the magnitude function squared into a loss function squared. When we take the square root of any function, we find two possible solutions. When we find the square root of the loss function, we will select only the stable roots (the roots to the left of the $j\omega$-axis). Let's examine how this is done.

We know that when we replace s with $j\Omega$ in the loss function, we will have a real part, R, and an imaginary part, I.

$$L(j\Omega) = R + jI \tag{F-3}$$

Using polar-to-rectangular conversion, we have

$$L(j\Omega) = \sqrt{R^2 + I^2} \; \underline{/\tan^{-1}(I/R)} \tag{F-4}$$

We are concerned only with the magnitude part and will therefore not put any conditions on the phase angle.

$$ML(j\Omega) = \sqrt{R^2 + I^2} \tag{F-5a}$$

Because Eq. (F-1) is based on the magnitude squared, we square Eq. (F-5a).

$$ML^2(j\Omega) = R^2 + I^2 \tag{F-5b}$$

We can factor this into complex conjugates.

$$ML^2(j\Omega) = (R + jI)(R - jI) \tag{F-5c}$$

The first factor is $L(j\Omega)$, as seen from Eq. (F-3). The second term is equal to $L(-j\Omega)$. We may then write Eq. (F-5c) as

$$ML^2(j\Omega) = L(j\Omega)L(-j\Omega) \tag{F-6}$$

We can show that $L(-j\Omega)$ is equal to $(R - jI)$ by example.

EXAMPLE F-1

Show that $L(-j\Omega)$ is the complex conjugate of $L(j\Omega)$.

Solution: We start with $L(s)$, which will be a numerator polynomial divided by a denominator polynomial.

$$L(s) = \frac{A_0 + A_1 s + A_2 s^2 + A_3 s^3 + \cdots}{B_0 + B_1 s + B_2 s^2 + B_3 s^3 + \cdots}$$

Then substituting $j\Omega$ for s,

$$L(j\Omega) = \frac{A_0 + jA_1\Omega - A_2\Omega^2 - jA_3\Omega^3 + \cdots}{B_0 + jB_1\Omega - B_2\Omega^2 - jB^3\Omega^3 + \cdots}$$

$$L(j\Omega) = \frac{(A_0 - A_2\Omega^2 + \cdots) + j(A_1\Omega - A_3\Omega^3 + \cdots)}{(B_0 - B_2\Omega^2 + \cdots) + j(B_1\Omega - B_3\Omega^3 + \cdots)}$$

We notice that all the odd powers of s will contain a j factor and the even powers of s do not. When we use $s = -j\Omega$, the even powers of s (the real part) will still have the same sign because a negative value to an even power is positive, but the odd powers of s will change sign.

$$L(-j\Omega) = \frac{A_0 - jA_1\Omega - A_2\Omega^2 + jA_3\Omega^3 + \cdots}{B_0 - jB_1\Omega - B_2\Omega^2 + jB_3\Omega^3 + \cdots}$$

$$L(-j\Omega) = \frac{(A_0 - A_2\Omega^2 + \cdots) - j(A_1\Omega - A_3\Omega^3 + \cdots)}{(B_0 - B_2\Omega^2 + \cdots) - j(B_1\Omega - B_3\Omega^3 + \cdots)}$$

Therefore, $L(j\Omega)$ and $L(-j\Omega)$ are complex conjugates.

Going back to Eq. (F-6), we can substitute s in for $j\Omega$.

$$ML^2(j\Omega)|_{j\Omega=s} = L(s)L(-s) \qquad \text{(F-7)}$$

When $j\Omega$ is replaced by s in the squared magnitude equation, $ML^2(j\Omega)$, we can solve for the roots. These roots will either be part of $L(s)$ or $L(-s)$. We use the roots to the left of the $j\omega$-axis in the s-plane that are the roots of $L(s)$. We then have the loss function.

The loss function is normally factored with complex conjugate roots paired and can be written as

$$L(s) = \frac{(s + a_0)^J \displaystyle\prod_{i=1}^{\text{int}(n/2)} (s^2 + a_i s + b_i)}{C_D \displaystyle\prod_{i=1}^{\text{int}(n/2)} (s^2 + c_i)} \qquad \text{(F-8)}$$

where n is the order of the filter
int$(n/2)$ is the integer portion of $n/2$
$J = 0$ for even n, and $j = 1$ for odd n

For the Butterworth and Chebyshev filters, there are no roots in the denominator.

We can find the denominator constant, C_D, if we know the value of $L(s)$ at a given frequency. For the Butterworth, Chebyshev, and elliptic filter equations, the magnitude of the loss function, Eq. (F-8), will be equal to $L_{\max}$ at $s = j$.

$$L_{\max} = 20 \log \left| \frac{(j + a_0)^J \displaystyle\prod_{i=1}^{\text{int}(n/2)} (-1 + a_i j + b_i)}{C_D \displaystyle\prod_{i=1}^{\text{int}(n/2)} (-1 + c_i)} \right|_{\text{mag}}$$

Solving for C_D gives

$$C_D = \left| \frac{(j + a_0)^J \displaystyle\prod_{i=1}^{\text{int}(n/2)} (-1 + a_i j + b_i)}{\displaystyle\prod_{i=1}^{\text{int}(n/2)} (-1 + c_i)} \right|_{\text{mag}} \left(\frac{1}{10^{L_{\max}/2}} \right) \qquad \text{(F-9)}$$

Notice that we need only the magnitude.

The following is a summary of the procedure for finding the filter equation.

1. Determine the value of ε using Eq. (F-2)
2. Select $F^2(j\Omega)$ in Eq. (F-1). This function will determine whether the filter is a Butterworth, Chebyshev, or elliptic.
3. Replace $j\Omega$ with s to form Eq. (F-7).
4. We then find the roots of Eq. (F-7). Different methods are used for this, depending on what is convenient. The rest of this appendix shows this step for each filter equation.
5. From the roots we choose the roots that are stable (to the left of the $j\omega$-axis).
6. Calculate the denominator constant using Eq. (F-9).

F-2 BUTTERWORTH FILTER EQUATION

For the Butterworth filter equation we start with an even-power polynomial for $F^2(j\Omega)$ in Eq. (F-1) and force it to have as few maximums and minimums as possible. (This filter is sometimes called the *maximally flat filter* because of this.) We therefore make, at $\Omega = 0$, as many derivatives equal to zero as possible. This polynomial is then substituted into Eq. (F-1) for $F^2(j\Omega)$. The magnitude squared for this function must contain only even powers.

$$F^2(j\Omega) = A_1\Omega^2 + A_2\Omega^4 + A_3\Omega^6 + \cdots + A_n\Omega^{2n} \qquad \textbf{(F-10)}$$

We can now evaluate the function by making, at $\Omega = 0$, as many derivatives equal to zero as possible.

$$\frac{d}{d\Omega} ML^2(j\Omega) = 2A_1\Omega + 4A_2\Omega^3 + 6A_3\Omega^5 + \cdots$$

Therefore,

$$\frac{d}{d\Omega} ML^2(j\Omega)\Big|_{\Omega=0} = 0$$

In this case the function is already zero for any value of Ω.
Finding the second derivative, we have

$$\frac{d^2}{d\Omega^2} ML^2(j\Omega) = 2A_1 + 12A_2\Omega^2 + 30A_3\Omega^4 + \cdots$$

Therefore,

$$\frac{d^2}{d\Omega^2} ML^2(j\Omega)\Big|_{\Omega=0} = 2A_1$$

For this to be zero, A_1 must equal zero.

As each derivative is set equal to zero, we will notice that odd derivatives are automatically equal to zero, and even derivatives will require another A coefficient to be zero.

We cannot let all the derivatives go to zero or our function will not be frequency dependent. The last derivative possible will have to be a constant times Ω^{2n}. We will choose A_n such that this constant will be equal to 1. Substituting this into Eq. (F-1) yields

$$ML^2(j\Omega) = 1 + \varepsilon^2\Omega^{2n} \tag{F-11}$$

We replace $j\Omega$ with s as shown in Eq. (F-7), but this equation does not have a $j\Omega$ term. We must rearrange the equation to have a $j\Omega$ term to substitute into.

$$1 + \varepsilon^2[(-1)(j)^2\Omega^2]^n$$

$$1 + \varepsilon^2[(-1)(j\Omega)^2]^n$$

Equation (F-7) therefore becomes

$$ML^2(j\Omega)|_{j\Omega=s} = L(s)L(-s) = 1 + \varepsilon^2[-(s)^2]^n \tag{F-12}$$

Eq. (F-2) can be used to find ε, therefore ε is a known value.

We then must factor this equation and solve for the stable roots. To solve the roots, we set the right side of the equation equal to zero.

$$1 + \varepsilon^2(-s^2)^n = 0 \tag{F-13}$$

We now need to focus our attention on solving the equation to find the $L(s)L(-s)$. First we need to list a few trigonometric identities that are required to solve Eq. (F-13).

$$-1 = \cos(-\pi) + j\sin(-\pi) \tag{F-14a}$$

$$= \cos(\pi) + j\sin(\pi)$$

$$e^{jx} = \cos(x) + j\sin(x) \tag{F-14b}$$

In Eq. (F-13) we move the 1 to the right side of the equation, which will make it -1. Using Eq. (F-14a), we can express this -1 as

$$-1 = \cos(-\pi) + j\sin(-\pi)$$

$$= \cos(-\pi + 2\pi k) + j\sin(-\pi + 2\pi k) \text{ where } k = 0, 1, 2, 3, \cdots$$

Using Eq. (F-14b), we can express this as an exponential function.

$$-1 = e^{j(2\pi k-\pi)}$$

Therefore, Eq. (F-13) becomes

$$\varepsilon^2(-s^2)^n = e^{j(2\pi k-\pi)}$$

or

$$\varepsilon^2(-1)^n(s^2)^n = e^{j(2\pi k - \pi)}$$

Because -1 and its reciprocal are the same value, we can move $(-1)^n$ to the right side of the equation.

$$\varepsilon^2(s^2)^n = e^{j(2\pi k - \pi)}(-1)^n$$

Using Eqs. (F-14a) and (F-14b), we can express $(-1)^n$ in an exponential form.

$$\varepsilon^2(s^2)^n = e^{j(2\pi k - \pi)}e^{j\pi n}$$

$$= e^{j(2\pi k - \pi + \pi n)}$$

Taking the $(2n)$th root of both sides gives

$$\varepsilon^{1/n}s = e^{j\pi(2k - 1 + n)/2n}$$

Solving for s and using Eq. (F-14b), we obtain

$$s = \varepsilon^{-1/n}\left\{\cos\left[\frac{\pi}{2}\left(\frac{2k - 1 + n}{n}\right)\right] + j\sin\left[\frac{\pi}{2}\left(\frac{2k - 1 + n}{n}\right)\right]\right\} \qquad \textbf{(F-15)}$$

Because s was to the $2n$ power, there are $2n$ roots. Therefore, $k = 0, 1, 2, 3, \cdots,$ $(2n - 1)$, but because the roots repeat after $2n - 1$, we may use $k = 1, 2, 3, \cdots, 2n$ for convenience.

Equation (F-15) will be the roots for both $L(s)$ and $L(-s)$ in Eq. (F-12). When we evaluate the roots, we must choose only the roots to the left of the $j\omega$-axis.

Summarizing how to find the roots for the Butterworth filter equations:

1. Select an L_{max} and filter order, n.
2. Calculate ε from Eq. (F-2).

$$\varepsilon = \sqrt{10^{0.1L_{max}} - 1}$$

3. Calculate the numerator roots from Eq. (F-15) using only the roots to the left of the $j\omega$-axis.

$$s = \varepsilon^{-1/n}\left\{\cos\left[\frac{\pi}{2}\left(\frac{2k - 1 + n}{n}\right)\right]\right.$$

$$\left. + j\sin\left[\frac{\pi}{2}\left(\frac{2k - 1 + n}{n}\right)\right]\right\} \qquad \text{where } k = 1, 2, 3, \cdots, 2n$$

4. Pair the complex conjugates roots to form Eq. (F-8). Notice that the Butterworth filter equation does not have denominator factors.

$$L(s) = \frac{(s + a_0)^J \displaystyle\prod_{i=1}^{\text{int}(n/2)} (s^2 + a_i s + b_i)}{C_D}$$

where n is the order of the filter
int($n/2$) is the integer portion of $n/2$
$J = 0$ for even n, and $J = 1$ for odd n

5. Determine the denominator constant using Eq. (F-9).

$$C_D = \left| \frac{(j + a_0)^J \displaystyle\prod_{i=1}^{\text{int}(n/2)} (-1 + a_i j + b_i)}{1} \right|_{\text{mag}} \left(\frac{1}{10^{L_{\text{max}}/2}} \right)$$

where n is the order of the filter
int($n/2$) is the integer portion of $n/2$
$J = 0$ for even n, and $J = 1$ for odd n

6. Complete the loss function by substituting the value of C_D into Eq. (F-8).

$$L(s) = \frac{(s + a_0)^J \displaystyle\prod_{i=1}^{\text{int}(n/2)} (s^2 + a_i s + b_i)}{C_D}$$

F-3 CHEBYSHEV FILTER EQUATION

The Chebyshev filter equation is made by using the Chebyshev function substituted in for $F(j\Omega)$ in Eq. (F-1). The Chebyshev function can be found in almost any filter textbook. The nth-order Chebyshev function is

$$C_n(\Omega) = \begin{cases} \cos(n \cos^{-1}\Omega) & 0 \le |\Omega| \le 1 \qquad \textbf{(F-16a)} \\ \cosh(n \cosh^{-1}\Omega) & |\Omega| > 1 \qquad \textbf{(F-16b)} \end{cases}$$

The Chebyshev function can also be expressed in a polynomial form. Because the polynomial will not be used to find the Chebyshev filter equation, it will not be presented here.

The Chebyshev function is substituted into Eq. (F-1) for $F(j\Omega)$.

$$ML^2(j\Omega) = 1 + \varepsilon^2 C_n^2(\Omega) \qquad \textbf{(F-17)}$$

We will set the right side of Eq. (F-17) equal to zero, substitute s in for $j\Omega$, and solve for the roots of s.

$$1 + \varepsilon^2 C_n^2(\Omega) = 0$$

$$C_n^2\left(\frac{j\Omega}{j}\right) = \frac{-1}{\varepsilon^2}$$

$$C_n\left(\frac{s}{j}\right) = \frac{\pm j}{\varepsilon} \qquad \textbf{(F-18)}$$

There are two possibilities for $C_n(s/j)$, as shown in Eqs. (F-16a) and (F-16b), so we need to solve Eq. (F-18) for both cases. However, we will only solve for the case of $0 \le |\Omega| \le 1$, because the solution using the other case will be similar. Therefore, we have

$$C_n\left(\frac{s}{j}\right) = \cos\left[n \cos^{-1}\left(\frac{s}{j}\right)\right] \tag{F-19}$$

The inverse cosine will have a real part, R_1, and an imaginary part, I_1.

$$R_1 + jI_1 = \cos^{-1}\left(\frac{s}{j}\right) \tag{F-20}$$

Substituting into Eq. (F-19) yields

$$C_n\left(\frac{s}{j}\right) = \cos[n(R_1 + jI_1)]$$

Substituting this into Eq. (F-18) gives

$$\cos[n(R_1 + jI_1)] = \frac{\pm j}{\varepsilon} \tag{F-21}$$

Using the following trigonometric identities,

$$\cos(x + y) = \cos(x)\cos(y) - \sin(x)\sin(y) \tag{F-22a}$$

$$\cos(jx) = \cosh(x) \tag{F-22b}$$

$$\sin(jx) = j\sinh(x) \tag{F-22c}$$

Equation (F-21) becomes

$$\cos(nR_1)\cosh(nI_1) - j\sin(nR_1)\sinh(nI_1) = \frac{\pm j}{\varepsilon}$$

We can associate the real part and imaginary part on the left side of the equation to the real part and imaginary part on the right side of the equation.

$$\cos(nR_1)\cosh(nI_1) = 0 \tag{F-23a}$$

$$-\sin(nR_1)\sinh(nI_1) = \frac{\pm 1}{\varepsilon} \tag{F-23b}$$

Because, in Eq. (F-23a), $\cosh(nI_1) \geq 1$ for all values of nI_1, then $\cos(nR_1)$ must equal zero. Therefore, we want nR_1 to be odd multiples of $\pi/2$. There are many ways to express this, but a convenient way for our purpose is

$$nR_1 = \left(k - \frac{1}{2}\right)\pi \qquad k = 1, 2, \cdots, 2n$$

$$R_1 = \left(\frac{2k-1}{2n}\right)\pi \qquad k = 1, 2, \cdots, 2n \qquad \textbf{(F-24)}$$

Looking at the imaginary part, we can substitute Eq. (F-24) in Eq. (F-23b).

$$-\sin\left[\left(\frac{2k-1}{2}\right)\pi\right]\sinh(nI_1) = \frac{\pm 1}{\varepsilon} \qquad k = 1, 2, \cdots, 2n$$

For the values of k, the sine part on the left will be either ± 1. We will only use positive ε, so we can simplify our equation to

$$\sinh(nI_1) = \frac{1}{\varepsilon}$$

$$I_1 = \frac{1}{n}\sinh^{-1}\left(\frac{1}{\varepsilon}\right) \qquad \textbf{(F-25)}$$

where

$$\sinh^{-1}(x) = \ln(x + \sqrt{x^2 + 1})$$

We can calculate the real part, R_1, and imaginary part, I_1, using Eqs. (F-24) and (F-25) and substitute back into Eq. (F-20).

$$\cos^{-1}\left(\frac{s}{j}\right) = R_1 + jI_1$$

Solving for s and applying Eqs. (F-22a), (F-22b), and (F-22c) gives

$$s = j\cos(R_1 + jI_1)$$

$$s = \sin(R_1)\sinh(I_1) + j\cos(R_1)\cosh(I_1) \qquad \textbf{(F-26)}$$

where R_1 and I_1 are defined by Eqs. (F-24) and (F-25).

With a similar procedure we could derive Eq. (F-17) using the Chebyshev function of Eq. (F-16b). For $|\Omega| > 1$, the resulting equation would be identical.

Following is a summary of how to find the roots for the Chebyshev filter equations:

1. Select an L_{max} and filter order, n.
2. Calculate ε from Eq. (F-2).

$$\varepsilon = \sqrt{10^{0.1L_{max}} - 1}$$

3. Calculate the numerator roots from Eqs. (F-26), (F-24), and (F-25) using only the roots to the left of the $j\omega$-axis.

$$s = \sin(R_1)\sinh(I_1) + j\cos(R_1)\cosh(I_1)$$

$$R_1 = \left(\frac{2k-1}{2n}\right)\pi \qquad k = 1, 2, \cdots, 2n$$

$$I_1 = \frac{1}{n}\sinh^{-1}\left(\frac{1}{\varepsilon}\right)$$

where

$$\sinh^{-1}(x) = \ln(x + \sqrt{x^2 + 1})$$

4. Pair the complex–conjugate roots to form Eq. (F-8). Notice that the Chebyshev filter equation does not have denominator factors.

$$L(s) = \frac{(s + a_0)^J \displaystyle\prod_{i=1}^{\mathrm{int}(n/2)} (s^2 + a_i s + b_i)}{C_D}$$

where n is the order of the filter
    $\mathrm{int}(n/2)$ is the integer portion of $n/2$
    $J = 0$ for even n, and $J = 1$ for odd n

5. Determine the denominator constant using Eq. (F-9).

$$C_D = \left| \frac{(j + a_0)^J \displaystyle\prod_{i=1}^{\mathrm{int}(n/2)} (-1 + a_i j + b_i)}{1} \right|_{\mathrm{mag}} \left(\frac{1}{10^{L_{\mathrm{max}}/2}}\right)$$

where n is the order of the filter
    $\mathrm{int}(n/2)$ is the integer portion of $n/2$
    $J = 0$ for even n, and $J = 1$ for odd n

6. Complete the loss function by substituting the value of C_D into Eq. (F-8).

$$L(s) = \frac{(s + a_0)^J \displaystyle\prod_{i=1}^{\mathrm{int}(n/2)} (s^2 + a_i s + b_i)}{C_D}$$

F-4 ELLIPTIC FILTER EQUATION

The elliptic filter equation uses the Chebyshev rational function for $F(j\Omega)$ in Eq. (F-1). Because this function is complex, we will concentrate on how to find the elliptic function equations. A complete discussion of the elliptic function equations can be found in

Daniels.[1] We will only summarize and make note of what is required to solve for the roots of the elliptic function equation.

The nth-order Chebyshev rational function can be written with $j\Omega$ replaced by s as

$$R_n(s) = C_R s^J \prod_{i=1}^{\text{int}(n/2)} \frac{s^2 - r_{zi}^2}{s^2 - r_{pi}^2} \qquad \text{(F-27)}$$

where $J = 0$ for even n, and $J = 1$ for odd n

$$C_R = \prod_{i=1}^{\text{int}(n/2)} \frac{1 - r_{pi}^2}{1 - r_{zi}^2}$$

int($n/2$) is the integer of $n/2$

This is substituted back into Eq. (F-1) to form the loss function.

$$F^2(s) = 1 + \varepsilon^2 R_n^2(s) \qquad \text{(F-28)}$$

It is important to notice that the denominator for the Chebyshev rational function will also be the denominator for the elliptic filter equation.

The roots of the Chebyshev rational function are found by

$$r_{zi}z = \text{sn}(y_i; \Omega_s^{-1}) \qquad \text{(F-29a)}$$

$$r_{pi} = \frac{1}{\Omega_s^{-1} r_{zi}} \qquad \text{(F-29b)}$$

where sn is the elliptic sine function
Ω_s is the normalized stop-band edge frequency
$i = 1, 2, 3, \cdots, \text{int}(n/2)$

The calculation of the elliptic sine function of Eq. (F-29a) is quite involved, and the procedure is as follows:

1. Evaluate the complete Jacobian elliptic function of the first kind, and then the complementary form.

    The complete Jacobian elliptic function is defined as

$$K(\Omega_s^{-1}) = \int_0^{\pi/2} \frac{d\theta}{[1 - \Omega_s^{-1} \sin^2(\theta)]^{1/2}} \qquad \text{(F-30)}$$

Using a binomial series expansion, we can evaluate this integral by

$$K(\Omega_s^{-1}) = \frac{\pi}{2} \left[1 + \left(\frac{1}{2}\right)^2 (\Omega_s^{-1})^2 + \left(\frac{1 \cdot 3}{2 \cdot 4}\right)^2 (\Omega_s^{-1})^4 \right.$$

$$\left. + \left(\frac{1 \cdot 3 \cdot 5}{2 \cdot 4 \cdot 6}\right)^2 (\Omega_s^{-1})^6 + \cdots \right] \qquad \text{(F-31)}$$

[1] R. W. Daniels, *Approximation Methods for Electronic Filter Design,* McGraw-Hill Book Company, New York, 1974.

The complementary form, $K'(\Omega_s^{-1})$, uses $\Omega_s^{-1'}$ substituted into Eqs. (F-30) and (F-31) for Ω_s^{-1}, where

$$\Omega_s^{-1'} = \sqrt{1 - (\Omega_s^{-1})^2} \qquad \text{(F-32)}$$

This changes Eq. (F-31) into

$$K'(\Omega_s^{-1}) = \frac{\pi}{2}\left[1 + \left(\frac{1}{2}\right)^2 (\Omega_s^{-1'})^2 + \left(\frac{1 \cdot 3}{2 \cdot 4}\right)^2 (\Omega_s^{-1'})^4 \right.$$

$$\left. + \left(\frac{1 \cdot 3 \cdot 5}{2 \cdot 4 \cdot 6}\right)^2 (\Omega_s^{-1'})^6 + \cdots \right] \qquad \text{(F-33)}$$

2. The values of $K(\Omega_s^{-1})$ and $K'(\Omega_s^{-1})$ are used to find the value of y in the Jacobian elliptic integral.

$$y_i = \int_0^{ui} \frac{d\theta}{[1 - \Omega_s^{-1}\sin^2(\theta)]^{1/2}} \qquad \text{(F-34a)}$$

where for odd n

$$y_i = i\,\frac{2K(\Omega_s^{-1})}{n} \qquad \text{(F-34b)}$$

and for even n,

$$y_i = (i - 0.5)\frac{2K(\Omega_s^{-1})}{n} \qquad i = 1, 2, 3, \cdots, \text{int}(n/2) \qquad \text{(F-34c)}$$

3. Knowing the specific value the Jacobian elliptic integral must be, we solve for the upper limit, u_i, required to obtain this specific value. Using a numerical process from the *Handbook of Mathematical Functions*,[2] we obtain

$$u_i = v + \sum_{m=1}^{\infty} \frac{2q^m \sin(2mv)}{m(1 + q^{2m})} \qquad \text{(F-35)}$$

where

$$v = \frac{\pi y_i}{2K(\Omega_s^{-1})}$$

$$q = e^{-\pi K'(\Omega_s^{-1})/K(\Omega_s^{-1})}$$

---

[2] L. M. Milne-Thomson, *Handbook of Mathematical Functions with Formulas, Graphs, and Mathematical Tables*. National Bureau of Standards Applied Mathematics Series 55, Section 17, p. 591, issued June 1964, third printing March 1965.

4. Last, we find the sine of the upper limit for each y_i. Therefore,

$$\text{sn}(y_i; \Omega_s^{-1}) = \sin(u_i) \qquad \text{(F-36)}$$

Once this is found, we can solve Eqs. (F-29a) and (F-29b), which are the roots of Eq. (F-27). At this point we have found the roots of the Chebyshev rational function, $R_n(s)$. We still must find the roots of the loss function, $L(s)$. Solving the roots of the loss function is not a trivial problem.

Daniels[3] points out that because the roots in the rational function are close together, they should be spread apart to increase the accuracy of finding the roots of the loss function. This transformation we will call Z-transformation. This is not the same as Z-transforms and should not be though of as being even remotely related.

We see in Eq. (F-28) that the rational function is squared. We will therefore define the Z-transformation of the Chebyshev rational function squared as

$$R_n^2(Z) = \frac{C_{ZN}^2}{C_{ZD}^2 (Z^2 - 1)^J} \prod_{i=1}^{\text{int}(n/2)} \frac{(Z^2 - r_{zzi}^2)^2}{(Z^2 - r_{zpi}^2)^2} \qquad \text{(F-37)}$$

where

$$C_{ZN}^2 = C_R^2 \prod_{i=1}^{\text{int}(n/2)} (r_{zi}^2)^2$$

C_R from Eq. (F–27)

$$C_{ZD}^2 = (-1)^J \prod_{i=1}^{\text{int}(n/2)} (r_{pi}^2)^2$$

$$r_{zzi}^2 = 1 - \frac{1}{r_{zi}^2}$$

$$r_{zpi}^2 = 1 - \frac{1}{r_{pi}^2}$$

$J = 0$ for even n, and $J = 1$ for odd n

The factors of Eq. (F-37) can be multiplied out to form a numerator polynomial, $N(Z)$, divided by a denominator polynomial, $D(Z)$.

$$R_n^2(Z) = \frac{N^2(Z)}{D^2(Z)} \qquad \text{(F-38)}$$

---

[3] R. W. Daniels, *Approximation Methods for Electronic Filter Design*, McGraw-Hill Book Company, New York, 1974.

This equation can be substituted into Eq. (F-28) to form a Z-transformation loss function squared.

$$F^2(s) = 1 + \varepsilon^2 \frac{N^2(Z)}{D^2(Z)} \tag{F-39}$$

which can be written as

$$F_n^2(Z) = \frac{D^2(Z) + \varepsilon^2 N^2(Z)}{D^2(Z)} \tag{F-40}$$

Here we see the numerator and denominator polynomials of the rational function must be multiplied out and combined into a single numerator polynomial. Because we already have the factored roots of the denominator in the s-domain, we only have to factor the numerator and transform it from the Z-transformation to the s-domain.

The numerator of Eq. (F-40) will often be larger than the fourth order. Numerical methods of factoring will therefore be required. One such method is Bairstow's method. A complete discussion of this method can be found in *Applied Numerical Methods*[4], we will only apply the method to our case.

This method is an iterative method that starts with an arbitrary quadratic, divides the polynomial by this quadratic, corrects the value of the quadratic, and then divides the polynomial by the corrected quadratic. This is repeated until the quadratic is corrected to the accuracy required. Because we are using a quadratic, we will not have complex numbers to deal with.

In our case there will be only even powers of the variable Z. Therefore,

$$D^2(Z) + \varepsilon^2 N^2(Z) = C(Z^{2I} + A_1 Z^{2(I-1)} + A_2 Z^{2(I-2)} + \cdots + A_1) \tag{F-41}$$

where I is the degree of the polynomial
notice that A_0 is equal to 1

We need to find an a_z and b_z such that

$$C(Z^4 + a_z Z^2 + b_z)(B_0 Z^{2(I-2)} + B_1 Z^{2(I-3)} + \cdots + B_{I-2})$$

$$= D^2(Z) + \varepsilon^2 N^2(Z) \tag{F-42}$$

We will then take the remaining B polynomial and factor out another quadratic until we have all the factors of the polynomial as

$$D^2(Z) + \varepsilon^2 N^2(Z) = C(Z^2 - a_{z0})^J \prod_{i=1}^{\text{int}(n/2)} (Z^4 + a_{zi} Z^2 + b_{zi}) \tag{F-43}$$

where C is defined in Eq. (F-41)
$J = 0$ for even n, and $J = 1$ for odd n

---

[4] M. L. James, G. M. Smith, and J. C. Wolford, *Applied Numerical Methods for Digital Computation with Fortran and CSMP*, 2nd ed., Harper & Row, Publishers, Inc., New York, 1977, pp. 141–146.

Using variables defined in Eqs. (F-41), (F-42), and (F-43), we can define Bairstow's method as follows:

1. Make sure that the polynomial is greater than the second order, $n > 2$, since Bairstow's method will not be required for $n \leq 2$.
2. Set M equal to the degree of the polynomial, n. Set i to an initial value of 1 to point to an a_{zi} and b_{zi}.
3. Factor the constant, C, out of polynomial A as shown in Eq. (F-41) so that the first coefficient is 1.
4. Choose a starting value for a_{zi} and b_{zi}. Values from -1 to 5 will usually give good results.
5. Set initial values for the starting value of a B polynomial and a C polynomial.

$$B_0 = 1 \qquad B_1 = A_1 - a_i$$

$$C_0 = 1 \qquad C_1 = B_1 - a_i$$

6. Initialize a counter variable. $K = 2$.
7. Find the next B coefficient.

$$B_K = A_K - B_{K-1}a_{zi} - B_{K-2}b_{zi}$$

8. If $K = M$, go to step 11; if not, continue.
9. Find the next C coefficient.

$$C_K = B_K - C_{K-1}a_{zi} - C_{K-2}b_{zi}$$

10. Increment the counter K ($K = K + 1$) and go back to step 7.
11. Calculate the correction factor for a_{zi} and b_{zi}.

$$D_a = \frac{B_n C_{n-3} - B_{n-1}C_{n-2}}{C_{n-1}C_{n-3} - C_{n-2}^2}$$

$$D_b = \frac{B_{n-1}C_{n-1} - B_n C_{n-2}}{C_{n-1}C_{n-3} - C_{n-2}^2}$$

Here we need to check if D_a's and D_b's magnitude is continuing to decrease (converge) or increase (diverge). If either one is diverging, we need to go back to step 4 and choose different starting a_{zi} and b_{zi} values.

12. Calculate the new a_{zi} and b_{zi}.

$$a_{zi} = a_{zi} + D_a$$

$$b_{zi} = b_{zi} + D_b$$

13. Determine of more accuracy is required. If $|D_a| + |D_b| >$ a desired small number, go back to step 5. Ideally, D_a and D_b should become zero.
14. The a_{zi} and b_{zi} at this point are coefficients in one of the factors.

$$Z^4 + a_{zi}Z^2 + b_{zi}$$

Or if $b_{zi} = 0$, we have found

$$Z^2 + a_{z0}$$

15. Set up to factor the remaining polynomial by moving the B coefficients to the A array, and reduce M by one factor.

$$A_j = B_j \text{ for } j = 1 \text{ to } n - 2$$

$$\text{if } b_{zi} <> 0, \text{ then } M = M - 2$$

$$\text{if } b_{zi} = 0, \text{ then } M = M - 1$$

16. Set i to point to the next a_{zi} and b_{zi} value ($i = i + 1$).
17. If $M > 2$, go to step 4 to start looking for the next a_{zi} and b_{zi} values.
18. If $M = 1$, then a_{zi} is a_{z0}.

When we finish Bairstow's method, we will have the factored Z-transformation shown in Eq. (F-43). This is the numerator of the loss function squared. We can find the inverse Z-transformation and the stable roots by the following procedure.

$$\sqrt{D^2(s) + \varepsilon^2 N^2(s)} = C(s - a_0)^J \prod_{i=1}^{\text{int}(n/2)} (s^2 + a_i s + b) \qquad \textbf{(F-44)}$$

where

$$a_i = \left(\frac{-2 - a_{zi} + 2\sqrt{1 + a_{zi} + b_{zi}}}{1 + a_{zi} + b_{zi}} \right)^{1/2}$$

$$b_i = \left(\frac{1}{1 + a_{zi} + b_{zi}} \right)^{1/2}$$

$$a_0 = \left(\frac{1}{a_{z0} - 1} \right)^{1/2}$$

$$J = 0 \text{ for even } n, \text{ and } J = 1 \text{ for odd } n$$

The constant, C, we will not evaluate because we will evaluate the loss functions constant using Eq. (F-9).

We now have the roots for the numerator and denominator of the loss function. With the elliptic function equations we need to find $L_{\min}$, because this function does not continue to increase in attenuation in the stop band. This value will be

$$L_{\min} = 10 \log \sqrt{1 + \varepsilon^2 X^2} \tag{F-45}$$

where

$$X = \frac{1}{(\Omega_s^{-1})^n \displaystyle\prod_{i=0}^{\mathrm{int}(n/2)-1} \mathrm{sn}^4(y_i; \Omega_s^{-1})}$$

The elliptic sine function to the fourth power is calculated by using Eqs. (F-35) and (F-36) with

$$y_i = \frac{(1 + 2i)K(\Omega_s^{-1})}{n}$$

In the summary of this procedure, the equations that specifically apply to the elliptic function equations will only have the equation numbers listed because they are so complex. This will facilitate the understanding of the order of using the equations.

Following is a summary of how to find the roots for the elliptic filter equations.

1. Select an $L_{\max}$ and filter order, n.
2. Calculate ε from Eq. (F-2).

$$\varepsilon = \sqrt{10^{0.1L_{\max}} - 1}$$

3. Find the Chebyshev rational function roots for Eq. (F-27).
   a. Find $K(\Omega^{-1})$ using Eq. (F-31).
   b. Find $K'(\Omega_s^{-1})$ using Eqs. (F-32) and (F-33).
   c. Find y_i using Eq. (F-34).
   d. Find u_i using Eq. (F-35).
   e. Find the r_{zi} roots of Eq. (F-29a) using Eq. (F-36).
   f. Find the r_{pi} roots of Eq. (F-29b) that are also the roots of the denominator of the loss function squared.
4. Convert the Chebyshev rational function to the squared Z-transformation Chebyshev rational function using Eq. (F-37).
5. Form the Z-transformation loss function numerator of Eq. (F-40).
6. Factor the numerator squared using Bairstow's method.
7. Find the stable roots of the loss function numerator using Eq. (F-44).
8. Write the elliptic function equation of Eq. (F-8) using the roots found.

$$L(s) = \frac{(s + a_0)^J \displaystyle\prod_{i=1}^{\mathrm{int}(n/2)} (s^2 + a_i s + b_i)}{C_D \displaystyle\prod_{i=1}^{\mathrm{int}(n/2)} (s^2 + c_i)}$$

9. Determine the denominator constant using Eq. (F-9).

$$C_D = \left| \frac{(j + a_0)^J \displaystyle\prod_{i=1}^{\text{int}(n/2)} (-1 + a_i j + b_i)}{\displaystyle\prod_{i=1}^{\text{int}(n/2)} (-1 + c_i)} \right|_{\text{mag}} \left(\frac{1}{10^{L_{max}/2}} \right)$$

where n is the order of the filter
int$(n/2)$ is the integer portion of $n/2$
$J = 0$ for even n, and $J = 1$ for odd n

10. Form the completed loss function by substituting the value of C_D into Eq. (F-8).
11. Find the pass-band loss, L_{min}, using Eq. (F-45).

Appendix G

Road Map for This Text

There are over 16,000 roads though this text due to the compartmentalization of the material. Because the compartmentalization is not absolute, their is a main road that must be taken in order to understand the fundamentals of analog signal processing necessary to designing active filters. In order to make it possible to navigate through the thousands of roads, a road map, or flow chart, is given at the end of this introduction. With this road map, the rigor and time frame for a course can be adjusted to the required curriculum and student's capabilities.

The road map uses a typical flow chart format. Therefore, when going from the beginning to the end, you must always move in the direction of the arrows. Each block contains both the title and the section number from the table of contents so that not only the material content is indicated but also the location of the material in the text is indicated.

In the map, the main road and optional side roads are clearly indicated. As an example, note Subsections 3-3-6 and 3-3-7 in the map for Chapter 3. Here the main road bypasses these two sections, but when a more rigorous understanding of Laplace transforms is desired, these sections can be covered. Notice how the flow of the arrows shows that either subsection could be used without the other, but if both are used, Subsection 3-3-6 should be covered first.

The MATLAB and TI-89 sections are not shown in the map. These two computing devices could be used in every section of the chapters, but it would be impractical to show each type of problem being worked and then a reference alongside of every block in the map. Therefore, only a few select examples are chosen to demonstrate how MATLAB and the TI-89 can be used, and these are not noted in the map in order to prevent clutter.

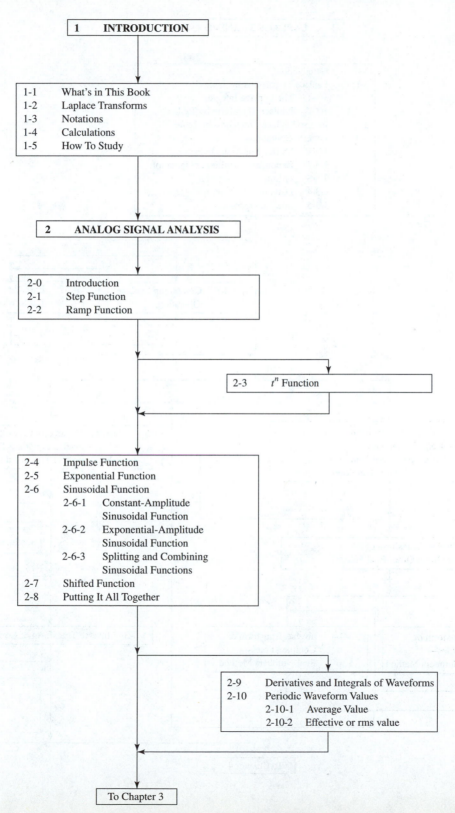

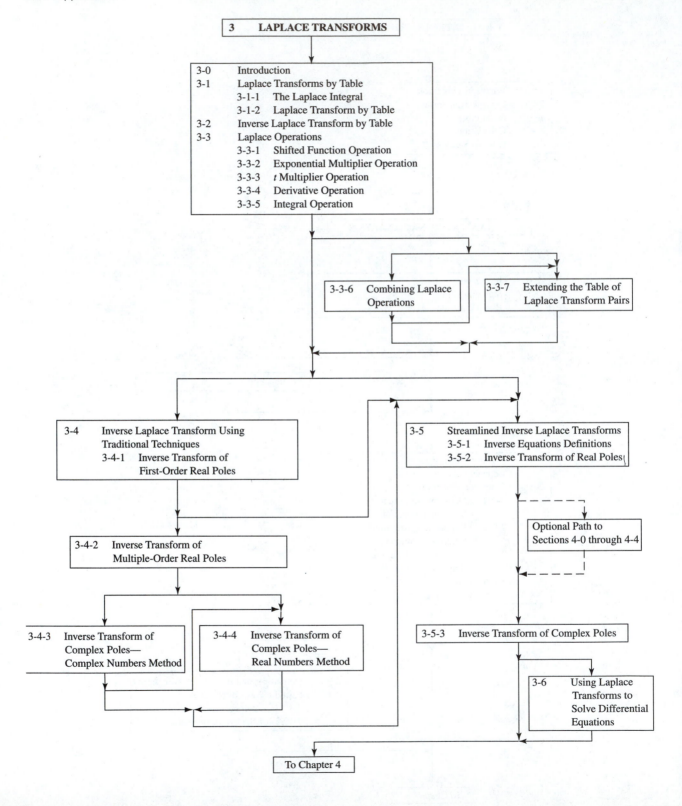

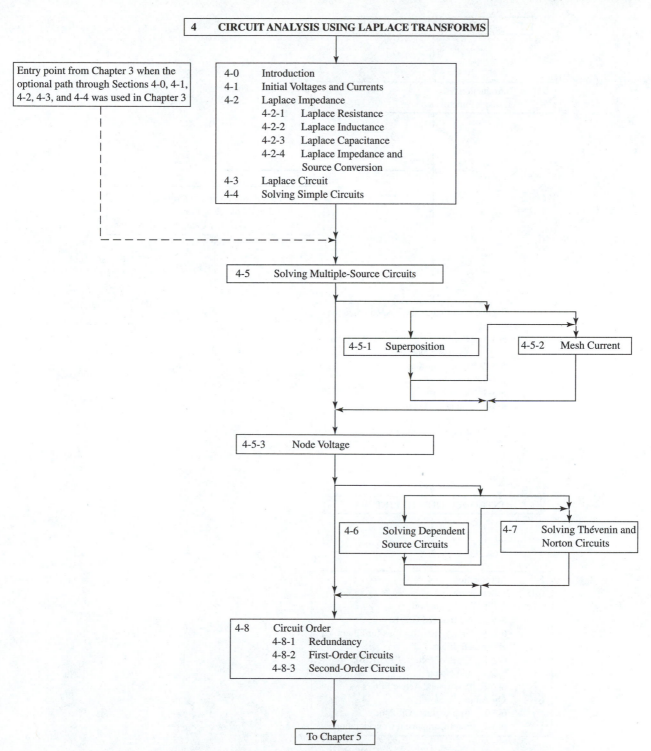

4   CIRCUIT ANALYSIS USING LAPLACE TRANSFORMS

Entry point from Chapter 3 when the optional path through Sections 4-0, 4-1, 4-2, 4-3, and 4-4 was used in Chapter 3

4-0   Introduction
4-1   Initial Voltages and Currents
4-2   Laplace Impedance
    4-2-1   Laplace Resistance
    4-2-2   Laplace Inductance
    4-2-3   Laplace Capacitance
    4-2-4   Laplace Impedance and
            Source Conversion
4-3   Laplace Circuit
4-4   Solving Simple Circuits

4-5   Solving Multiple-Source Circuits

4-5-1   Superposition

4-5-2   Mesh Current

4-5-3   Node Voltage

4-6   Solving Dependent
    Source Circuits

4-7   Solving Thévenin and
    Norton Circuits

4-8   Circuit Order
    4-8-1   Redundancy
    4-8-2   First-Order Circuits
    4-8-3   Second-Order Circuits

To Chapter 5

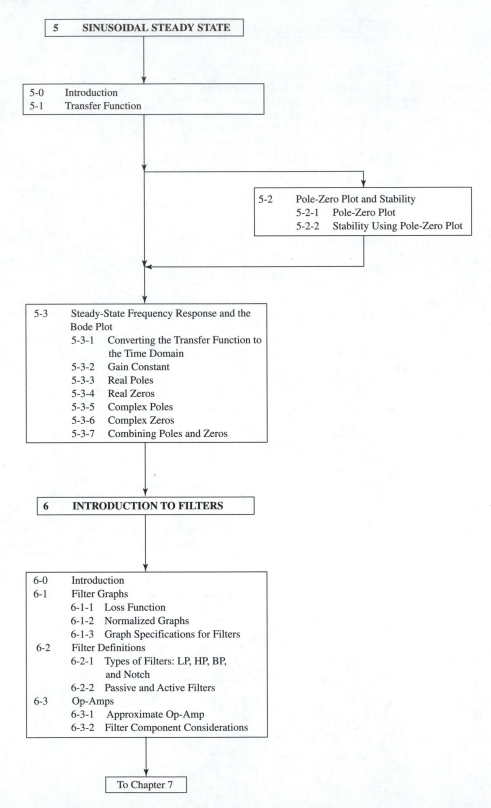

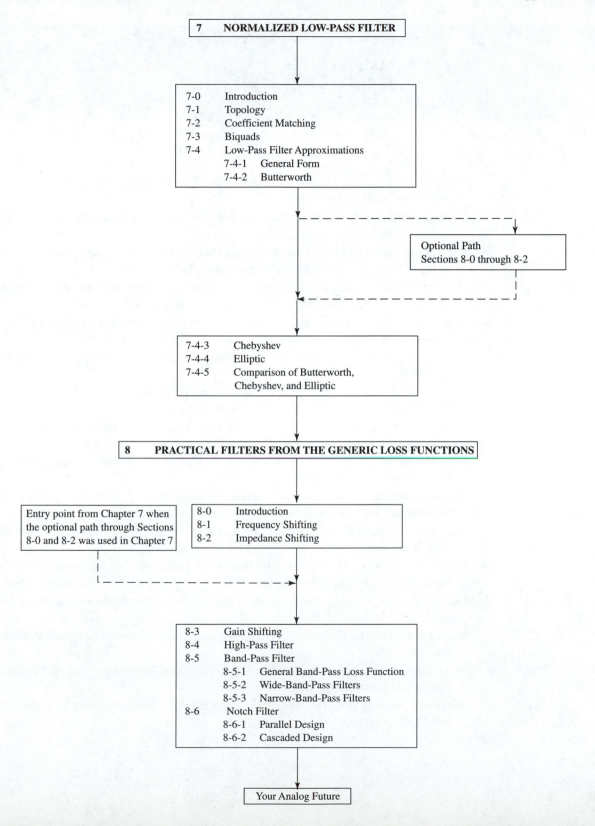

Bibliography

CHRISTIAN, ERICH, and EGON EISENMANN. *Filter Design Tables and Graphs.* New York: John Wiley & Sons, Inc., 1966.

DANIELS, RICHARD W. *Approximation Methods for Electronic Filter Design.* New York: McGraw-Hill Book Company, 1974.

DARYANANI, GOBIND. *Principles of Active Network Synthesis and Design.* New York: John Wiley & Sons, Inc., 1976, by Bell Laboratories, Inc.

HUELSMAN, LAWRENCE P., and PHILLIP E. ALLEN. *Introduction to the Theory and Design of Active Filters.* New York: McGraw-Hill Book Company, 1980.

JAMES, M. L., G. M. SMITH, and J. C. WOLFORD. *Applied Numerical Methods for Digital Computation,* 2nd ed. New York: Harper & Row, Publishers, Inc., 1977.

JOHNSON, DAVID E. *Introduction to Filter Theory.* Englewood Cliffs, N.J.: Prentice-Hall, Inc., 1976.

LACROIX, ARILD, and KARL-HEINZ WITTE. *Design Tables for Discrete Time Normalized Low-Pass Filters.* Dedham, Mass.: Artech House, Inc., 1986.

LAGO, GLADWYN, and LLOYD M. BENNINGFIELD. *Circuit and System Theory.* New York: John Wiley & Sons, Inc., 1979.

LAM, HARRY Y-F. *Analog and Digital Filters: Design and Realization.* Englewood Cliffs, N.J.: Prentice-Hall, Inc., 1979.

MILNE-THOMSON, L. M. "Elliptic Integrals," in *Handbook of Mathematical Functions.* Washington, D.C.: National Bureau of Standards, March 1965.

MOSCHYTZ, GEORGE SAMSON, and P. HORN. *Active Filter Design Handbook.* New York: John Wiley & Sons, Inc., 1981.

NATARAJAN, SUNDARAM. *Theory and Design of Linear Active Networks.* New York: Macmillan Publishing Company, Inc., 1987.

RICHMOND, A. E. *Calculus for Electronics,* 2nd ed. New York: McGraw-Hill Book Company, 1972.

STANLEY, WILLIAM D. *Network Analysis with Applications.* Reston, Va.: Reston Publishing Co., Inc., 1985.

——. *Transform Circuit Analysis for Engineering and Technology,* 2nd ed. Englewood Cliffs, N.J.: Prentice-Hall, Inc., 1989.

WILLIAMS, ARTHUR BERNARD. *Electronic Filter Design Handbook.* New York: McGraw-Hill Book Company, 1981.

Answers to Odd-Numbered Problems

CHAPTER 2

1. Fig. P2-1

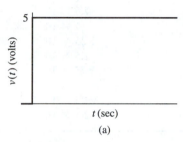
(a)

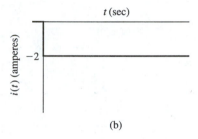
(b)

Figure P2-1

3. $i(t) = 6 \times 10^{-3} u(t)$

5. Fig. P2-2

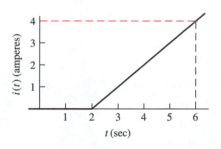

Figure P2-2

9. Fig. P2-3

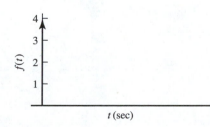

Figure P2-3

11. (a) $\tau = 5 \times 10^{-4}$
    (b) $\alpha = 2 \times 10^3$
    (c) Table P2-1
    (d) Fig. P2-4

Table P2-1

t (time constants in seconds)	t (seconds)	$v(t)$ $5e^{-2000t} u(t)$
0	0	5
$0.5\,\tau$	2.5×10^{-4}	3.0327
$1\,\tau$	5×10^{-4}	1.8394
$1.5\,\tau$	7.5×10^{-4}	1.1157
$2\,\tau$	1×10^{-3}	6.7668×10^{-1}
$2.5\,\tau$	1.25×10^{-3}	4.1042×10^{-1}
$3\,\tau$	1.5×10^{-3}	2.4894×10^{-1}
$3.5\,\tau$	1.75×10^{-3}	1.5099×10^{-1}
$4\,\tau$	2×10^{-3}	9.1578×10^{-2}
$4.5\,\tau$	2.25×10^{-3}	5.5545×10^{-2}
$5\,\tau$	2.5×10^{-3}	0

Figure P2-4

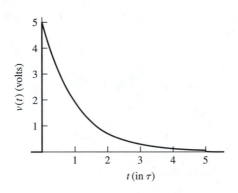

13. (a) $14 \sin(6t + \pi/6)$ $14 \sin[(1080/\pi)t° + 30°]$
    (b) $9 \cos(5t + \pi/12)$ $9 \cos[(900/\pi)t° + 15°]$

15. (a) 2.5 sec
    (b) 0 A
    (c) Table P2-2
    (d) Fig. P2-5

Table P2-2

t in seconds	$i(t) = 4e^{-2t} \sin(5t)\, u(t)$
-3.9026×10^{-1}	0 since $u(t) = 0$
0^+	0
2.3806×10^{-1}	2.3071
6.2832×10^{-1}	0
8.6638×10^{-1}	-0.65661×10^{-1}
1.2567	0
1.4947	1.86880×10^{-1}
1.8850	0
2.1230	-5.31870×10^{-2}
2.5133	0 since we are past 5 τ
2.7513	0 since we are past 5 τ

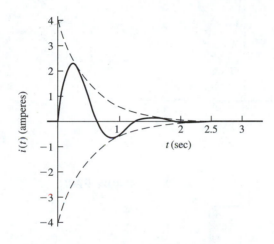

Figure P2-5

17. $5.4464 \sin(4\pi t) + 8.3867 \cos(4\pi t)$

19. $-2.5712 \sin(3\pi t) + 3.0642 \cos(3\pi t)$

21. $7.8102 e^{-7t} \sin(4\pi t + 39.806°)$

23. $4(t - 2)\, u(t - 2)$

25. $7e^{-2(t-3\times10^{-6})} \sin[3(t - 3 \times 10^{-6}) + 40°] u(t - 3 \times 10^{-6})$

27. $4e^{-2t} \sin(4t + 10°)\, u(t)$

29. Fig. P2-6

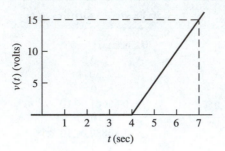

Figure P2-6

31. Fig. P2-7

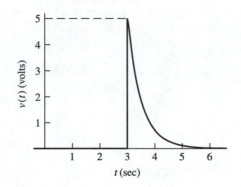

Figure P2-7

33. Fig. P2-8

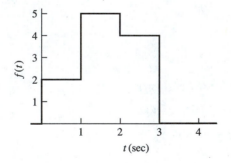

Figure P2-8

35. Fig. P2-9

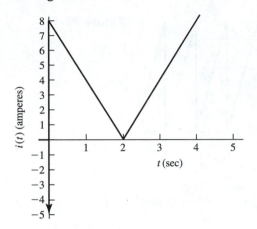

Figure P2-9

37. Fig. P2-10

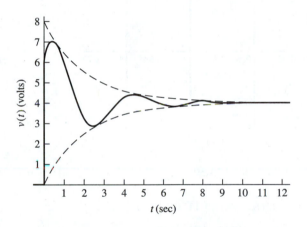

Figure P2-10

39. $i(t) = 4\,\delta(t) + 8\,u(t - 2) - 8\,u(t - 3)$

41. $4t\,u(t) + 2\,u(t - 1) - 7(t - 1)\,u(t - 1) + 3(t - 2)\,u(t - 2) - 3\,u(t - 3)$

43. $f(t) = 2\,1u(t) + 3.25935e^{-0.55112t}\sin(2.0944t - 1.2566)\,1u(t)$

45. Fig. P2-11

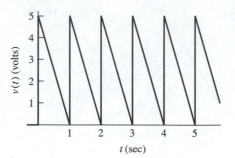

Figure P2-11

47. $20\,u(t) + \sum\limits_{n=1}^{\infty} (-1)^n\, 40\, u(t - n(0.5 \times 10^{-3}))$

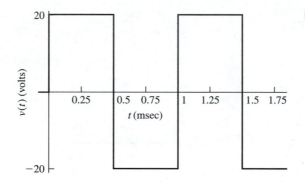

Figure P2-12

49. $v(t) = 8.9095 \sin(120\pi t) + \sum\limits_{n=1}^{\infty} 17.819 \sin[120\pi(t - n/120)]\, u(t - n/120)$

51. $u(t) + 7t\,u(t) - 4(t - 1)^2\, u(t - 1)$

53. $\delta(t) + 10(t - 2)\, u(t - 2)$

55. $5(t - 1)\, u(t - 1) + 0.2e^{-2(t-2)}\, u(t - 2) + 0.22361 e^{-2(t-2)} \sin[4(t - 2) - 116.57°]\, u(t - 2)$

57. $6\,\delta(t) - 18e^{-3t}\, u(t) + 0.96593\,\delta(t - 3) + 4 \sin[4(t - 3) + 165°]\, u(t - 3)$

59. $5 \times 10^{-3}\,\delta(t) + 2 \times 10^{-2}\, u(t - 3)$

61. $-t^2\, u(t) + \sum\limits_{n=0}^{\infty} 4(t - 2n)\, u(t - 2n)$

63. $8t\,u(t) - 8(t - 2)\, u(t - 2) + 8(t - 3)\, u(t - 3) - 12(t - 5)\, u(t - 5) + (t - 5)^2\, u(t - 5) - (t - 7)^2\, u(t - 7)$

65. 4.3333

67. 6.3662

69. (a) $5e^{-(t-5)} u(t-5)$
    (b) $15(t-5) u(t-5) + 30 u(t-5)$

71. (a) 1 amperes rms
    (b) 20 watts

CHAPTER 3

1. See Appendix B, Section B-1-3

3. (a) $V(s) = \dfrac{3s}{s^2 + 144}$

    (b) $I(s) = \dfrac{31.416}{s^2 + 39.478}$

    (c) $I(s) = \dfrac{3.4641 \, (s - 1.7321)}{s^2 + 9}$

5. $v(t) = 14 \cos(4t) \, u(t)$

7. $5.6569 \sin(2t + 45°) \, u(t)$

9. (a) $\dfrac{-4(s-1)e^{-2s}}{s^2}$

    (b) $\dfrac{0.57940(s - 2.8134)e^{-s}}{s^2 + 4}$

11. $i(t) = \dfrac{10.3923(s + 3.7321)}{[(s+2)^2 + 9]}$

13. $i(t) = \dfrac{(s+3)(s-3)}{(s^2+9)^2}$

15. (a) $\dfrac{-24}{s+6}$

    (b) $\dfrac{4s}{s+6}$

17. (a) $\dfrac{12}{s(s+4)}$

    (b) $\dfrac{10}{s}$

19. $V(s) = \dfrac{240(s+10)}{[(s+10)^2 + 900]}$

21. $V(s) = \dfrac{s(s+4)e^{-9s}}{[(s+4)^2 + (3\pi)^2]}$

23. (a) $\dfrac{\omega}{[(s + \alpha)^2 + \omega^2]}$

(b) $\dfrac{(s + \alpha)}{[(s + \alpha)^2 + \omega^2]}$

25. $f(t) = 5e^{-8t}\,u(t) - 5e^{-10t}\,u(t)$

27. $i(t) = 2\,u(t) + 3e^{-10t}\,u(t)$

29. $f(t) = 3\,\delta(t) + 30e^{-4t}\,u(t) - 12e^{-6t}\,u(t)$

31. $f(t) = 70te^{-6t}\,u(t) - 7e^{-6t}\,u(t) + 14e^{-8t}\,u(t)$

33. $i(t) = 2t^3\,u(t) + 8t^2\,u(t) + 7t\,u(t) + u(t)$

35. $f(t) = -6e^{-6.5t}\,u(t) + 20e^{-6t}\,\sin(4t + 126.87°)\,u(t)$

37. $f(t) = 80\,u(t) + 75e^{-2.5t}\,\sin(2t + 126.87°)\,u(t)$

39. $\dfrac{s(s + 8)}{(s + 6.5)[(s + 6)^2 + (4)^2]}$

41. $R_{-5} = 4$, $R_{-6} =$ does not exist because it is not a pole, and $R_{-4+j2} = 32\ \angle 0°$

43. $e^{-7t}\,u(t) - e^{-10t}\,u(t)$

45. $5\,u(t) - 4e^{-t}\,u(t)$

47. $\delta(t) - 30e^{-9t}\,u(t) + 51e^{-6t}\,u(t)$

49. $2.5e^{-6t}\,u(t) + [13t - 1.5]e^{-4t}\,u(t)$

51. $[42t^2 + 23t^2 + 2t]\,u(t)$

53. $-2e^{-5t}\,u(t) + 3e^{-6t}\,\sin(3t + 90°)\,u(t)$

55. $2.1e^{-2t}\,u(t) + 3.5\,\sin(4t - 36.870°)\,u(t)$

57. $[-60t - 28]e^{-4t}\,u(t) + 85e^{-2t}\,\sin(4t + 53.130°)\,u(t)$

59. $v(t) = [0.625t^2\,\sin(t + 126.87°) + 1.7002t\,\sin(t - 107.10°) + 1.625\,\sin(t)]e^{-2t}\,u(t)$

61. $i(t) = 2.5e^{-t} - 2.5e^{-5t}$

63. $p(t) = 16 + 15e^{-5t}\,\sin(4t - 126.87°)$

CHAPTER 4

1. Fig. P4-1

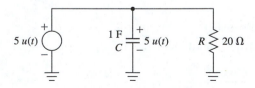

Figure P4-1

3. Fig. P4-2

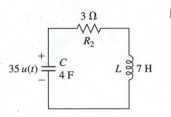

Figure P4-2

5. Fig. P4-3

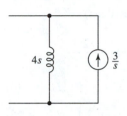
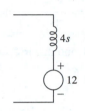

Figure P4-3

7. Figure P4-4

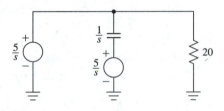

Figure P4-4

9. Figure P4-5

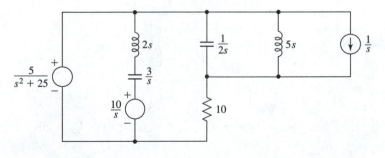

Figure P4-5

11. (a) $\dfrac{8}{s(s+4)}$

(b) $2\,u(t) - 2e^{-4t}\,u(t)$

(c) transient term $= -2e^{-4t}\,u(t)$, steady-state term $= 2\,u(t)$

13. (a) $\dfrac{-2.4}{s(s + 0.2)}$

 (b) $-12\,u(t) + 12e^{-0.2t}\,u(t)$

 (c) -12 volts

15. (a) $\dfrac{20s}{(s + 0.5)}$

 (b) $20\,\delta(t) - 10e^{-0.5t}\,u(t)$

17. $-4\,u(t) + 1.5e^{-5t}\,u(t)$

 $I(s)$ due to $4\,u(t) \Rightarrow \dfrac{-20}{s(s + 5)}$, due to $10\,u(t) \Rightarrow 0$,

 due to $2\,u(t) \Rightarrow 0$, and due to $5\,\delta(t) \Rightarrow \dfrac{-2.5}{(s + 5)}$

19. (a) $(2 + s + 3)$ I_1 $- (3)$ $\qquad\qquad$ I_2 $- (0)$ $\qquad$ $I_3 = 2$

 $\qquad\quad -(3)$ $\quad I_1$ $+ \left(3 + 6 + \dfrac{3}{s} + 2\right) I_2$ $- (6)$ $\qquad$ $I_3 = \dfrac{-5}{s}$

 $\qquad\quad -(0)$ $\qquad I_1$ $- (6)$ $\qquad\qquad$ I_2 $+ (6 + 2s)\ \ I_3 = \dfrac{12}{s}$

 (b) $\qquad\quad (s + 5)$ $\ \ I_1$ $- (3)$ $\qquad I_2$ $- (0)$ $\qquad I_3 = 2$

 $\qquad\qquad -(3s)$ $\quad I_1$ $+ (11s + 3)$ $\ I_2$ $- (6s)$ $\qquad I_3 = -5$

 $\qquad\qquad -(0)$ $\qquad I_1$ $- (6s)$ $\qquad I_2$ $+ s(2s + 6)\ \ I_3 = 12$

21. (a) $I_R(s) = \dfrac{-2(s^2 + 16.5)}{(s + 3)(s^2 + 16)}$ $\qquad$ $I_C(s) = \dfrac{2s}{(s^2 + 16)}$

 (b) $I_S(s) = \dfrac{4[(s + 0.75)^2 + (2.7726)^2]}{(s + 3)(s^2 + 16)}$

23. (a) $\left(\dfrac{1}{2} + \dfrac{1}{2} + \dfrac{1}{4s}\right) V_1$ $- \left(\dfrac{1}{4s}\right)$ $\qquad$ $V_2 = \dfrac{5}{s^2 + 4}$

 $\qquad -\left(\dfrac{1}{4s}\right)$ $\qquad V_1$ $+ \left(\dfrac{1}{4s} + 1\right) V_2 = \dfrac{-12}{s}$

 (b) $(4s + 1)(s^2 + 4)$ $\ V_1$ $- (s^2 + 4)$ $\quad V_2 = 20s$

 $\qquad\qquad -1$ $\qquad\qquad V_1$ $+ (4s + 1)$ $\quad V_2 = -48$

 (c) $\qquad\qquad \dfrac{2(s - 2.1568)(s + 2.7818)}{s(s^2 + 4)(s + 0.5)}$

25. Same answers as Problem 23

27. (a) $8I_1(s)$

 (b) $\dfrac{0.4}{s} V_1(s)$

 (c) $0.8I_1(s)$

29. $\dfrac{20(s - 0.16667)}{s[(s - 0.33333)^2 + (1.7951)^2]}$

31. (a) $\dfrac{4[s + 0.1)^2 + 0.49]}{(s + 0.2)}$

    (b) $\dfrac{-1.2}{(s + 4)(s + 0.2)}$

    (c) Fig. P4-6

Figure P4-6

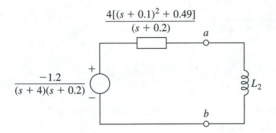

33. (a) $\dfrac{4[(s + 0.1)^2 + 0.49]}{(s + 0.2)}$

    (b) $\dfrac{-0.3}{(s + 4)[(s + 0.1)^2 + 0.49]}$

    (c) Fig. P4-7

    (d) See Fig. P4-6 in Problem 31

Figure P4-7

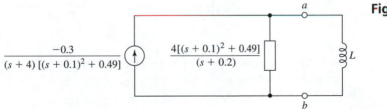

35. (a) 3rd order
    (b) 5th order

37. $-0.5\,\delta(t) + e^{-2t}\,u(t)$

39. (a) $\zeta = 1.45$, $\omega_n = 10$ rad/sec, $\alpha = 4$ and 25, $\omega_d =$ does not exist
    (b) $\zeta = 0$, $\omega_n = 3$ rad/sec, $\alpha = 0$, $\omega_d = 3$ rad/sec

41. $C > \dfrac{1}{2}$ for underdamped, $C = \dfrac{1}{2}$ for critically damped, $C < \dfrac{1}{2}$ for overdamped

CHAPTER 5

1. (a) $\dfrac{s}{(s + 0.5)}$

   (b) $e^{-0.5t}\,u(t);\ \delta(t) - 0.5e^{-0.5t}\,u(t)$

3. (a) $\dfrac{-0.2}{(s + 0.02)}$

   (b) $-10\,u(t) + 10e^{-0.02t}\,u(t)$

5. (a) $\dfrac{1}{[(s + 0.5)^2 + 1.75]}$

   (b) $\dfrac{5e^{-s}}{(s + 2)[(s + 0.5)^2 + 1.75]}$

7. (a) Fig. P5-1
   (b) Transient: $s = -3, -1 \pm j6$, steady state: $s = 0$
   (c) Marginally stable
   (d) $s = -3, -1 \pm j6$

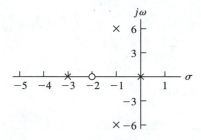

Figure P5-1

9. (a) Fig. P5-2
   (b) Transient: $s = -1, -4, -4$, steady state: none
   (c) Unstable
   (d) $s = -4, -4, -1$

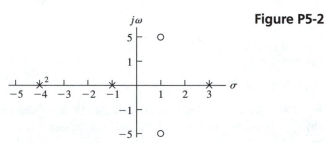

Figure P5-2

11. $$\dfrac{1.3889\,s^2\left(\dfrac{s}{10}+1\right)}{\left(\dfrac{s}{2}+1\right)\left(\dfrac{s^2}{18}+\dfrac{1}{3}s+1\right)}$$

13. (a) Fig. P5-3a
    (b) Fig. P5-3b

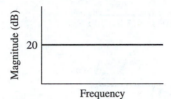
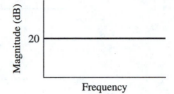
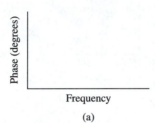
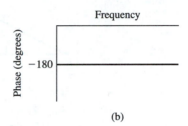

(a)

(b)

Figure P5-3

15. (a) Fig. P5-4
    (b) -1.4451 dB, $-32.142°$

Figure P5-4

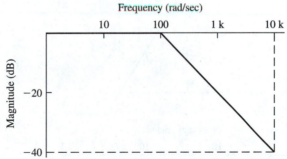
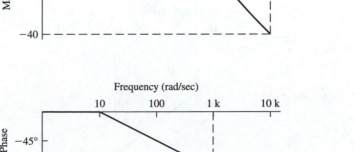

17. Fig. P5-5

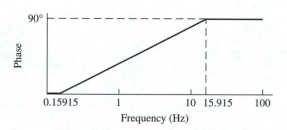

Figure P5-5

19. (a) Fig. P5-6
    (b) 2.6954 dB, 16.4924 rad/sec

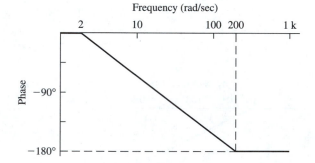

Figure P5-6

21. (a) Fig. P5-7
    (b) $\zeta > 0.70711$; therefore, none to find

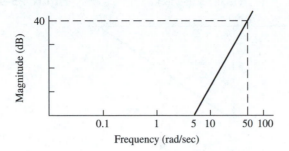

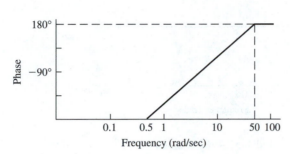

Figure P5-7

23. (a) Fig. P5-8

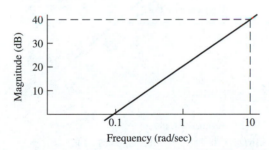

Figure P5-8

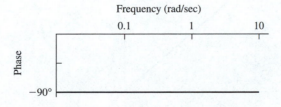

(b) Fig. P5-9

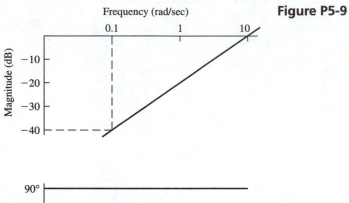

Figure P5-9

25. (a) 31.956 dB, 31.4960 rad/sec
Fig. P5-10

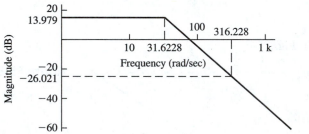

Figure P5-10

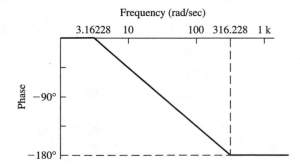

(b) −24.481 dB, 1.9799 *K* rad/sec
   Fig. P5-11

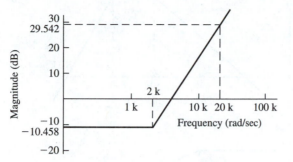

Figure P5-11

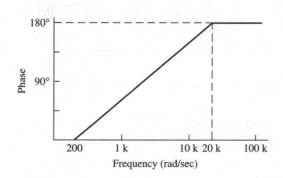

27. (a) Fig. P5-12
    (b) 0.079577 Hz, 3.1831 kHz
    (c) 14.6 dB, −57.427°

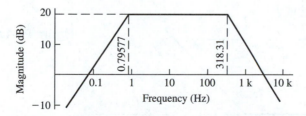

Figure P5-12

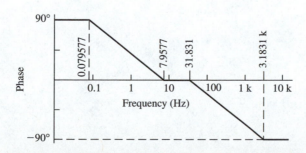

29. Fig. P5-13

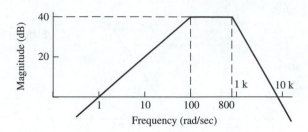

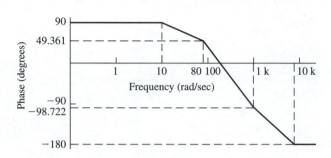

Figure P5-13

31. $\dfrac{6.2421 \times 10^{13}}{(s + 31.416)(s + 6.2832 \times 10^{6})}$

CHAPTER 6

1. (a) Amplified
   (b) +5 dB

3. (a) Fig. P6-1a

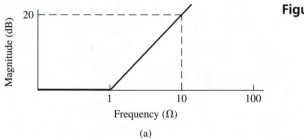

Figure P6-1a

(a)

(b) 10.458 dB, 3.3333

(c) Fig. P6-1b

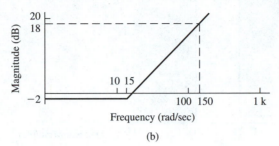

Figure P6-1b

(b)

5. (a) Fig. P6-2a

(b) 3.8764 dB, 6.25

(c) Fig. P6-2b

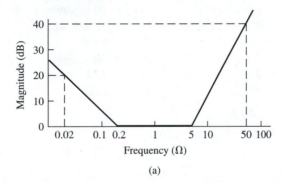

Figure P6-2

(a)

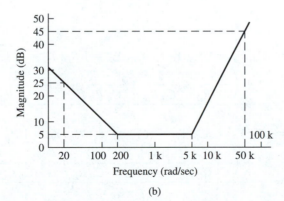

(b)

7. Low pass; Fig. P6-3

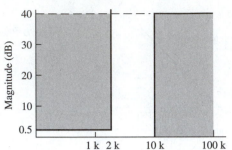

Figure P6-3

9. $\omega_{s1} = 2$ krad/sec $\qquad \omega_{s2} = 12$ krad/sec
$\omega_{p1} = 2.4$ krad/sec $\qquad \omega_{p2} = 10$ krad/sec
$L_{min} = 40$ dB $\qquad\qquad L_{max} = 3$ dB

11. (a) $\omega_{s1} = 5$ krad/sec $\qquad\qquad \omega_{s2} = 10$ krad/sec
$\qquad\omega_{p1} = 4.1667$ krad/sec $\qquad \omega_{p2} = 12$ krad/sec
$\qquad L_{min} = 40$ dB $\qquad\qquad\quad L_{max} = 2$ dB
(b) $\omega_{s1} = 2.4$ krad/sec $\qquad\qquad \omega_{s2} = 10$ krad/sec
$\qquad\omega_{p1} = 2$ krad/sec $\qquad\qquad\; \omega_{p2} = 12$ krad/sec
$\qquad L_{max} = 40$ dB $\qquad\qquad\quad L_{min} = 2$ dB

13. $\dfrac{4.7045 \times 10^{13}}{(s + 50.265)(s + 6.2832 \times 10^{6})}$

15. $\dfrac{V_o}{V_{in}} = \dfrac{-R_2}{R_1}$

CHAPTER 7

5. (a) $Y_A = sC_A$, $Y_B = 1/R_B$, $Y_C = sC_C$, $Y_D = 1/R_D$, $Y_E = 0$

$$\frac{V_{in}}{V_o} = \frac{s^2 + \left(\dfrac{1}{C_C R_D} + \dfrac{1}{C_A R_D} + \dfrac{1-K}{C_A R_B}\right)s + \dfrac{1}{C_A C_C R_B R_D}}{K s^2}$$

(b) $Y_A = sC_A$, $Y_B = sC_B$, $Y_C = sC_C$, $Y_D = 1/R_D$, $Y_E = 1/R_E$

$$\frac{V_{in}}{V_o} = \frac{s^2 + \left(\dfrac{C_A}{C_B C_C R_D} + \dfrac{1}{C_C R_D} + \dfrac{1}{C_B R_D}\right)s + \dfrac{1}{C_B C_C R_D R_E}}{\left(\dfrac{-C_A}{C_B}\right)s^2}$$

7. $R_2 = R_A = R_C = 1\ \Omega$, $C_B = C_D = 1$ F, $R_1 = 0.5858\ \Omega$, and $h = 1.5858$

9. (a) Second-order low pass
(b) Second-order band pass
(c) First-order high pass

11. 16.454 dB

13. (a) $C_{D1} = 0.1$ F, $C_{E1} = 1$ F
$R_{A1} = 0.73664$ Ω, $R_{B1} = R_{C1} = 2.8309$ Ω
$C_{D2} = 0.1$ F, $C_{E2} = 1$ F
$R_{A2} = 6.7360$ Ω, $R_{B2} = R_{C2} = 2.8309$ Ω

(b) $C_{B1} = C_{D1} = 1$ F
$R_{A1} = R_{C1} = 0.89522$ Ω, $R_{11} = 0.1523$ Ω, $R_{21} = 1$ Ω
$C_{B2} = C_{D2} = 1$ F
$R_{A2} = R_{C2} = 0.89522$ Ω, $R_{11} = 1.2346$ Ω, $R_{22} = 1$ Ω
$R_{F3} = 0.38833$ Ω, $R_{13} = 1$ Ω (inverting amplifier)

15. (a) $C_D = 0.1$ F, $C_E = 1$ F, $h = 0.70711$
$R_A = 8.9478$ Ω, $R_B = R_C = 3.7606$ Ω

(b) $C_D = C_B = 1$ F, $h = 1.5801$
$R_A = R_C = 1.1892$ Ω, $R_1 = 1.2346$ Ω, $R_2 = 1$ Ω

17. (a) $1.0264 \, / -152.29°$ or 0.22618 dB
(b) $6.1506 \, / -42.395°$ or 15.778 dB

19. $C_{B1} = 1$ F, $R_{A1} = 66.839$ Ω, $R_{B1} = 3.3183$ Ω
$C_{A2} = C_{E2} = 1$ F, $C_{C2} = 2$ F, $C_{G2} = 20$ F, $R_{12} = 19.975$ Ω, $R_{22} = 1$ Ω
$R_{B2} = R_{F2} = 0.17364$ Ω, $R_{D2} = 0.086822$ Ω, $R_{G2} = 8.4064$ Ω

CHAPTER 8

1. $C_1 = 2$ F, $C_2 = 0.2$ F
$R_1 = 7.9577 \times 10^{-5}$ Ω, $R_2 = 1.5915 \times 10^{-4}$ Ω

3. $C_1 = 0.002$ F, $C_2 = 0.02$ F, $R_1 = 3$ Ω, $R_2 = 2.2$ Ω

5. (a) $C_1 = 10$ μF, $C_2 = 1$ μF, $R_1 = 4$ MΩ, $R_2 = 200$ kΩ
(b) $C_1 = 100$ μF, $C_2 = 10$ μF, $R_1 = 400$ kΩ, $R_2 = 20$ kΩ

7. $C_B = C_D = 0.001$ μF, $R_2 = R_A = R_C = 18.696$ kΩ, $R_1 = 10.9518$ kΩ,
and $h = 6.5145$

9. (a) $\dfrac{s + 1.5569}{2.7686}$

(b) $\dfrac{s + 1.5569}{0.87551}$

11. (a) $R_\beta = 9.1127$ kΩ; output between R_α and R_β
(b) $R_\beta = 7.3375$ kΩ; feedback between R_α and R_β

13. (a) $R_\beta = 99.206$ kΩ; feedback point is between R_α and R_β
(b) $R_\beta = 23.182$ kΩ; output is between R_α and R_β

15. $\dfrac{(s + 2.3803)(s^2 + 0.45344s + 1.0793)}{s^3}$

17. $C_A = C_C = 0.22$ μF, $R_B = R_D = 20.939$ kΩ, $R_1 = 12.265$ kΩ, $R_\beta = 5.4055$ kΩ;
feedback between R_α and R_β

19. $Q = 2.2222$, $BW_{R/S} = 1.35$ krad/sec, $BW_{Hz} = 214.86$ Hz, $f_o = 477.47$ Hz, and $\omega_o = 3$ krad/sec

21. First stage: $C_{A1} = C_{B1} = C_{C1} = 0.022\ \mu F$, $R_{D1} = 30.242$ kΩ, $R_{E1} = 3.0887$ kΩ
    $R_{\beta 1} = 1.8851$ kΩ, output to next stage between $R_{\alpha 1}$ and $R_{\beta 1}$

   Second stage: $C_{D2} = 0.0022\ \mu F$, $C_{E2} = 0.022\ \mu F$, $R_{A2} = 5.1225$ kΩ,
    $R_{B2} = R_{C2} = 5.2852$ kΩ
    $R_{\beta 2} = 1.8851$ kΩ, final output from between $R_{\alpha 2}$ and $R_{\beta 2}$

23. $$\frac{(s^2 + 0.11590s + 1)(s^2 + 5.5044 \times 10^{-2}s + 0.90450)(s^2 + 6.0856 \times 10^{-2}s + 1.1056)}{1.5569 \times 10^{-3}s^3}$$

25. $$\frac{(s^2 + 0.18496s + 0.52641)(s^2 + 0.35136s + 1.89965)}{0.34722s^2}$$

27. $C_B = C_C = 1\ \mu F$, $R_A = R_E = 20.541$ kΩ, $R_D = 2.4342$ kΩ
    $R_F = 100$ kΩ, $R_1 = 3.3321$ kΩ, $R_2 = 56.234$ kΩ

29. $$\frac{(s^2 + 0.67105s + 1)(s^2 + 3.9337 \times 10^{-2}s + 1.24)(s^2 + 3.1724 \times 10^{-2}s + 0.80645)}{0.99998(s^2 + 1)^3}$$

31. First stage: $C_{A1} = C_{E1} = 0.1\ \mu F$, $C_{C1} = 0.2\ \mu F$, $C_{G1} = $ none
    $R_{B1} = R_{F1} = 2.3216$ kΩ, $R_{D1} = 1.1608$ kΩ, $R_{G1} = 44.969$ kΩ
    $R_{11} = 2.3740$ kΩ, $R_{21} = 2.3216$ kΩ

   Second stage: $C_{A2} = C_{E2} = 0.1\ \mu F$ $C_{C2} = 0.2\ \mu F$, $C_{G2} = 0.1\ \mu F$
    $R_{B2} = R_{F2} = 2.4419$ kΩ, $R_{D2} = 1.2209$ kΩ, $R_{G2} = 2.8404$ kΩ
    $R_{12} = 6.7902$ kΩ, $R_{22} = 2.4419$ kΩ

Index

A

Active filters, 1, 306–7, 307
Amplification, 296
Analog filter, 307
Analog signal analysis, 7–79
   combining, 35–50
   derivatives and integrals of
     waveforms, 51–59
   exponential function, 18–22
   impulse function, 15–18
   MATLAB in, 68–74
   periodic waveform values, 59–61
     average, 62–63
     effective or rms, 63–68
   ramp function, 12–14
   shifted function, 33–35
   sinusoidal function, 22–23
   step function, 8–12
   TI-89 in, 74–79
   t^n function, 14–15
Approximate op-amp, 307–13
Argument, 8
Attenuation, 296
Average value of waveform, 62

B

Band-pass filter, 302, 303–4, 389–99
   general functions, 389–90
   narrow, 394–99
   wide, 390–93
Biquads, 329
Bode magnitude function, 252
Bode phase function, 252
Bode plots, 246–78, 323, 329–32
Break frequency, 248
Butterworth band-pass filter, 389
Butterworth filter equation, 470–73
Butterworth filters, 334–41, 378, 383,
   384, 385
   comparison of Chebyshev filters and
     elliptic filters with, 356–57

C

Calculations, 2
Capacitors, 51, 187
   values, 376
Cascaded design for notch filter, 403–5
Cauer approximation, 348
Chebyshev band-pass filter response,
   394–99
Chebyshev filter equations, 473–76
Chebyshev filters, 341–48, 378, 384,
   385, 395
   comparison of elliptic filters and
     Butterworth filters with, 356–57
Chebyshev loss function, 351
Circuit analysis using Laplace
   transforms, 179–229
   circuit, 189–91
   impedance, 183–89
   initial voltages and currents, 180–83
   MATLAB in, 223–26
   order, 216–23
   solving dependent source circuits,
     207–10
   solving multiple-source circuits,
     197–207
   solving simple circuits, 191–97
   solving Thévenin and Norton circuits,
     210–16
   TI-89 in, 226–29
Circuit order, 216–23
   first, 217–18
   redundancy, 217
   second, 218–23
Circuits
   first-order, 217–18
   normalized, 326
   second-order, 218–23
   solving dependent source, 207–10
   solving multiple-source, 197–207
   solving Norton, 210–16
   solving simple, 191–97

   solving Thévenin, 210–16
Coefficient matching, 320, 323–29
Complex-numbers method, 135–39
   inverse transform of complex-poles,
     135–39
Complex poles, 257–59
   deriving formula, 430–34
   inverse transform of, 153–59
     complex-numbers method, 135–39
     real-numbers method, 139–42
Complex waveforms, 1
Complex zeros, 259–62
Constant-amplitude sinusoidal function,
   22–26
Convolution, 238
Corner frequency, 248
Cover-up method, 127, 147
Current-controlled current source
   (CCCS), 208
Current-controlled voltage source
   (CCVS), 208
Current divider rule, 438

D

Damped resonant frequency, 220
Damping constant, 21
DC circuit equations, 435–41
   current divider rule, 438
   identifying parallel circuits, 435
   identifying series circuits, 435
   Kirchhoff's voltage law, 438–40
   Kitchhoff's current law, 440–41
   Ohm's law, 436
   parallel current sources, 436
   series voltage sources, 435
   voltage and current measurements,
     436–37
   voltage divider rule, 437–38
Decade, 251
Dependent source circuits, solving,
   207–10

Derivative operation, 108–12
Derivatives of waveforms, 51–59
Differential equations, Laplace
    transforms in solving, 160–63
Digital filter, 307
Dual-feedback topologies, 321, 322
  inverting, 461–64
  noninverting, 457–60

E

Effective value, 63–68
Elliptic filters, 348–56
  comparison of Chebyshev filters and
    Butterworth filters with, 356–57
  equation, 476–84
Elliptic loss function, 378, 384, 385
Equiripple approximation, 341
Euler's formula, 135, 138
Exponential amplitude sinusoidal
    function, 26–31
Exponential function, 18–22
Exponential multiplier operation, 104–6
Extended transform pairs, 421

F

Filter graphs, 295–302
  loss function, 296
  normalized graphs, 297–301
  specifications for, 301–2
Filters, 295–314. *See also* Normalized
    low-pass filter
  active, 306–7, 307
  analog, 307
  band-pass, 302, 303–4
  Butterworth, 334–41
  Chebyshev, 341–48
  component considerations, 313–14
  digital, 307
  elliptic, 348–56
  graph specifications for, 301–2
  high-pass, 302
  low-pass, 302
  normalized, 319
  notch, 302–3
  op-amps, 307–14
  passive, 306–7
  practical, from generic loss functions,
    373–414
    band-pass, 389–99
    frequency shifting, 373–75
    gain shifting, 380–84
    high-pass, 384, 385

impedance shifting, 376–80
    MATLAB in, 405–11
    notch, 399–405
    TI-90 in, 411–14
Filter table calculations, 467
  Butterworth filter equation, 470–73
  Chebyshev filter equations, 473–76
  elliptic filter equation, 476–84
  general procedure, 467–70
First-order circuits, 217–18
First-order real poles, inverse transform
    of, 121–28
Fixed-value component, 314
Frequency shifting, 373–75
Function notation, 2

G

Gain, 238
Gain constant, 248–49
Gain function, 296
Gain shifting, 380–84
General band-pass loss functions,
    389–90
Generic approximations, 333
Generic loss functions, 333
  practical filters from, 373–414
    band-pass filter, 389–99
    frequency shifting, 373–75
    gain shifting, 380–84
    high-pass filter, 384–88
    impedance shifting, 375–80
    MATLAB in, 405–11
    notch filter, 398–404
    TI-89 in, 411–14
Graphs
  filter, 295–302
  normalized, 297–301
  semilog, 443–52
Graph specifications for filters, 301–2

H

High-pass filter, 302, 384–88
High-pass function, conversion of low-
    pass function into
    MATLAB in, 405–11
    TI-89 in, 411–14

I

Impedance shifting, 376–80
Impulse function, 15–18
Impulse response, 217, 238
Inductors, 51, 185

Integral-differential equation, 112
Integral operation, 112–13
Integrals of waveforms, 51–59
Inverse equations definitions, 142–46
Inverse Laplace transforms
  of complex poles, 153–59
    complex-numbers method, 135–39
    real-numbers method, 139–42
  of first-order real poles, 121–28
  of multiple-order real poles, 129–35
  of real poles, 146–53
  streamlined, 142–59
  by table, 97–98
Inverting band-pass filter, 399
Inverting dual-feedback topology,
    461–64
  band pass, 463–64
  high pass, 462–63
  low pass, 461–62
Inverting single-feedback topology,
    455–57
  high pass, 456–57
  low pass, 455–56

K

Kirchhoff's current law, 196, 440–41
Kirchhoff's voltage law, 182, 197, 212,
    214, 215, 438–40

L

Laplace capacitance, 186–87
Laplace circuit, 189–91
Laplace derivations
  complex poles formula, 430–34
  Laplace transform operations, 426–30
  Laplace transform pairs, 423–26
Laplace impedance, 183–89
  source conversion and, 187–89
Laplace inductance, 185–86
Laplace integral, 94–96
Laplace operations, 98–121
  combining transforms, 113–17
  derivative, 108–12
  exponential multiplier, 104–6
  extending table of Laplace transform
    pairs, 117–21
  integral, 112–13
  shifted function, 99–104
  t multiplier, 106–8
Laplace resistance, 184
Laplace transform pairs
  deriving, 423–26

extending table of, 117–21
Laplace transforms, 93–171
  circuit analysis using, 179–229
    circuit, 189–91
    impedance, 183–89
    initial voltages and currents,
      180–83
    MATLAB in, 223–26
    order, 216–23
    solving dependent source circuits,
      207–10
    solving multiple-source circuits,
      197–207
    solving simple circuits, 191–97
    solving Thévenin and Norton
      circuits, 210–16
    TI-89 in, 226–29
  defined, 1, 94
  deriving, 426–30
  inverse
    by table, 97–98
    traditional techniques in, 121–42
  MATLAB in, 164–67
  operations, 98–121
  in solving differential equations,
    160–63
  streamlined inverse, 142–59
  by table, 94–97
  TI-89 in, 167–71
Log scale
  calculating distances on, 444–47
  reading, 443–44
Loss functions, 296, 319
Low-pass filters, 302
  approximations, 332–57
  normalized, 319–69
    approximations, 332–57
    biquads, 329–32
    coefficient matching, 320, 323–29
    MATLAB in, 358–63
    TI-89 in, 362–69
    topology, 320–23
Low-pass function, conversion of high-
    pass function into
    MATLAB in, 405–11
    TI-89 in, 411–14

M

Magnitude function, 247, 250
Mathematical expression, graphing, 35
MATLAB, 3–5
  for analog signal analysis, 68–74
  to connect low-pass function into
    high-pass function, 405–11
  for current analysis using Laplace
    transforms, 223–26
  for Laplace transforms, 164–67
  for normalized low-pass filter, 358–63
  for sinusoidal steady state, 278–82
Maximally flat approximation, 334
Maximally flat filter, 470
Mesh current method, 200–204
Multiple-order real poles, inverse
    transform of, 129–35
Multiple-source circuits, solving,
    197–207
Multiplying constant, 21

N

Narrow-band-pass filters, 394–99
Natural equivalent circuits, 184
  for the capacitor, 187
  for the inductor, 186
Node voltage method, 204–7, 206
Noninverting dual-feedback topology,
    457–60
  band pass, 460
  high pass, 459
  low pass, 458
Noninverting single-feedback topology,
    453–55
  high pass, 454–55
  low pass, 453–54
Nonzero poles, 250
Normalized circuit, 326
Normalized filter, 319
Normalized first-order Butterworth
    notch filter, 399
Normalized graphs, 297–301
Normalized low-pass filters, 319–69
  approximations, 332–57
  biquads, 329–32
  coefficient matching, 320, 323–29
  MATLAB in, 358–63
  TI-89 in, 362–69
  topology, 320–23
Norton equivalent, 188
Norton equivalent circuits, 213
  solving, 210–16
Norton equivalent impedance, 211
Notch filters, 302, 398–404
  cascaded design, 403–5
  parallel design, 399–403
  twin-T topology in designing, 403

O

Ohm's law, 184, 188, 195, 196, 198, 208,
    212, 222, 436
Op-amps, 307–14
  approximate, 307–13
  topologies, 453–66
    inverting dual-feedback, 461–64
    inverting single-feedback, 455–57
    noninverting dual-feedback,
      457–60
    noninverting single-feedback,
      453–55
    twin-T, 464–66

P

Parallel circuits, identifying, 435
Parallel current sources, 436
Parallel design for notch filter, 399–403
Partial fraction expansion, 121
Pascal's triangle, 154
Pass-band loss, 380
Passive filter, 306–7
Periodic waveform values, 59–61
  average, 62–63
  effective or rms, 63–68
Phase function, 247, 250
Polar-rectangular conversion, 254
Polar-to-rectangular conversion, 31
Poles, 121–22
  combining zeros and, 262–78
  complex, 257–59
    deriving, 430–34
  real, 250–55
Pole-zero plot, 241–42
  stability using, 242–46
Prime notation, 162

Q

Quadratic equation, 94

R

Ramp function, 12–14
RC product, 374
Reactive components, 180
Real-numbers method, 139–42
  inverse transform of complex-poles,
    139–42
Real poles, 250–55
  inverse transform of, 146–53
    multiple-order, 129–35
Real zeros, 254–57
Rectangular-to-polar conversion, 31, 251

Redundancy in circuit order, 217
Removing impulse, 122
Residue, 123, 144
Resistors, 180, 184
  tolerance, 314
rms value, 63–68
Roll-off, 251
Roll-off rates, calculating, on semilog
  graphs, 447–50

S

Second-order circuits, 218–23
Semilog graph paper, construction of,
  450–52
Semilog graphs, 443–52
  calculating distances on, 444–47
  calculating roll-off rates on, 447–50
  reading, 443–44
Sensitivity, 314
Series circuits, identifying, 435
Series voltage sources, 435
Shifted function operation, 33–35,
  99–104
Single-feedback topologies, 321
  inverting, 455–57
  noninverting, 453–55
Sinusoidal functions, 22–33
  constant-amplitude, 22–26
  exponential amplitude, 26–31
  splitting and combining, 31–33
Sinusoidal steady state, 237–88
  frequency response and bode plot,
    246–78
  MATLAB in, 278–82
  pole-zero plot, 241–42
  stability using pole-zero plot, 242–46
  TI-89 in, 282–88
  transfer function, 238–41
Source conversion, Laplace impedance
  and, 187–89
S-plane, 241
Stability using pole-zero plot, 242–46
Steady-state frequency and bode plot,
  246–78
Steady-state response, 191, 192
Steady-state terms, 191, 192

Step function, 8–12
  multiplication of, by another function,
    10
  rules for, 9
Step response, 217, 238
Streamlined inverse Laplace transforms,
  142–59
  inverse equations defined, 142–46
  inverse transforms of complex poles,
    153–59
  inverse transforms of real poles,
    146–53
Superposition, 197–200
Switch function, 10. *See also* Step
  function

T

Table of Laplace transforms pairs,
  extending, 117–21
Tables. *See also* Transform tables
  inverse Laplace transforms by, 97–98
  Laplace transforms by, 94–97
Thévenin equivalent circuits, 188, 212
  solving, 210–16
Thévenin equivalent impedance, 211
TI-89, 3–5
  for analog signal analysis, 74–79
  to connect low-pass function into
    high-pass function, 411–14
  for current analysis using Laplace
    transforms, 226–29
  for Laplace transform, 167–71
  for normalized low-pass filter, 362–69
  for sinusoidal steady state, 282–88
Time-domain function, 94–95, 96, 109
  converting transfer function to,
    247–49
Time-shifted function, 99
Time-shifting operation, 99
T multiplier operation, 106–8
T^n function, 14–15
Topology of normalized low-pass filter,
  320–23
Transfer function, 238–41
  converting, to time domain, 247–49
Transform identities, 420

Transform operations, 419
  combining, 113–17
  deriving Laplace, 426–30
Transform pairs, 94, 96, 419
  deriving Laplace, 423–26
  extended, 421
Transform tables
  extended transform pairs, 421
  transform operations, 419
  transform pairs, 419
Transient responses, 191, 192
Transient terms, 191
Twin-T topology, 322–23, 464–66
  band pass, 465–66
  in designing notch filters, 403
  high pass, 465–66
  low pass, 465–66

U

Unit step function, 8

V

Voltage and current measurements,
  436–37
Voltage-controlled current source
  (VCCS), 208
Voltage-controlled voltage source
  (VCVS), 207–8, 308
Voltage divider rule, 437–38

W

Waveforms
  average value of, 62–63
  complex, 1
  derivatives of, 51–59
  integrals of, 51–59
  periodic values, 59–68
Wide-band-pass filters, 390–93

Z

Zero crossings, 27, 45, 363–69
Zero poles, 250
Zeros, 121–22
  combining poles and, 262–78
  complex, 259–62
  real, 254–57
Zero-to-one transition, 9